About the Authors

Patricia Barnes-Svarney is a science and science fiction writer. Over the past few decades, she has written or coauthored close to three dozen books, including *When the Earth Moves: Rogue Earthquakes, Tremors, and Aftershocks* and the award-winning *New York Public Library Science Desk Reference*. Thomas E. Svarney is a scientist who has written extensively about the natural world. His books, with Patricia Barnes-Svarney, include Visible Ink Press' *The Handy Dinosaur Answer Book, The Handy Math Answer Book*, and *The Handy Ocean Answer Book,* as well as *Skies of Fury: Weather Weirdness around the World* and *The Oryx Guide to Natural History*. You can read more about their work and writing at www.pattybarnes.net.

THE
HANDY
BIOLOGY
ANSWER
BOOK

THE
HANDY
BIOLOGY
ANSWER
BOOK

SECOND EDITION

Patricia Barnes-Svarney and Thomas E. Svarney

VISIBLE
INK
PRESS

Detroit

THE HANDY BIOLOGY ANSWER BOOK

Visible Ink Press®
43311 Joy Rd., #414
Canton, MI 48187–2075

Visible Ink Press is a registered trademark of Visible Ink Press LLC.

Most Visible Ink Press books are available at special quantity discounts when purchased in bulk by corporations, organizations, or groups. Customized printings, special imprints, messages, and excerpts can be produced to meet your needs. For more information, contact Special Markets Director, Visible Ink Press, www.visibleink.com, or 734–667–3211.

Managing Editor: Kevin S. Hile
Art Director: Mary Claire Krzewinski
Typesetting: Marco Di Vita
Proofreaders: Shoshana Hurwitz and Aarti Stephens
Indexer: Larry Baker

Cover images: Shutterstock.

ISBN: 978-1-57859-490-0 (paperback)
ISBN: 978-1-57859-524-2 (pdf ebook)
ISBN: 978-1-57859-526-6 (Kindle ebook)
ISBN: 978-1-57859-525-9 (ePub ebook)

Library of Congress Cataloging-in-Publication Data
Barnes-Svarney, Patricia L.
  The handy biology answer book / by Patricia Barnes-Svarney and Thomas E. Svarney. – 2nd edition.
    pages cm
  Previous edition: The handy biology answer book (Detroit : Visible Ink Press, 2004).
  Includes bibliographical references.
  ISBN 978-1-57859-490-0 (pbk. : alk. paper)
1. Biology–Miscellanea. I. Svarney, Thomas E. II. Title.
  QH349.H36 2015
  570–dc23
                                                                    2014009127

10 9 8 7 6 5 4 3 2 1

Contents

Acknowledgements

We are indebted to the authors of the first edition of *The Handy Biology Answer Book*—James Bobick, Naomi Balaban, Sandra Bobick, and Laurel Roberts. Their knowledge and research made our job of revising the book that much easier. And there are also the people behind the scenes, as always. We'd like to thank Roger Jänecke for all his help, and especially for asking us to revise the first edition; also thanks to Mary Claire Krzewinski for page and cover design, Marco Di Vita of the Graphix Group for typesetting, Shoshana Hurwitz for indexing, and Aarti Stephens and Shoshana (again) for proofreading. An extra special thank you to Kevin Hile, our understanding editor for many "Handy Answer" books—who, as an editor and writer *exceptionnel*, is truly surpassed by few. We'd also like to thank our wonderful agent, Agnes Birnbaum, for her help, patience, and above all, her friendship over the years.

We'd also like to acknowledge all those biologists—from botanists and bacteriologists to environmentalists and geneticists, past, present, and future—who have, are, and will try to solve the mysteries of life on our planet and beyond. It's almost impossible to comprehend the number and types of organisms that exist; and though we may never know all the answers, here's to everyone who tries to comprehend just how the many organisms—including humans—fit in the grand scheme of life.

Photo Credits

Electronic Illustrators Group: pp. 18, 46.

Barfooz: p. 92.

Kelvinsong: p. 56.

Nova: p. 153.

Public domain: pp. 36, 48, 57, 66, 74, 85, 88, 99, 132, 161, 297, 322, 325, 326, 331.

Shutterstock: 6, 7, 10, 13, 15, 25, 27, 27, 28, 38, 42, 50, 59, 63, 67, 69, 71, 78, 80, 94, 95, 108, 109, 115, 117, 125, 137, 142, 146, 153 (bottom), 156, 159, 164, 168, 178, 183, 188, 190, 192, 194, 196, 201, 202, 204, 205, 211, 218, 220, 225, 229, 231, 235, 242, 244, 249, 251, 252, 258, 260, 266, 270, 273, 276, 277, 279, 281, 283, 285, 288, 289, 291, 294, 310, 314, 318, 333, 341, 344, 346, 352, 353, 356, 361, 367, 370, 382, 385, 393, 395, 401, 403, 406, 409, 415, 418, 420, 422.

Yassine Mrabet: p. 4.

Introduction

Biology is a grand, glorious field; and life is an even bigger subject. To present all the information known about life on Earth in one book would be a lifetime's work—and a huge book. What we offer you is a condensed version of some of the most interesting and up-to-date information in the biology world.

We both consider ourselves naturalists for a reason: We spend most of our time outside in nature—surrounded by birds, fungi, animals, and plants, along with sundry protists and bacteria we can't see. These natural subjects are all there to watch and learn from—showing us the necessary phases of our world's organisms, such as birth, dormancy, hibernation, and, of course, death. In fact, you can't point to anything "out there" without awe, wonder, and an appreciation of all the other life forms that have lived on this planet for just over a billion years.

Here are some of the biological highlights of that past history—and of the present and future. We are indebted to the original authors of the first edition—Naomi Balaban, James Bobick, Sandra Bobick, and Laurel Bridges Roberts—and have kept many of their questions and answers throughout the book. And because a great deal has happened in the ten or so years since the publication of the first edition, we updated some of the original queries—and added many of our own.

We hope you will enjoy this step into the world of biology, and that you'll use this book as a platform to discover other books or Internet sites about your favorite natural topics. And above all, we hope this book will inspire you to walk, run, gaze, sit, or stand in this beautiful living world. Our planet teems with life—go out there and enjoy it!

Patricia Barnes-Svarney
and Thomas E. Svarney

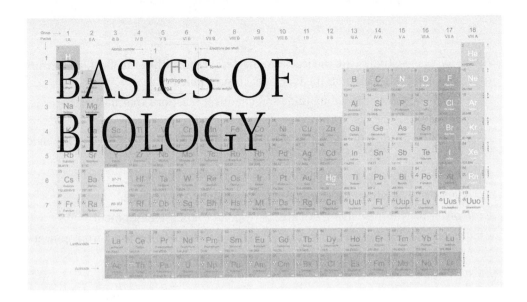

BIOLOGY AND LIFE

What is biology?

Biology is often called the science of life in studies that include everything from an organism's conception to its death. It is mainly concerned with the study of living systems—from animal to plant and everything in between—and includes the study of various organisms' cells, metabolism, reproduction, growth, activity of systems, and response to the stimuli in their environment.

Who coined the term "biology"?

French biologist Jean-Baptiste Pierre Antoine de Monet de Lamarck (1744–1829) is credited with coining the term "biology" (from the Greek terms *bios*, meaning "life," and *logy*, meaning "study of") in 1802 to describe the science of life. He was also the first to publish a version of an evolutionary tree that described the ancestral relationships among species (an early classification system), first to distinguish between vertebrates and invertebrates—and is often considered one of the first evolutionists.

What are some studies within the field of biology?

Numerous studies are within the field of biology. The following lists some of the most familiar biologically oriented scientific divisions and their relevant studies:

Anatomist—Studies the structures of living organisms (other divisions exist within this field, such as a comparative anatomist who studies the similarities and differences in animal body structures).

Astrobiologist—Studies the possibility of life—or the formation and/or possible distribution of life—on early Earth and throughout the solar system and universe.

Bacteriologist—Studies the intricacies of bacteria (and within this field, numerous other divisions exist based on the type of bacteria studied).

Biochemist—Studies the compounds and chemical reactions that take place in living organisms.

Biophysicist—Studies living things using the techniques and tools used in the field of physics.

Botanist—Studies the world of plants.

Cryobiologist—Studies how extreme cold affects living organisms.

Ecologist—Studies how living organisms respond to their environment.

Embryologist—Studies the formation and development of organisms from conception to adulthood.

Entomologist—Studies the structure, function, and behavior of insects.

Ethologist—Studies certain animal behavior under natural conditions.

Exobiologist—Studies the possibility of life elsewhere in the universe and how that life could come about.

Geneticist—Studies the field of heredity and genetics.

Gnotobioticist—Studies how organisms grow in a germ-free environment or studies organisms that grow in environments with certain specific germs.

Histologist—Studies the tissues of living organisms.

Ichthyologist—Studies fish (usually specific types, such as freshwater or ocean fish).

Lepidopterist—Studies organisms that live in freshwater areas.

Marine biologist—Studies life in the ocean (usually specific organisms, such as squid or sharks).

Molecular biologist—Studies the molecular processes that occur in the cells of organisms.

Mycologist—Studies the intricacies of fungi.

Oologist—Studies bird eggs, including the development of eggs from certain types of birds.

Organic chemist—Studies the compounds from living organisms.

Ornithologist—Studies the structure, function, and behavior of birds.

Paleontologist—Studies prehistoric life (although this is actually a field of geology, many paleontologists have an extensive background in biological studies).

Parasitologist—Studies the life cycle of parasites.

Taxonomist—Studies the classification of organisms.

Virologist—Studies the development of viruses and how they affect other organisms.

Zoologist—Studies the structure, function, development, and/or behavior of animals (usually in specific regions, such as desert or tundra animals, or specific animals, such as polar bears or grizzly bears).

What is life?

The definition of "life" is the most controversial subject—just mention the word to scientists would undoubtedly be a heated debate. It affects every branch of biology—from life on Earth to the possibility of life in outer space. But some general, often agreed-upon criteria exist for the definition of life (although some creatures exist that are contrary to the rules): Living organisms are usually complex and highly organized (with exceptions); most creatures respond to external stimuli (for example, plants that recoil on touch, and for higher level organisms, the ability to learn from the stimulus); the majority of organisms try to sustain internal homeostasis (a relative balance of an organism's internal systems, such as maintaining its temperature); most tend to take their energy from the surrounding environment and use it for their growth and reproduction; and most organisms reproduce (asexually or sexually—or even both), with their offspring evolving over time. Of course, these definitions do not take into consideration alternate forms of organisms—such as possible extraterrestrial life that could upset our Earth-centric view of life!

What was the Oparin-Haldane hypothesis?

In the 1920s, while working independently, Russian biochemist Aleksandr Oparin (1894–1980) and British geneticist and biochemist John Burdon Sanderson Haldane (1892–1964) both proposed scenarios for the "prebiotic" conditions on Earth (the conditions that would have allowed organic life to evolve). Although they differed on details, both models described an early Earth with an atmosphere containing ammonia and water vapor. Both also surmised that the assemblage of organic molecules began in the atmosphere and then moved into the seas. The Oparin-Haldane model includes the idea that organic molecules—including amino acids and nucleotides—were synthesized without living cells (or abiotically); then the organic building blocks in the prebiotic soup were assembled into polymers of proteins and nucleic acids; and finally, the biological polymers were assembled into self-replicating organisms that fed on the existing organic molecules. (For more about nucleic acids, see the chapter "DNA, RNA, Chromosomes, and Genes.")

What was the Miller-Urey Synthesis experiment?

In 1953, American chemist and biologist Stanley Lloyd Miller (1930–2007) and American physical chemist Harold Clayton Urey (1893–1981) designed an experiment—called the Miller-Urey Synthesis—to understand the conditions on early Earth and to test the Oparin-Haldane hypothesis. Simulating what was thought to be the atmospheric conditions on Earth about four billion years ago—a hot environment filled with simple organic chemical substances such as water (H_2O), ammonia (NH_3), hydrogen gas (H_2), methane (CH_4), and other mineral salts—the scientists subjected the mix to a continual electrical discharge (essentially to simulate lightning strikes). After about a week into the experiment, four major organic molecules in their simplest forms were gener-

3

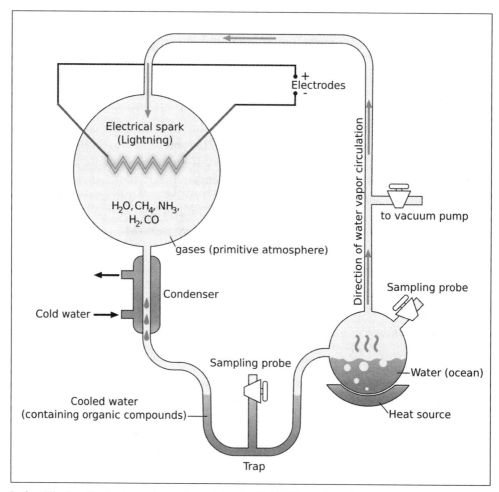

In the Miller-Urey Synthesis experiment chemists Stanley Lloyd Miller and Harold Clayton Urey simulated what conditions on Earth might have been like four billion years ago. The result was that chemical substances essential for the formation of life were created.

ated: nucleotides, sugars, fatty acids, and a total of five amino acids—all thought to be the precursors to life.

What did scientists eventually discover about Miller's experiment?

After Stanley Miller died in 2007, scientists who inherited the original experiment looked even closer at Miller and Urey's results—thanks to advances in analytical tools. They found that far more organic molecules existed than Miller reported, with fourteen amino acids and five amines (a class of organic compounds derived from ammonia). The scientists also uncovered two additional experiments that were never published. One produced a lower diversity of organic molecules, while the other produced a much wider variety. In the latter experiment, Miller included conditions similar to those of volcanic eruptions—something that scientists believe was quite prevalent on the early Earth—with the experiment producing twenty-two amino acids, five amines, and many hydroxylated molecules. These and other experiments suggest that the early Earth's volcanic activity may have been instrumental in producing the precursors to life.

What is the heterotroph hypothesis?

The heterotroph hypothesis suggests that the first primitive life forms on early Earth—evolving about 3.5 billion years ago—could not manufacture their own food (thus, they were heterotrophic). Because of the lack of oxygen in the early atmosphere, they were anaerobic (did not need oxygen to survive) and probably absorbed the primordial soup's organic molecules as nutrients.

What possible mechanisms helped early cells to group together and self-replicate?

The main criteria for living cells are a membrane capable of separating the inside of the cell from its surroundings, genetic material capable of being reproduced, and the ability to acquire and use energy (metabolism). But how did those early single cells "come together" to form organic compounds and eventually self-replicate?

The mechanism(s) that eventually helped to form organic compounds is still a highly debated subject. One suggestion is that the first cells collected together and eventually self-replicated in ocean foam. Another theory states that the clay may have contributed its own energy (clay can store, transform, and release chemical energy) to encourage the growth of cells. British-born American theoretical physicist Freeman Dyson (1923–) hypothesized the "double origin theory," in which "two separate kinds of creatures [exist], one kind capable of metabolism without exact replication and the other kind capable of replication without metabolism." And still another idea is that larger molecules called polymers (proteins bonded together) somehow connected together and eventually became self-replicating.

What is panspermia?

Panspermia, meaning "all-seeding," is the idea that organic molecules are in space and that microorganisms, spores, or bacteria attached to tiny particles of matter can travel through space, and in theory, eventually land on a suitable planet and initiate the rise of life there. The first known mention of the term was by the Greek philosopher Anaxagoras (c. 5 B.C.E.). The idea was revived in the nineteenth century by several scientists of the time, including the British scientist Lord Kelvin (1824–1907), who suggested that life may have arrived here from outer space, perhaps carried by meteorites. In 1903, the Swedish chemist Svante Arrhenius (1859–1927) put forward the more complex idea that life on Earth was "seeded" by means of extraterrestrial spores, bacteria, and microorganisms coming here on tiny bits of cosmic matter. In 1974, British astronomer Sir Fred Hoyle (1915–2001) and Sri Lankan-born British mathematician Chandra Wickramasinghe (1939–) proposed that dust in interstellar space contained carbon, noting

5

that even today, life forms continue to enter the Earth's atmosphere (they also said these organics may be responsible for new diseases or epidemic outbreaks).

The theory further suggests that, based on life forms scientists have discovered on Earth that can withstand the rigors of extreme environments, life such as bacteria could travel dormant in space for an extended period. They could eventually collide with planets or intermingle with protoplanetary disks (broken-up chunks of rock and debris that eventually form a solar system), with the bacteria (or other life) eventually becoming active. But note: panspermia is not meant to mean how life began, but is the method that may cause its ability to survive and spread.

What space-borne organic molecules have been discovered?

Scientists continue to search for possible organic molecules in the solar system and throughout the universe. In recent years, improved technology has allowed scientists to discover a multitude of organic molecules and structures. The following lists only a few of these discoveries:

Meteorites—In 2008, analysis of the Murchison meteorite found in Antarctica indicated that the organic compounds in the rock were not terrestrial (in other words, contaminated by Earth organics), but from nonterrestrial origins. Since this rock is thought to have originated in our solar system, some scientists believe it may be evidence that organic compounds were around when the Earth formed—and may have played a role in the development of life on Earth. And in 2011, scientists examining meteorites on Earth suggested that building blocks of DNA (deoxyribonucleic acid) may have formed in outer space.

Comets—In 2009, scientists identified the amino acid glycine in a comet for the first time. By 2013, scientists discovered that comets could be breeding grounds for creating complex dipeptides—linked pairs of amino acids that indicate life.

In space—In 2011, astronomers reported that cosmic dust contains complex organic matter; they suggested that the organics were created naturally—and quite rapidly—by stars. In 2012, scientists discovered a sugar molecule called glycolaldehyde—needed to form ribonucleic acid (RNA)—in a distant star system. In 2013, scientists studying a giant gas cloud around 25,000 light years from Earth found a molecule thought to be a precursor to a key component of DNA, and another, called cyanomethanimine, may be one of the key steps in the processing of adenine, an amino acid (for more information about RNA and DNA, see the chapter "DNA, RNA, Chromosomes, and Genes").

Could life on Earth have been based on silicon instead of carbon?

Yes, technically, life on Earth could have been based on silicon instead of carbon because the element has the same bonding properties as carbon. But silicon is second only to carbon in its presence on Earth, thus carbon-based life evolved. (Note: Silicon is never found alone in nature, but always exists as silica [silicon dioxide] or silicates [made up of a compound made of silicon, oxygen, and at least one metal].) But that does not mean no organisms exist that contain silica. For example, a plant called horsetail has one of the highest contents of silica in the plant kingdom. Called a "living fossil," it is the descendant of plants that lived over a hundred million years ago.

Why is water so important to life?

We are all aqueous creatures, whether because of living in a watery environment or because of the significant amount of water contained within living organisms. Therefore, all chemical reactions in living organisms take place in an aqueous environment. Water is important to all living organisms due to its unique molecular structure (H_2O), which is V-shaped, with hydrogen atoms at the points of the V and an oxygen atom at the apex of the V. In the covalent bond (for more about covalent bonds, see ahead in this chapter) between oxygen and hydrogen, the electrons spend more time closer to the oxygen nucleus than to the hydrogen nucleus. This uneven or unequal sharing of electrons results in a water molecule with a slightly negative pole and a slightly positive pole.

Water is the universal solvent in biological systems, so what does this mean for living organisms?

A solvent is a substance that can dissolve other matter; because all chemical reactions that support life occur in water, water is known as the universal solvent. In

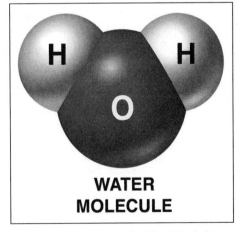

WATER MOLECULE

A water molecule is essential to life on Earth. Its slightly positive and negative poles encourages other molecules to organize themselves in aqueous solutions.

Pure, liquid water is most dense at 39.2°F (3.98°C) and decreases in density as it freezes. The water molecules in ice are held in a relatively rigid geometric pattern by their hydrogen bonds, producing an open, porous structure. Liquid water has fewer bonds; therefore, more molecules can occupy the same space, making liquid water denser than ice.

fact, it is the polar nature of the water molecule (it contains both positive and negative poles) that causes it to act as a solvent—and any substance with an electric charge will be attracted to one end of the molecule. (If a molecule is attracted to water, it is termed hydrophilic; if it is repelled by water, it is termed hydrophobic.)

How many organisms have lived on Earth since life began?

How many organisms have lived on the Earth since life began continues to be a very controversial subject. Some scientists believe more than two billion species have lived on our Earth over time, including those living today. In fact, some scientists estimate that about 90 to 99.9 percent of all animal and plant species that have ever lived on our world are now extinct. There are reasons why this number is difficult to pin down, including the fact that much of early life—especially those with soft bodies—left no trace. In addition, many of the fossils that exist are buried deep into the ground or have been weathered away by natural physical processes (for example, glacial or water erosion).

How did different forms of life evolve on Earth?

No one really knows how life evolved on Earth. One of the reasons is the minute size (single cells) of the first organisms, which makes it difficult to detect them in ancient rocks. In addition, most of the oldest rocks have been exposed to the heat and pressure of geologic activity over time, making detection impossible by erasing all traces of that life. The following is only one interpretation of how early life on Earth developed (all years are approximations):

- 3.6 billion years ago, simple cells (prokaryotes) evolved.
- 3.4 billion years ago, stromatolites began the process of photosynthesis.
- 2 billion years ago, complex cells (eukaryotes) developed.
- 1 billion years ago, multicellular life began.
- 600 million years ago, simple animals evolved in the oceans.
- 570 million years ago, arthropods (ancestors of insects, arachnids, and crustaceans) began to become more widespread.
- 550 million years ago, complex animals began to evolve.

What was the Cambrian Explosion?

The Cambrian Explosion occurred, logically, at the beginning of the Cambrian period, about 544 million years ago (on the geologic time scale, it also marked the end of the Precambrian era and the beginning of the Paleozoic era). At this time, a huge explosion of life occurred in the oceans—most of them similar to modern marine animal groups—with a rapid diversification between the different groups. It took about another one hundred million years—around 440 million years ago—before the first animals crawled on land and a second burst of animal growth occurred.

What is the earliest evidence of life found thus far?

In 2013, scientists studying some of the oldest rocks in the world found traces of life that date back 3.49 billion years. Located in the Pilbara region of Western Australia, the area contains a collection of well-preserved, ancient sedimentary rocks. The region was originally a sandy coastal plain; the sands were eventually built up into microbial mats by microbes. Over millions of years, the sand turned into rock, preserving the bacterial mats. Although no fossils remain in the ancient rock, the researchers found that the rock's mats contained weblike patterns and textures—called Microbially Induced Sedimentary Structures (MISS)—probably created by an ecosystem of different ancient bacteria.

Today, microbial mats still form in places such as the Pilbara—mainly in the form of stromatolites (for more about stromatolites, see this chapter). Cyanobacteria (and other bacteria) live in the mats, which produce oxygen through photosynthesis. This is probably the same process that occurred around 2.4 billion years ago, when it is thought that cyanobacteria produced an abundant amount of oxygen, setting the stage for our oxygen-rich atmosphere and oxygen-dependent organisms.

Where was life recently found in an unlikely place on Earth?

Most people don't think of "life" thousands of feet under the icy continent of Antarctica. But in 2011, living bacteria were found in core samples from Antarctica's Lake Vostok—waters lying 12,100 feet (3,700 meters) below the ice. In 2013, other evidence of life was found 2,624.7 feet (800 meters) under the ice sheet that covers Lake Whillans in Antarctica. Scientists found cells containing DNA (deoxyribonucleic acid) in the subglacial lake—cells that were actively using oxygen. Although some scientists believe the cells were from contamination by the surrounding ice, the scientists who discovered the cells cited two main reasons to support their claim: the water contained cell concentrations about one hundred times higher than the cell count in the glacier's meltwater, and the minerals in the lake water were at least one hundred times higher than the glacier's meltwater. The scientists also estimated that the water in the subglacial lake—and thus, the cells—had probably been cut off from the surface for 100,000 years.

Archaebacteria can live in extreme environments where other organisms would quickly perish, such as this sulphurous hot spring in Yellowstone National Park.

What is an extremophile?

An extremophile is an organism capable of surviving extreme environments. In fact, scientists continue to discover that life can inhabit many zones—from beyond the boiling point of water, below freezing, under extreme radiation, around 2.5 miles (4 kilometers) underground, and over 6 miles (11 kilometers) below sea level. For example, in 1977, scientists aboard the research submarine Alvin discovered life far below the ocean surface where no light penetrates. It was shown that the volcanic vents supplied enough nutrients for life to thrive without sunlight in a process called chemosynthesis (the ability to convert chemicals into food). These extremophiles—some bacteria and animals—thrive in temperatures above 212°F (around 100°C), the boiling point of water; some bacteria can survive even higher temperatures. Still other "extreme" bacteria can survive in oceanic pressures 6.84 miles (11 kilometers) under water, while still others survive arid, frigid, or even acidic environments. Bacteria are not the only extremophiles—in 2012, scientists mimicking Martian conditions in the Mars Simulation Laboratory in Germany found that lichens could survive for at least thirty-four days (the length of the simulation) on the Red Planet.

Have any signs of life been found in our solar system?

Although no definitive life has been found in our solar system, scientists are still searching for evidence. For example, Mars may be smaller than the Earth, and farther away from the Sun than Earth, but some scientists believe the planet may have once had small organisms living on its surface. The latest probe, the rover Curiosity, tested the surface rock and soil in 2012 and 2013, hoping to prove that organic material is present on the planet and is actually from the planet, not contamination from meteorites or the rover itself. And scientists are also suggesting, based on living bacteria found in Antarctica's Lake Vostok (12,100 feet [3,700 meters] deep), that the frozen satellites (moons) of the outer planets—such as Jupiter's Europa and Neptune's Triton—may harbor bacteria in or under their ice. In fact, some scientists believe that the pull of Jupiter causes Europa to have "tides," allowing the ice to melt under the planetary ice coating and create a watery ocean that may harbor life.

CLASSIFICATION OF LIFE

What is systematics?

Systematics is the area of biology devoted to the classification of organisms. Originally introduced by Swedish naturalist Carolus Linnaeus (Carl von Linné, 1707–1778), who based his classification system on physical traits, systematics now includes the similarities of DNA, RNA, and proteins across species as criteria for classification.

How has the classification of organisms changed throughout history?

A long list of scientists exists who have tried to classify organisms, and even today, not one single classification system has been agreed upon. Initially, from Aristotle (384–322 B.C.E.) to Carolus Linnaeus, scientists who proposed the earliest classification systems divided living organisms into two kingdoms—plants and animals. The following lists some of the other highlights of classification:

- During the nineteenth century, German zoologist Ernst Haeckel (1834–1919) proposed establishing a third kingdom—Protista—for simple organisms that did not appear to fit in either the plant or animal kingdom.

- In 1969, American plant ecologist Robert Harding Whitaker (1920–1980) proposed a system of classification based on five different kingdoms. The groups Whitaker suggested were the bacteria group Prokaryotae (originally called Monera), Protista, Fungi (for multicellular forms of nonphotosynthetic heterotrophs and single-celled yeasts), Plantae, and Animalia. (This classification system is still widely accepted.)

- A six-kingdom system of classification was proposed in 1977 by American microbiologist and biophysicist Carl Woese (1928–2012), including Archaebacteria and Eubacteria (both for bacteria), Protista, Fungi, Plantae, and Animalia. And in 1981, Woese further proposed a classification system based on three domains (a level of classification higher than kingdom): Bacteria, Archaea, and Eukarya. The domain Eukarya is further subdivided into four kingdoms: Protista, Fungi, Plantae, and Animalia.

Until recently, what were some ways to classify living organisms?

Like many things in science, a certain subject cannot always be explained one way—and the classification of living organisms is no exception. Until about the mid-1990s, the following represented one of the most commonly used classifications of organisms and their respective characteristics (it is still used in some literature):

Kingdom	Cell Type	Characteristics
Monera* (Bacterial and Archaean Kingdoms)	Prokaryotic	Single cells lacking distinct nuclei and other membranous organelles
Protista	Eukaryotic	Mainly unicellular or simple multicellular, some containing chloroplasts. Includes protozoa, algae, and slime molds
Fungi	Eukaryotic	Single-celled or multicellular, yeasts, not capable of photosynthesis
Plantae	Eukaryotic	Multicellular organisms with chloroplasts capable of photosynthesis
Animalia	Eukaryotic	Multicellular organisms, many with complex organ systems

*This division name is no longer in use.

What is the latest (to date) way to classify organisms?

One of the latest classification systems is based on DNA analysis—a much more accurate way to reflect the evolutionary history and interconnections between organisms. This is the three-domain system, which includes Bacteria, Archaea, and Eukarya. The domains Bacteria (Eubacteria or "true" bacteria) and Archae (Archaebacteria or "ancient" bacteria) consist of unicellular organisms with prokaryotic cells. The domain Eukarya consists of four kingdoms: Protista, Fungi, Plantae, and Animalia; organisms in these groups have eukaryotic cells (for more about eukaryotic and prokaryotic cells, see the chapter "Cellular Basics").

Do other ways to classify living organisms exist?

Yes, seemingly a plethora of other classification listings exist—all depending on various criteria. For example, more informally, animals are often classified as the Metazoa subkingdom in the traditional two-kingdom system of classification (animals and plants). Thus, the Metazoa subkingdom is often considered to be synonymous with the Animalia kingdom. This subkingdom includes all animals except the protozoa (for more about protozoa, see the chapter "Bacteria, Viruses, and Protists").

In yet another example, some biologists divide the Animalia kingdom into two subkingdoms: the parazoa (from the Greek *para*, meaning "alongside," and *zoa*, meaning "animal"), which includes multicellular animals with a digestive tract (all animals except Porifera, or sponges) and the eumetazoa (from the Greek *eu*, meaning "true"; *meta*, meaning "later"; and *zoa*, meaning "animal"), which includes multicellular organisms with less specialized cells than the eumetazoa and includes the single phylum of Porifera. (For more about animals, see the chapter "Aquatic and Land Animal Diversity.")

What organisms are included in the kingdom Fungi?

Of the bewildering variety of organisms that live on the planet Earth—and perhaps the most unusual and peculiarly different from human beings—are fungi. Members of the kingdom Fungi range from single-celled yeasts to *Armillaria ostoyea*, a species that covers 2,200 acres (890 hectares)! Also included are mushrooms that are commonly consumed, the black mold that forms on stale bread, the mildew that grows on damp shower curtains, rusts, smuts, puffballs, toadstools, shelf fungi, and the death cap mushroom, *Amanita phalloides*. Fungi are able to rot timber, attack living plants, spoil food, and

Mushrooms like these are a type of fungi, which are neither plants nor animals but, rather, constitute their own kingdom.

afflict humans with athlete's foot or even worse maladies. Fungi also decompose dead organisms, fallen leaves, and other organic materials. In addition—and on the bright side—fungi produce antibiotics and other drugs, make bread rise, and ferment beer and wine. (For more about fungi, see the chapter "Fungi.")

How many different species of fungi are on Earth?

According to scientific reports in 2011, an estimated 8,700,000 (give or take 1.3 million) total species are on Earth—from microorganisms and plants to animals. Overall, around 6.5 million are terrestrial (land-based) and 2.2 million (about 25 percent) are in the oceans. Of those organisms described and catalogued by 2011, just over 953,000 species are animals, about 215,000 species are plants, around 43,000 species are fungi, and just over 8,000 species are protozoa. Many scientists agree that many more species have yet to be uncovered, with estimates of about 86 percent of all species on land and 91 percent in the oceans yet to be discovered, described, and catalogued.

Who first proposed the kingdom Protista?

The German zoologist Ernst Haeckel (1834–1919) first proposed the kingdom Protista in 1866 for the newly discovered organisms that were neither plant nor animal. The term "protist" is derived from the Greek term *protistos*, meaning "the very first." (For more about protists, see the chapter "Bacteria, Viruses, and Protists.")

What is thought to be the most primitive group of animals?

Sponges—from the phylum Porifera (Latin for *porus*, meaning "pore," and *fera*, meaning "bearing") are thought to represent the most primitive animals. These organisms are collections of specialized cells without true tissues or organs, and their bodies are not symmetrical. They have a specialized way of gathering nutrients from waters and are known as filter feeders. (For more about sponges, see the chapter "Aquatic and Land Animal Diversity.")

BASIC CHEMISTRY FOR BIOLOGY

What is biochemistry?

As a field of scientific study, chemistry may be divided into various subgroups. One major subgroup is organic chemistry, a field that refers to the study of carbon-based compounds, including carbohydrates and hydrocarbons such as methane and butane. When this discipline further focuses on the study of the organic molecules that are important to living organisms, it is known as biochemistry.

What is an atom?

An atom is the smallest unit of an element, containing the unique chemical properties of that element. Atoms are very small—several million atoms could fit in the period at the end of this sentence.

Parts of an Atom

Subatomic Particle	Charge	Mass	Location
Proton	Positive	1.7×10^{-24} g	Nucleus
Neutron	Neutral	1.7×10^{-24} g	Nucleus
Electron	Negative	9.1×10^{-28} g	Orbits around nucleus

How does the nucleus of an atom differ from the nucleus of an organism's cell?

The English word "nucleus" is derived from the Latin word *nucula*, meaning "kernel" or "core." The nucleus of an atom is an enclosed, positively charged center, containing protons and neutrons. The nucleus of an organism's cell is a membrane-enclosed feature (called an organelle) that contains the genetic material of that cell (for more information about cells, see the chapter "Cellular Basics").

What is the Periodic Table of the Elements?

The Periodic Table of the Elements is a listing of all the known chemicals and their symbols. The first ninety-two elements occur in nature (with a few exceptions); the remaining have been artificially created in laboratory particle accelerators. Many of these elements are important to organic chemistry—in particular, the bonding of certain elements resulting in the formation of organics, such as hydrocarbons and polymers.

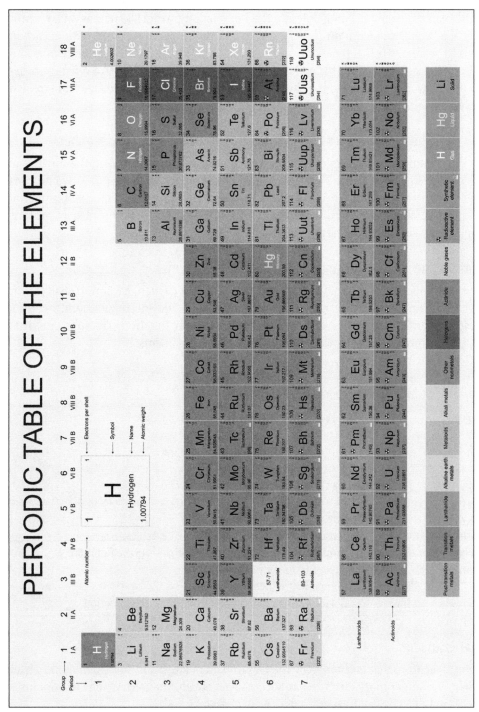

The periodic table of all the chemical elements currently known to science.

Why is carbon an important element?

Carbon is what makes life as we know it exist. It makes up 18 percent of the weight of the human body, and all molecules in the body (except water)—such as sugars, proteins, fats, and DNA—contain carbon. Due to its unique electron configuration, carbon needs to share electrons. It can form four covalent bonds with other carbon atoms or a variety of other elements, forming long chains of molecules, each with a different property. In addition, it forms bonds with many other molecules, from hydrogen and oxygen to even some metals.

How does the mass number of an element differ from the atomic number?

The mass number is the sum of the number of protons and neutrons in the nucleus of an element. For example, the mass number of helium is 4, because it has two protons and two neutrons in its nucleus. Since it has only two protons, the atomic number of helium is 2. When the atomic number changes (for example, the number of protons change), the result is a different element.

What are the most important elements in living systems?

The most important elements in living systems include oxygen, carbon, hydrogen, nitrogen, calcium, phosphorus, potassium, sulfur, sodium, chlorine, magnesium, and iron. These elements are essential to life due to their cellular function. The following lists the most common and important elements in living organisms:

Element	Percent of Humans by Weight	Functions in Life
Oxygen	65	Part of water and most organic molecules; molecular oxygen
Carbon	18	Backbone of organic molecules
Hydrogen	10	Part of all organic molecules and water
Nitrogen	3	Component of proteins and nucleic acids
Calcium	2	Part of bone; essential for nerves and muscles
Phosphorus	1	Part of cell membranes and energy storage molecules; part of bone
Potassium	0.3	Important for nerve function
Sulfur	0.2	Structural component of some proteins
Sodium	0.1	Primary ion in body fluids; essential for nerve function
Chlorine	0.1	Major ion in body fluids
Iron	Trace	Component of hemoglobin
Magnesium	Trace	Cofactor for enzymes; important to muscle function

What is an ion?

An ion is an atom that is charged by the loss or gain of electrons. For example, when an atom gains one or more electrons, it becomes negatively charged. When an atom loses one or more electrons, it becomes positively charged.

What is a chemical bond?

A chemical bond is an attraction between the electrons present in the outermost energy level or shell of a particular atom. This outermost energy level is known as the valence shell. Atoms with an unfilled outer shell are less stable and tend to share, accept, or donate electrons. When this happens, a chemical bond is formed. In living systems, chemical reactions—with help from enzymes—link atoms together to form molecules.

What are the major types of bonds?

Three major types of chemical bonds exist: covalent, ionic, and hydrogen. The form of bond that is established is determined by a specific arrangement between the electrons. Ionic bonds are formed when electrons are exchanged between two atoms and the resulting bond is relatively weak. For example, salt is held together by ionic bonds between sodium (Na^+) and chloride (Cl^-) ions. Covalent bonds occur when electrons are shared between atoms; this form of bond is strongest and is found in both energy-rich molecules and molecules essential to life. For example, hydrogen and oxygen molecules in water are held together by covalent bonds. Hydrogen bonds are temporary, but they are important because they are crucial to the shape of a particular protein and have the ability to be rapidly formed and reformed, as in the case of muscle contraction. The following chart summarizes the three types of chemical bonds and their characteristics:

Type	Strength	Description	Examples
Covalent	Strong	Sharing of electrons results in each atom having a filled outermost shell of electrons	Bonds between hydrogen and oxygen in a molecule of water
Hydrogen	Weak	Bond between oppositely charged regions of molecules that have covalently bonded hydrogen atoms	Bonds between molecules of water
Ionic	Moderate	Bond between two oppositely charged atoms that were formed by the permanent transfer of one or more electrons	Bond between Na^+ and Cl^- in salt

What determines the type of bond that forms between atoms?

The electron structure of an atom is the best predictor of its chemical behavior. Atoms with electron-filled outer shells tend not to form bonds. However, those atoms with one, two, six, or seven electrons in the outer shell tend to become ions and form ionic bonds. Atoms with greater than two or less than six electrons tend to form covalent bonds.

What is an isotope?

Atoms of an element that have different numbers of neutrons are isotopes of the same element. Isotopes of an element have the same atomic number but different mass numbers. Common examples are the isotopes of carbon: ^{12}C and ^{14}C. ^{12}C has six protons, six electrons, and six neutrons, while ^{14}C has six protons, six electrons, and eight neutrons. Some isotopes are physically stable, while others, known as radioisotopes, are unstable. Radioisotopes undergo radioactive decay, emitting both particles and energy. If the decay leads to a change in the number of protons, the atomic number changes, transforming the isotopes into a different element.

How does one prepare a 1:10 dilution?

To dilute means to weaken or reduce the intensity, strength, or purity of a substance, or to make more fluid by adding a liquid. For example, a 1:10 dilution means one part in a total of ten parts. Three different ways exist to prepare a 1:10 dilution: 1) the weight-to-weight (w:w) method, 2) the weight-to-volume (w:v) method, and 3) the volume-to-volume (v:v) method. In the weight-to-weight method, 1.0 gram of a solute (a substance dissolved in a solution or mixture of some type) is dissolved in 9.0 grams of solvent (a substance having the ability to dissolve another substance), yielding a total of ten parts by weight, one of which is solute. In the weight-to-volume method, enough solvent is added to 1.0 gram of solute to make a total volume of 10 millileters. In this method, one part (by weight) is dispersed in ten total parts (by volume).

Since most biological solutions are very dilute, most research does not use the weight-to-volume method. The weight-to-weight method is used more often and overall, the volume-to-volume method is preferred when the solute is a liquid used to change the concentration of a solution. For example, one milliliter of solute, such as ethanol, added to 9.0 milliliters of water yields a ten-part solution, one part of which is the solute.

What are isomers?

Isomers are compounds with the same molecular formula but differing atomic structure within their molecules. Three major isomers exist: structural isomers differ in their connections, geometric isomers differ in their symmetry about a double bond, and optical isomers are mirror images of each other. (For more information about molecules, see ahead in this chapter.)

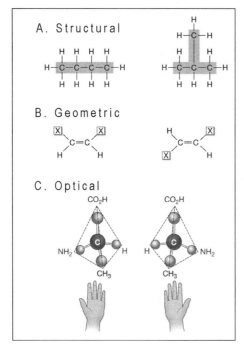

The three types of isomers are A) structural, which are connected in different ways. In this example, butane and isobutane (called an isomer of butane) differ in covalent partners; B) geometric, which differ in arrangement about a double bond (in these diagrams X represents an atom or group of atoms attached to a double-bonded carbon; and C) optical (or enantiomers), which are mirror images of each other, like left and right hands—but they cannot be superimposed on each other.

What food can determine if a solution is acidic or basic?

One easy-to-find food can be used to determine if a solution is acidic or basic—the red cabbage. This vegetable contains a water-soluble pigment called flavin—also found in plums, apple skins, and grapes—which is also called an anthocyanin. If you chop some red cabbage into small pieces, cover them with boiling water, and allow the mixture to sit for about ten minutes, you can use the cabbage juice to discover the pH of a solution. Basic solutions will turn the anthocyanin in the cabbage juice a greenish-yellow, neutral solutions will turn purple, and acidic solutions will turn red.

What is meant by pH?

The term "pH" is taken from the French phrase *l'puissance d'hydrogen*, meaning "the power of hydrogen." Scientifically, pH refers to the -log of the H^+ (positive hydrogen). The mathematical equation to determine pH is usually written as follows: $pH = -\log [H^+]$. For example, if the hydrogen ion concentration in, say, a solution is 1/10,000,000 or 10^{-7}, then the pH value is 7.

The composition of water can also be used to understand the concept of pH: Water is composed of two hydrogen atoms bonded covalently to an oxygen atom. In a solution of water, some water molecules (H_2O) will break apart into the component ions—H^+ and OH^- ions; it is the balance of these two ions that determines pH. When more H^+ ions than OH^- ions exist, the solution is an acid, and when more OH^- ions than H^+ ions exist, the solution is a base.

Why is pH so important to life?

The concentration of hydrogen ions in water influences the chemical reactions of other molecules. An increase in the concentration of electrically charged ions can interfere with or influence the ability of molecules (specifically proteins) to chemically interact. In general, most living systems function at an internal pH close to 7, but biologically active molecules vary in pH levels depending on the molecule and where it functions.

What is the pH scale?

The pH scale is the measurement of the H^+ concentration (hydrogen ions) in an aqueous solution and is used to measure the acidity or alkalinity of that solution. The pH scale ranges from 0 to 14. A neutral solution has a pH of 7; a solution with a pH greater than 7 is basic (or alkaline), and a solution with a pH less than 7 is acidic. In other words, the lower the pH number, the more acidic the solution; the higher the pH number, the more basic the solution. As the pH scale is logarithmic, each whole number drop on the scale represents a tenfold increase in acidity (meaning the concentration of H^+ increases tenfold), and of course, each whole number rise on the scale represents a tenfold increase in alkalinity.

Scale of pH Values

pH Value	Examples of Solutions
0	hydrochloric acid (HCl), battery acid
1	stomach acid (1.0–3.0)
2	lemon juice (2.3)
3	vinegar, wine, soft drinks, beer, orange juice, some acid rain
4	tomatoes, grapes, bananas (4.6)
5	black coffee, most shaving lotions, bread, normal rainwater
6	urine (5–7), milk (6.6), saliva (6.2–7.4)
7	pure water, blood (7.3–7.5)
8	egg white (8), seawater (7.8–8.3)
9	baking soda, phosphate detergents, Clorox
10	soap solutions, milk of magnesia
11	household ammonia (10.5–11.9), nonphosphate detergents
12	washing soda (sodium carbonate)
13	hair remover, oven cleaner

What is the SI system of measurement?

French scientists as far back as the seventeenth and eighteenth centuries questioned the hodgepodge of the many illogical and imprecise standards used for measurement. Thus, they began a crusade to make a comprehensive, logical, precise, and universal measurement system called Système Internationale d'Unités, or SI for short. The SI uses the metric system as its base. Since all the units are in multiples of ten, calculations are simplified. Today, all countries except the United States, Myanmar (formerly Burma), and Liberia use this system. However, some elements within American society do use SI—scientists, exporting and importing industries, and federal agencies.

What are the SI units of measurement?

The SI or metric system has seven fundamental standards: the meter (for length), the kilogram (for mass), the second (for time), the ampere (for electric current), the kelvin (for temperature), the candela (for luminous intensity), and the mole (for amount of substance). In addition, two supplementary units—the radian (plane angle) and steradian (solid angle)—and a large number of derived units compose the current system, which is still evolving. Some derived units, which use special names, are the hertz, newton, pascal, joule, watt, coulomb, volt, farad, ohm, siemens, weber, tesla, henry, lumen, lux, becquerel, gray, and sievert.

Very large or small dimensions are expressed through a series of prefixes, which increase or decrease in multiples of ten. For example, a decimeter is 1/10 of a meter, a centimeter is 1/100 of a meter, and a millimeter is 1/1000 of a meter. A dekameter is 10 meters, a hectometer is 100 meters, and a kilometer is 1,000 meters. The use of these prefixes enables the system to express these units in an orderly way and avoid inventing new names and new relationships.

What is scientific notation?

Scientific notation allows scientists to manipulate very large or small numbers. It is based on the fact that all numbers can be expressed as the product of two numbers, one of which is the power of the number ten (written as the small superscript next to the number ten and called the exponent). Positive exponents indicate how many times the number must be multiplied by ten, while negative exponents indicate how many times a number must be divided by ten. The following lists how to interpret scientific notation:

$$1,000,000,000 = 1 \times 10^9 \qquad 0.000000001 = 1 \times 10^{-9}$$
$$100,000,000 = 1 \times 10^8 \qquad 0.00000001 = 1 \times 10^{-8}$$
$$10,000,000 = 1 \times 10^7 \qquad 0.0000001 = 1 \times 10^{-7}$$
$$1,000,000 = 1 \times 10^6 \qquad 0.000001 = 1 \times 10^{-6}$$
$$100,000 = 1 \times 10^5 \qquad 0.00001 = 1 \times 10^{-5}$$
$$10,000 = 1 \times 10^4 \qquad 0.0001 = 1 \times 10^{-4}$$
$$1,000 = 1 \times 10^3 \qquad 0.001 = 1 \times 10^{-3}$$
$$100 = 1 \times 10^2 \qquad 0.01 = 1 \times 10^{-2}$$
$$10 = 1 \times 10^1 \qquad 0.1 = 1 \times 10^{-1}$$
$$1 = 1 \times 10^0 \qquad 1 = 1 \times 10^0$$

How are Celsius temperatures converted into Fahrenheit temperatures?

Temperature is the level of heat in a gas, liquid, or solid. The freezing and boiling points of water are used as standard reference levels in both the metric (Celsius or, less common, Centigrade) and the English system (Fahrenheit). In the metric system, the difference between freezing and boiling is divided into one hundred equal intervals each called a degree Celsius (°C); in the English system, the intervals are divided into 180 units, with one unit called a degree Fahrenheit (°F).

The formula for converting Celsius temperatures into Fahrenheit is °F = (°C × 9/5) + 32. The formula for converting Fahrenheit temperatures into Celsius is °C = (°F − 32) × 5/9. Some comparisons between the two scales are as follows:

Temperature	Fahrenheit	Centigrade
Absolute zero	−459.67	−273.15
Point of equality	−40	−40
0°F	0	−17.8
Freezing point of water	32	0
Normal human blood temperature	98.4	36.9
100°F	100	37.8
Boiling point of water (at standard pressure)	212	100

What is the Kelvin temperature scale?

Temperature can be measured from absolute zero (no heat, no motion). The resulting temperature scale is the Kelvin temperature scale, named after its inventor, Belfast-born

British mathematical physicist and engineer William Thomson, First Baron Kelvin (also known as Lord Kelvin; 1824–1907), who devised it in 1848. The Kelvin (symbol K) has the same magnitude as the degree Celsius (the difference between freezing and boiling water is 100 degrees), but the two temperatures differ by 273.15 degrees (absolute zero, which is -273.15 degrees on the Celsius scale). For example, the normal human body temperature of 98.6°F is equal to 37°C and 310.15 K.

MOLECULES AND BIOLOGY

What are molecules and why are they important to living organisms?

Molecules are made of specific combinations of atoms. For example, carbon dioxide is made of one carbon atom and two oxygen atoms; water is made of two hydrogen atoms and one oxygen atom—with all the atoms joined by chemical bonds. Complex molecules such as starch may have hundreds of various atoms linked together in a specific pattern. Four molecules are referred to as bioorganic because they are essential to living organisms and contain carbon: nucleic acids, proteins, carbohydrates, and lipids. These molecules are all large, and they are formed by a specific type of smaller molecule, known as a monomer.

What role do bonds have in bioorganic molecules?

Bonds are important to the structure of many bioorganic molecules. Because chemical reactions involve electron activity at the subatomic level, a molecule's shape often determines function. For example, morphine has a shape similar to an endorphin, a naturally occurring molecule in the brain. Endorphins are pain suppressant molecules; thus, morphine essentially mimics the function of endorphins and can be used as a potent pain reliever.

What is a "mole" (or mol)?

A mole (mol) is a fundamental unit of measure for molecules; it refers to either the gram atomic weight or the gram molecular weight of a substance. A mole is equal to the quantity of a substance that contains 6.02×10^{23} atoms, molecules, or formula units. This number is also called Avogadro's number, named after Amedeo Avogadro (Lorenzo Romano Amedeo Carlo, Count of Quaregna and Cerreto; 1776–1856; he is also considered to be one of the founders of modern physical chemistry).

What are functional groups?

Numerous patterns of atoms and bonds exist in organic compounds. These configurations of atoms are called functional groups, as each has specific, predictable properties. For example, the carboxyl functional group (symbolized as COOH) has both a carbonyl and a hydroxyl group attached to the same carbon atom, resulting in new properties. Hydroxyl functional groups have one hydrogen linked to one oxygen atom (symbolized as -OH). These groups readily form hydrogen bonds (which is why certain molecules are soluble in water); for example, alcohols and sugars are full of hydroxyl groups.

What is a polymer?

A polymer is any of two or more compounds that are formed by the process of polymerization, or the process of changing the molecular arrangement of a compound to form new compounds. These new compounds have the same percent composition as the original, but they have a greater molecular weight and different properties. For example, acetylene and benzene are have the same chemical composition, but different weights and properties.

What is a macromolecule?

Macromolecules are literally "giant" polymers made from the chemical linking of smaller units called monomers. To be considered a macromolecule, a molecule has to have a molecular weight greater than 1,000 daltons. (A dalton is a standard unit of measurement and refers to the mass of a proton; it can be used interchangeably with the words "atomic mass unit" [amu]).

How are macromolecules built?

The four types of very large molecules—carbohydrates, nucleic acids, lipids, and proteins—are important to life and are quite diverse in terms of structure, size, and function. But overall, the same mechanisms build and break them down. To understand this, the following applies to these large molecules:

- All are comprised of single units linked together to create a chain, similar to a freight train with many cars.
- All the monomers, or single units, contain carbon.
- All monomers are linked together through a process known as dehydration synthesis, which literally means "building by removing water." A hydrogen atom (H) is removed from one monomer, and a hydroxide (OH) group is removed from the next monomer in line. Atoms on the ends of the two monomers will then form a covalent bond to fill their electron shells, thereby building a polymer.
- All polymers are broken down by the same method, hydrolysis, or "breaking with water." By adding H_2O (which contains hydrogen and hydroxide groups) back to the monomers, the bond is broken and the macromolecule separates into smaller pieces.

What are the most common macromolecules used as energy sources by cells?

Cells use a variety of macromolecules as energy sources. Carbohydrates, lipids, and even proteins can be metabolized for energy; ATP and related compounds are also used for temporary energy storage. (For more about cells, see the chapter "Cellular Basics"; for more about the various human energy sources for cells, see the chapter "Biology and You.")

What are some common energy sources for cells?

Various molecules have certain energy, and those can be used by cells (the kilocalorie/gram is also expressed as kilogram calorie, or the amount of heat required to raise the temperature of water one degree Celsius). The following chart lists the energy source and the kilocalorie per gram energy yield for the cells:

Energy source	Energy yield
Carbohydrate	4 kilocalories/gram
Fat (lipids)	9 kilocalories/gram
Protein	4 kilocalories/gram

MOLECULES AND ENERGY

What is a metabolic pathway?

A metabolic pathway is a series of interconnected reactions that share common mechanisms. Each reaction is dependent on a specific precursor: a chemical, an enzyme, or the transfer of energy.

How do plant and animal cells store energy?

Plants and animals use glucose as their main energy source, but the way this molecule is stored differs. Animals store their glucose subunits in the form of glycogen, a series of long, branched chains of glucose. Plants store their glucose as starch, formed by long, unbranched chains of glucose molecules. Both glycogen and starch are formed through

---

What is the difference between aerobic and anaerobic organisms?

Aerobic refers to organisms that require oxygen to exist; for example, most living organisms need oxygen to stay alive. As humans, our cells get our energy by using oxygen to fuel our metabolism. Anaerobic refers to organisms that need little or no oxygen to exist; it often refers to bacteria, such as those found in the human small intestines.

the chemical reaction of dehydration synthesis, and both are broken down through the process of hydrolysis.

What is ATP?

ATP (adenosine triphosphate) is the universal energy currency of a cell for both plants and animals. Its secret lies in its structure: ATP contains three negatively charged phosphate groups. When the bond between the outermost two phosphate groups is broken, ATP becomes ADP (adenosine diphosphate). This reaction releases 7.3 kcal/mole of ATP, which is a great deal of energy by cell standards.

All cells need the ATP in order to survive. For example, in humans, ATP is used for a large range of biological actions, with each cell in the body estimated to use between one to two billion ATPs per minute, from muscle contractions to providing the energy needed to move the "tail" of a sperm cell in order to reach the female's egg cell. In plants, ATP is not only used in photosynthesis, but in the plant's root hair cells, which need ATP to absorb the essential mineral ions from the soil in order to grow.

What is the difference between catabolic and anabolic reactions?

Catabolic and anabolic reactions are metabolic processes. Both the capture and use of energy by organisms involves a series of thousands of reactions (what we call metabolism). A catabolic reaction is one that breaks down large molecules to produce energy; for example, digestion is a catabolic reaction. An anabolic reaction is one that involves creating large molecules out of smaller molecules, for example, when your body makes fat out of the extra nutrients you eat.

What is the Krebs cycle?

The Krebs cycle (also referred to as the citric acid cycle) is central to aerobic metabolism and is extremely important in the respiration cycle of mammals (specifically why

A diagram of the molecular structure of ATP (adenosine triphosphate), which contains negatively charged phosphate groups. When these bonds are broken, they produce energy for the cell.

we exhale carbon dioxide) and other multicellular organisms. In general, the cycle is an adaptation that allows cells to gain energy from glucose. The process is essential to obtaining high-energy electrons during the final breakdown of the glucose molecule; the essential by-products of this cycle are carbon dioxide and water. It is named after the German chemist Hans Krebs (1900–1981), who received the 1953 Nobel Prize in Physiology or Medicine for his discovery.

What is the Calvin cycle?

Photosynthesis in plants creates reactions that are both light dependent and light independent. The Calvin cycle is part of the reaction that is light independent and occurs in the chloroplast. This cycle allows the plant to capture carbon dioxide, thus leading to the formation of sugar ($C_6H_{12}O_6$). It is named in honor of American chemist Melvin Calvin (1911–1997), who received the 1961 Nobel Prize in Chemistry for unraveling the process of glucose biosynthesis.

Why would most organisms die without oxygen?

Most living organisms are aerobic—specifically, they require oxygen to complete the total breakdown of glucose. In particular, as many as thirty-six ATP are produced through aerobic metabolism of one glucose molecule. Without oxygen, cells do not synthesize enough ATP to maintain a multicellular organism.

Most people think that we need oxygen to breathe. Although this is true, we also need oxygen to recycle the spent electrons and hydrogen ions (H^+) produced as by-products of aerobic respiration. With these by-products, oxygen combines and forms what is often referred to as "metabolic" water.

Why do we release carbon dioxide when we exhale?

The carbon dioxide we exhale is a result of the breakdown of glucose. This occurs during the energy-obtaining phase of our aerobic cells' respiration. During this process, all of the carbon atoms (from the $C_6H_{12}O_6$) are released as carbon dioxide molecules. Of the six carbon dioxide molecules generated, four are released via the Krebs cycle (see above for more about the Krebs cycle).

FERMENTATION

What is fermentation?

Scientists have theorized that fermentation was the process through which energy was first gathered from organic compounds. Fermentation evolved before Earth's atmosphere contained free oxygen, which occurred over 2.5 billion years ago. Fermentation is thus an ancient process and occurs normally in microorganisms that live in the absence of oxygen.

How did Louis Pasteur's theory of fermentation differ from the accepted concept of fermentation?

French chemist and microbiologist Louis Pasteur (1822–1895)—one of the founders of medical microbiology—proposed that fermentation is a process carried out by what he referred to as "living ferments." Other renowned chemists of the time believed that fermentation was a purely chemical process in which microorganisms were a by-product, not the cause.

Grapes sit in a bin awaiting processing in fermentation vats, where yeast will digest them and turn sugars into alcohol and carbon dioxide.

What are some examples of fermented foods?

Fermented foods are foods produced by microorganisms. A combination of bacteria and fungi are used in the fermentation process of many food products. Examples of fermented foods are cheese, sour cream, yogurt, sauerkraut, vinegar, certain baked products, olives, pickles, soy sauce, miso, chocolate, coffee, and most alcoholic beverages. (For more about many of these fermentation products, see the chapter "Biology and You.")

Can cells be used as factories?

Yes, humans have been using cells as factories for millennia—especially for the production of cheese, yogurt, beer, and wine—all of which relies on the ability of individual cells to produce specific products, such as lactic acid and ethanol. More recently, scientists have been able to manipulate cell genes, so that cells will produce substances that have little relation to their normal function. Examples of products from bioengineered cell factories include the production of human insulin to treat diabetics and Factor VIII (a naturally produced clotting factor in humans) for hemophiliacs.

A woman enjoys a frozen yogurt treat. Yogurt is an example of a fermented food. It is made when lactose in milk is fermented through bacterial action. These are good bacteria, though, and are helpful to digestion.

When did brewing first develop?

No one knows the exact date brewing came into practice—but it would be correct to say that beer brewing has had a long history. To date, chemical evidence from ancient pottery shows that beer was produced in the area now known as Iran about 7,000 years ago. Historians studying Sumerian clay tablets also suggest that this culture discovered fermentation about 6,000 years ago (the tablets depict people drinking). Chinese recorded brewing a beer called *kui* about 5,000 years ago; about 4,300 years ago, the Babylonians were brewing, with many women as brewers as well as priestesses (certain beers were used in religious ceremonies).

Without yeast, there would be no such thing as beer! Yeast is essential for the fermentation process.

How is alcohol produced?

The type of alcohol found in alcoholic beverages is known as ethanol—produced by yeast through the process known as alcoholic fermentation. During the breakdown of glucose to harvest energy, cells generate ethanol as a way to recycle a molecule that is crucial to their metabolism. Interestingly, yeast can only survive in media (like beer or wine) that contains less than 10 percent alcohol by volume. (For more about yeast, see the chapter "Fungi.") This means that beverages with an alcohol content greater than 10 percent have been fortified with additional alcohol or have been distilled. (Distillation involves the boiling off and then the condensation of the ethanol, which evaporates at a lower temperature than water.)

What is the role of yeast in beer production?

Beer is made by fermenting water, malt, sugar, hops, yeast (species *Saccaromyces* spp., used for the fermentation process), salt, and citric acid. Each ingredient has a specific role in the creation of beer. For example, malt is produced from a grain—usually barley—that has sprouted, been dried in a kiln, and ground into a powder, which gives beer its characteristic body and flavor. Hops is made from the fruit that grows on the herb *Humulus lupulus* (a member of the mulberry family). The fruit is picked when ripe and is then dried; this ingredient gives beer its slightly bitter flavor.

How is beer made?

Making beer is a somewhat complex process. One mostly commercial method begins by mixing and mashing malted barley with a cooked cereal grain such as corn. This mixture, called "wort," is filtered before hops are added to it. The wort is then heated until it is completely soluble. The hops are removed, and after the mixture has cooled, yeast is added. The beer ferments for eight to eleven days at temperatures that range from 50 to 70°F (10–21°C). The beer is then stored and kept at a state that is close to freezing. During the next few months, the liquid takes on its final character before carbon dioxide is added for effervescence. The beer is then refrigerated, filtered, and pasteurized in preparation for bottling or canning.

Is the same strain of yeast used to make lager and ale beers?

No, not all yeast is created equal for making beer! Two common strains of yeast are used to ferment beer. *Saccharomyces carlsbergensis*, also known as bottom yeast, sinks to the bottom of the fermentation vat. Strains of bottom yeast ferment best between temperatures of 42.8 to 53.6°F (6–12°C) and take eight to fourteen days to produce a lager beer. *Saccharomyces cerevisiae*, also known as top-fermenting yeast, is distributed throughout wort and is carried to the top of the fermenting vat by carbon dioxide (CO_2). Top-fermenting yeast ferments at a higher temperature—57.2 to 73.4°F (14–23°C) over only five to seven days—and produces ales, porter, and stout beers.

What bacteria are essential in the production of various food products?

Numerous bacteria are essential not only to our health, but in the production of many food products. The following lists familiar foods and the microorganisms that are responsible for their respective characteristics (for more about some of these fermented foods, see the chapter "Biology and You"):

Food	Microorganism
Buttermilk and sour cream	*Streptococcus cremoris, Leuconostoc citrovorum*
Pickles	*Enterobacter aerogenes, Leuconostoc* spp.*, *Lactobacillus brevis*
Sauerkraut	*Leuconostoc* spp., *Lactobacillus brevis*
Swiss cheese	*Lactobacillus* spp., *Propionibacterium* spp.
Vinegar	*Acetobacter aceti*
Yogurt	*Streptococcus thermophilus, Lactobacillus bulgaricus*

*spp. = unspecified species

What is the difference between regular and sweet acidophilus milk?

Both acidophilus milk and sweet acidophilus milk are inoculated with the bacterium *Lactobacillus acidophilus*. Many health practitioners believe that this bacterium, a normal member of the human intestinal flora, aids in digestion. Regular acidophilus milk is produced by adding the bacterium to vats of skim milk as part of the fermentation process; it has a characteristic sour taste. When the bacterium is added to pasteurized milk and packaged without fermentation, sweet acidophilus milk, which lacks the characteristic sour taste, is produced.

ENZYMES–AND PROTEINS–AT WORK

What is an enzyme?

Enzymes are proteins that act as biological catalysts. They decrease the amount of energy needed (called activation energy) to start a metabolic reaction. Without enzymes, you would not be able to obtain energy and nutrients from your food.

For example, when a person has lactose intolerance, their system cannot produce lactase, the enzyme that breaks down milk sugar (lactose) in most dairy products. Because of this, if any dairy is eaten, the milk sugar affects digestion, resulting in bloating, gas pains, and, if severe, vomiting and/or diarrhea. While this condition is not life-threatening for most adults, it can have severe consequences for infants, children, and the elderly. (For more about lactose intolerance, see the chapter "Biology and You.")

Who first used the term "enzyme," and how was it used?

Around 1876, German physiologist Wilhelm Kühne (1837–1900) proposed that the term "enzyme" be used to denote phenomena that occurred during fermentation. The word itself means "in yeast" and is derived from the Greek *en*, meaning "in," and *zyme*, meaning "yeast."

Enzymatic reactions can build up or break down specific molecules. The specific molecule an enzyme works on is the substrate; the molecule that results from the reaction is the product. For example, for people who are not lactose intolerant, lactose is the substrate, lactase is the enzyme, and the products are glucose and galactose.

How is thermodynamics used in terms of cellular studies?

Thermodynamics is the study of the relationships between energy and the activity of a cell. Overall, the laws of thermodynamics govern the way in which cells transform chemical compounds—in other words, how living systems transform one form of energy to another so they can carry out essential life functions.

How do I know my enzymes are working?

Obviously, it is difficult to track the activities of all the enzymes required by your body! However, in the case of the enzyme amylase, you can easily check to see if it is working. Amylase is an enzyme in your saliva that starts the breakdown of complex carbohydrates into simple sugars (glucose and maltose). A plain cracker held in the mouth long enough will begin this breakdown, and you will actually begin to taste the sweetness of the enzyme products. Also, lysozyme is an enzyme present in your respiratory tract secretions and tears. It prevents invasion by bacteria—which explains why the warm, moist, open environment of our eyes manages to remain relatively infection-free.

As in physics, these laws can apply to cells. The first law of thermodynamics states that the total energy of a system and its surroundings will always remain constant (energy cannot be created or destroyed, only transferred). The second law of thermodynamics states that systems tend to become disordered and it usually arises in the form of heat. Another term for this disorder is entropy—a term that refers to the disorder that tends to disrupt cells, or in physics, even the universe.

Why is enzyme shape so important?

Shape is critical to the function of all molecules, but especially enzymes, which are three-dimensional. The "active site" of an enzyme is the area where the substrate binds and the reaction takes place. How an enzyme reacts with its substrate is similar to how a ship docks; minor bonds form between the enzyme and substrate until docking is complete. Thus, anything affecting a protein's shape would have an effect on its "docking"— in other words, its ability to react with the substrate.

What effect does aspirin have on enzymes?

Aspirin is probably the most common over-the-counter medication, as it blocks the production of cyclooxygenase (COX) enzymes, COX-1 and COX-2. These two important enzymes have different functions: COX-1 catalyzes the biosynthesis of hormones important in maintaining the stomach lining; COX-2 catalyzes the biosynthesis of chemicals that promote inflammation, fever, and pain when an injury occurs. The positive effects of aspirin (pain and inflammation reduction) are due to the blocking of COX-2 enzymes; however, the negative aspects of aspirin (stomach problems) are due to the blocking of COX-1 enzymes.

How many enzymes are in the human body?

To date, approximately 75,000 enzymes are thought to exist in the human body—all divided into three classes: metabolic enzymes that run our bodies, digestive enzymes that digest our food, and food enzymes from raw foods that start our food digestion. A reason exists for so many enzymes: various metabolic functions may require a whole complex of enzymes to complete hundreds of reactions. In general, individual enzymes are named by adding the suffix "-ase" to the name of the substrate with which the enzyme reacts; for example, the enzyme amylase controls the breakdown of amylose (starch).

Do enzymes only work in specific environments?

Yes, enzymes do seem picky when it comes to their environment. This is because changes in temperature and pH can cause the structure of a protein to change. Thus, every enzyme has criteria that must be met in order for it to perform its function. For example, the amylase that is active in the mouth cannot function in the acidic environment of the stomach, and pepsin, which breaks down proteins in the stomach, cannot function in the mouth.

What are proteins, and how do they work?

Proteins do, in a word, everything. Proteins allow life to exist as we know it. They are the enzymes that are required for all metabolic reactions. They are also important to structures like muscles and act as both transporters and signal receptors. The following chart lists the types of functions of proteins:

Type of protein	Examples of functions
Defensive	Antibodies that respond to invasion
Enzymatic	Increase the rate of reactions, build and break down molecules
Hormonal	Insulin and glucagon, which control blood sugar
Receptor	Cell surface molecules that cause cells to respond to signals
Storage	Store amino acids for use in metabolic processes
Structural	Major components of muscles, skin, hair, horns, spider webs
Transport	Hemoglobin carries oxygen from lungs to cells

Why do you need protein in your diet?

Of the twenty amino acids used by humans as building blocks for proteins, eight are essential. In other words, our bodies cannot synthesize them, nor can we survive without them. Luckily, all of the essential amino acids can be acquired by eating animal meat (complete proteins) and/or certain plant sources (complementary proteins). Combining grains, such as rice and legumes (for example, beans or peas) in the correct proportion and amount, is usually sufficient for a person's daily requirement of protein.

What are some other problems associated with an insufficient intake of protein?

For most people—adults and children—an inadequate intake of protein can lead to several body-oriented problems. For example, because hair is made up primarily of protein molecules, a lack of protein in the diet will essentially cut off the supply, causing hairs to fall out or even slow down growth in order to conserve protein. Another problem with low protein intake is a decrease in lean body mass and compromised muscles and muscular strength. For example, with less protein consumed, it will take much longer for a person's muscles to recover after strength-training exercises.

What happens to the proteins present in an egg when it is cooked?

The "white" of an egg is rich in the protein albumin. When subjected to high heat, the bonds that form the three-dimensional structure of albumin are irreversibly changed. This causes the clear, jellylike consistency of the egg to become firm and white in a process known as protein denaturation.

What happens to proteins during a hair perm?

During a hair perm, the bonds that form the structure in the protein keratin (the major protein of hair) are chemically broken and then reformed; hydrogen bonds are primarily involved in this process. The same principle, breaking and forming bonds, is used to make hair either very curly or straight.

What happens to proteins in our skin as we age?

A variety of molecular changes typify aging. For example, production of the proteins—collectively known as collagen—in the skin slows down. As the amount of collagen decreases, the skin loses elasticity and begins to sag, bag, and wrinkle. The loss of elasticity can be demonstrated by gently pinching a fold of skin on the back of the hand: The skin of young children will immediately return to form when released, while that of an older person will take longer.

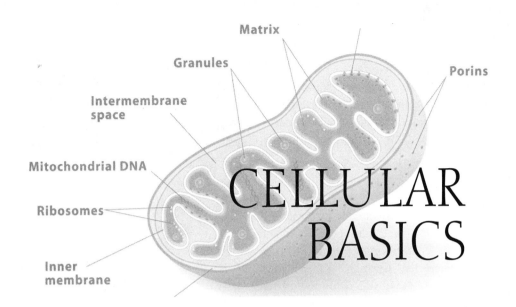

Matrix

Granules

Porins

Intermembrane
space

Mitochondrial DNA

Ribosomes

Inner
membrane

CELLULAR BASICS

HISTORICAL VIEWS OF CELLS

What is a cell?

In general, a cell is the basic unit of all forms of life. Cells are considered to be special-ized depending on their function, such as tissue cells, cells of the various animal organs, or even the cells that create a tomato plant. They can range in size from microscopic to the size of a chicken egg. In addition, an organism can be only one cell (single-celled) or multicelled, such as a human—an organism with more than 100,000,000,000,000 cells. (For more about humans and cells, see the chapter "Biology and You.")

What is the cell theory?

The cell theory is the concept that all living things are made up of essential units called cells and that they are the fundamental components of all life. The cell is the simplest collection of matter that can live. They represent diverse forms of life, existing as sin-gle-celled organisms or more complex organisms, such as multicellular plants and an-imals. They also include diverse specialized cells that cannot survive for long on their own. In addition, everything an organism does occurs fundamentally at the cellular level. Finally, all cells come from preexisting cells, and they are related by division to ear-lier cells—all of which have been modified in various ways during the long evolution-ary history of life on Earth.

What is the origin of the term "cell"?

In 1665, British physicist Robert Hooke first used the term "cell" to describe the divi-sions he observed in a slice of cork. Using a microscope that magnified thirty times, Hooke identified little chambers or compartments in the cork that he called *cellulae*—a Latin term meaning "little rooms"—because they reminded him of the small monastery cells inhabited by monks. (He further calculated that one square inch [6.45 square centimeters] of cork would contain 1,259,712,000 of these tiny chambers or "cells.") This word evolved into the modern term "cell."

What is cytology?

Cytology is the study of cellular structure based on microscopic techniques. Cytology became a separate branch of biology in 1892, when the German embryologist Oscar Hertwig (1849–1922) proposed that organic processes are reflections of cellular processes.

What were some important early discoveries associated with the cell?

Many discoveries happened along the way in the modern study of cells. The following lists a few of the major ones:

German embryologist Oscar Hertwig is considered the founder of cytology.

- In 1665, British physicist Robert Hooke (1635–1703), curator of instruments for the Royal Society of London, was the first to see a cell, initially in a section of cork and then in bones and plants.

- In 1824, French physiologist Henri Dutrochet (René-joachim-henri Dutrochet, 1776–1847) proposed that animals and plants had similar cell structures.

- In 1831, British botanist Robert Brown (1773–1858) discovered and named the cell nucleus in plant cells; German botanist Matthias Schleiden (1804–1881) named the nucleolus (the structure within the nucleus now known to be involved in the production of ribosomes) around that same time.

- Working independently, Schleiden and German physiologist Theodor Schwann (1810–1882) described preliminary forms of the general cell theory in 1839, the former stating that cells were the basic unit of plants and Schwann extending the idea to animals. (Schwann also discovered the first animal enzyme, pepsin, in 1836—an enzyme essential to digestion.)

- In 1855, Polish/German embryologist, physiologist, and neurologist Robert Remak (1815–1865) became the first to describe cell division. Shortly after Remak's discovery, German doctor, anthropologist, pathologist, biologist, and politician Rudolph Virchow (1821–1902) stated that all cells come from preexisting cells. The work of Schleiden, Schwann, and Virchow firmly established the cell theory.

- In 1868, German zoologist Ernst Haeckel (1834–1919) proposed that the nucleus was responsible for heredity.
- In 1888, German anatomist Wilhelm von Waldeyer-Hartz (Heinrich Wilhelm Gottfried von Waldeyer-Hartz, 1836–1921) named and observed chromosomes in the nucleus of a cell.
- In 1882, German biologist Walther Flemming (1843–1905) published his work *Zellsubstanz, Kern und Zelltheilung* (*Cell Substance, Nucleus and Cell Division*), becoming the first scientist to explain the entire process of cell division (he is also considered the "father of cytogenetics").

PROKARYOTIC AND EUKARYOTIC CELLS

What are the differences between prokaryotic and eukaryotic cells?

In 1937, the French marine biologist Edouard Chatton (1883–1947) first proposed the terms *procariotique* and *eucariotique* (French for prokaryotic and eukaryotic, respectively). Prokaryotic, meaning "before nucleus," was used to describe bacteria, while eukaryotic, meaning "true nucleus," was used to describe all other cells. Today, the terms are more well defined: eukaryotic cells are much more complex than prokaryotic cells, having compartmentalized interiors and membrane-contained organelles (small structures within cells that perform dedicated functions; for more about organelles, see this chapter) within their cytoplasm. The major feature of a eukaryotic cell is its membrane-bound nucleus—the active part of the cell that contains genetic information; prokaryotic cells do not have a nuclear membrane (the membrane that surrounds the nucleus of the cell).

The cells also differ in size: Eukaryotic cells are generally much larger and more complex than prokaryotic cells. In fact, most eukaryotic cells are one hundred to 1,000 times the volume of typical prokaryotic cells. The following lists the general differences between these cells (for more about the characteristics of cells, see this chapter):

Comparison of Prokaryotic and Eukaryotic Cells

Characteristic	Prokaryotic Cells	Eukaryotic Cells
Organisms	Eubacteria and Archaebacteria	Protista, Fungi, Plantae, Animalia
Cell size	Usually 1–10 MGRm across	Usually 10–100 MGRm across
Membrane-bound organelles	No	Yes
Ribosomes	Yes, very small	Yes, larger
Type of cell division	Cell fission	Mitosis and meiosis
DNA location	Nucleoid	Nucleus
Membranes	No internal	Many
Cytoskeleton	No	Yes
Metabolism	Anaerobic or aerobic	Aerobic

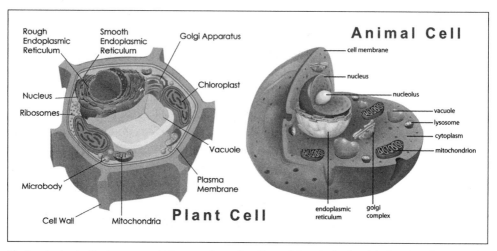

Plant and animal cells share some attribues, such as cell walls, nuclei, and mitochondria, but they also have differences, such as the chloroplasts that only exist in plant cells.

How big are animal cells?

Most animal cells are very small, and most cannot be seen without a microscope; in fact, most cells are smaller than the period at the end of this sentence. Two exceptions are bird and frog egg cells—both are larger cells readily observable with the unaided eye. For example, the chicken egg is actually one single cell—the nucleus, cytoplasm, and cellular membrane, which contains an abundance of yolk and albumin—the nutrients needed for the developing chick (embryo)—making this cell much larger than the normally functioning cells within the chicken.

How do the cells of bacteria, plants, and animals compare to each other?

Numerous differences exist between the cells of bacteria, plants, and animals—all of which make these organisms unique. The following table lists some of those major differences and even a few similarities:

	Bacterium (Prokaryote)	Plant (Eukaryote)	Animal (Eukaryote)
Exterior structures			
Cell wall	Present (protein polysaccharide)	Present (cellulose)	Absent
Plasma membrane	Present	Present	Present
Interior structures			
Endoplasmic reticulum	Absent	Usually present	Usually present
Ribosome	Present	Present	Present
Microtubule	Absent	Present	Present
Centriole	Absent	Absent	Present
Golgi apparatus	Absent	Present	Present
Cytoskeleton	Absent	Present	Present

	Bacterium (Prokaryote)	Plant (Eukaryote)	Animal (Eukaryote)
Other organelles			
Nucleus	Absent	Present	Present
Mitochondrion	Absent	Present	Present
Chloroplast	Absent	Present	Absent
Nucleolus	Absent	Present	Present
Chromosome	A single circle of naked DNA	Multiple; DNA-protein complex	Multiple; DNA-protein complex
Lysosome	Absent	Absent	Present
Vacuole	Absent	Usually a large single vacuole	Absent

Why are lipids important to animal and plant cells?

Lipids are another word for "fats." More formally, they are substances such as fats, oils, or waxes that dissolve in alcohol, but not in water; for example, cholesterol and triglycerides are lipids. Along with proteins and carbohydrates, lipids are the main constituents of plant and animal cells, serving as a source of food for the cell.

Do cells get their energy from only one source?

No, although many cells use glucose as their primary energy source, lipids and proteins can also be broken down to provide energy as well. Lipids are broken down into simpler and more stable substances (catabolized) into monomers (a structural unit of a polymer), glycerol, and fatty acids, which are then metabolized during cell respiration. Proteins are also catabolized to their amino acid building blocks, which are then fed into the process of glycolysis, also known as the Krebs cycle (for more about the Krebs cycle, see the chapter "Basics of Biology").

What is the average life span of a cell?

The life span of a cell varies greatly. In general, most living cells do not live longer than a month. Even cells that live longer, such as liver and brain cells, constantly renew their components so that no part of the cell is really more than a month old. Even plasma cells—that fight off antigens a body encounters immediately—are short-lived, dying off in about ten to seventeen days. Certain cells, such as memory cells, can survive much longer.

Do all cells require oxygen?

No, not all cells require oxygen; in fact, some cells use a metabolic process called fermentation to produce energy, not oxygen. For example, nonoxygen-dependent (anaerobic) organisms include yeast and bacteria that are able to live in an environment with low levels of oxygen.

What are the oldest, living cultured human cells?

The oldest, living cultured human cells are the HeLa cell line, or population of cells. They were the first line of human cells to survive in a test tube (called *in vitro*) and have been a standard for understanding many human biological processes. All HeLa cells are derived from Henrietta Lacks, a thirty-one-year-old woman from Baltimore, Maryland, who died of cervical cancer in 1951. This culture led to advancements in many areas, such as cancer, HIV/AIDS, and even helped in the development of a polio vaccine in the 1950s. In addition, using this cell line, scientists discovered that 80 to 90 percent of cervical cancers (carcinomas) contain human papillomavirus DNA.

In 2013, scientists did the first detailed sequence of the genome of a HeLa line. Along with discovering many widespread abnormalities in the structure and number of chromosomes, they found other unexpected results. For example, they discovered that regions of the chromosomes in each cell were arranged in the wrong order and had either extra or too few copies of genes. This was a sign of a newly discovered phenomenon called chromosome shattering—one that is associated with about 2 to 3 percent of all cancers.

Overall, the majority of organisms are oxygen dependent (aerobic) because of the high ATP—the main source of energy for cells—yield that oxygen provides. Under some circumstances, oxygen-dependent cells can gather energy from fermentation for short periods of time—but this shortcut eventually results in a buildup of lactic acid, a toxic waste product, and produces little ATP. (For more about fermentation, see the chapter "Basics of Biology.")

What is the chemical composition of a typical mammalian cell?

A typical mammalian cell—such as those for humans, apes, deer, and a plethora of other mammals—can be broken down into its chemical composition. On the average, the following chart lists those chemicals:

Molecular component	Percent of total cell weight
Water	70
Proteins	18
Phospholipids and other lipids	5
Miscellaneous small metabolites	3
Polysaccharides	2
Inorganic ions (sodium, potassium, magnesium, calcium, chlorine, etc.)	1
RNA	1.1
DNA	0.25

What are some sizes of cells?

Cells come in a variety of sizes. Bacteria are among the smallest (0.0079–0.012 inches [0.2–0.3 millimeters] in diameter), while cells of plants and animals are generally larger (0.39–1.97 inches [10–50 millimeters] in diameter). Cellular size is determined by the surface-to-volume ratio needed for substances and enzymes to complete their functions within a cell.

What are some of the most specialized cells in mammals?

Depending on the criteria you choose, several types of cells could be nominated as the most specialized cell in the mammalian body. The top candidate for this honor, of course, would be the cells that produce gametes, the mammalian sperm and eggs responsible for the continuation of the specific species of mammal!

The second choice is the red blood cells that carry oxygen and carbon dioxide in the blood—perhaps the most highly specialized cells in mammals. They live for roughly 120 days, but during that time they may travel more than 500 miles (800 kilometers) through various organs and blood vessels. Red blood cells lack a nucleus, so they are unable to reproduce; thus, new red blood cells form in the bone marrow.

What are the different types of blood cells?

The three main types of blood cells, all produced within the bone marrow, include the red blood cells that carry oxygen throughout the body, the white blood cells that help protect the body from infection, and the platelets that help your blood to clot normally. (For more about blood and the human body, see the chapters "Physiology: Animals Inside" and "Biology and You.")

What is keratin?

All of the surface cells of the body (except those in the eyes) contain a molecule called keratin, a fibrous protein that is particularly well suited to withstand abrasions. Keratin includes hair, nails, horns, hooves, wool, feathers, and the epithelial cells in the skin's outermost layer. In general, specific bonds in the keratin make this protein impervious to attack by certain enzymes and completely insoluble in hot or cold water.

Certain cells, like those of the nails and hair, have increased amounts of keratin that provide extra strength, helping them maintain their shape. Whether it is in the claws on the toes of your cat, the horns on the head of your favorite cow, or the five million hairs that cover your body, keratin strengthens and protects organisms from everyday wear and tear.

What are T cells?

Everyone should thank their T (and B) cells—the cells that keep many bacteria and viruses at bay. T cells are produced in the bone marrow; they later move to the thymus (where they mature) and are divided into two types:

Helper T cells—These cells help to regulate the immune defense of the body, activating B cells and Killer T cells. They become activated when a macrophage (white blood cells within tissues) consumes an invading antigen and makes it to the nearest lymph node; from there, the cell passes on the information to the T cell, where the receptor of the Helper T cell becomes activated. The Helper T cell then starts to divide, producing the proteins that activate the B and Killer T cells.

Killer T cells—These cells do as they are labeled—actually attacking the cells of the body infected by the viruses (it is also often heard in reference to attacking cancer cells). It "inspects" each cell it meets with its receptors—and if a cell is infected (it can detect small traces of the antigen), it kills off the cell.

Thus, along with B cells called memory cells (see below), the T cells can keep the intruding antigen from invading the body—and both in tandem help the immune system to activate much faster.

What are B cells?

B cells (also called B lymphocyte cells) look for antigens to match their receptors—and once they find them, a signal inside the B cell is set off. With the help of proteins from the T cells, they becomes active, dividing and producing clones. This produces two cell types:

Plasma—These produce a certain protein we all know called an antibody—and it matches the antigen that the B cell receptor matched. Thus, when the body is attacked by an antigen—for example, a virus—the plasma cells fight off the invaders (this is called the primary immune response), producing and releasing antibodies at an amazing rate.

Memory cells—After the plasma cells do their duty, the memory cells activate. These cells have a longer life span and can survive for several decades—and a small number can remain circulating in the blood for a lifetime. They provide the secondary

What's different about chicken pox?

The virus that causes chicken pox can cause the human body to produce memory cells: After having the chicken pox once, a re-exposure to the virus is fought off by the memory cells, preventing a person from contracting the disease more than once. But in this case, you may not get the "chicken pox," but another problem: After a chicken pox infection, the virus remains dormant in the body's nerve tissues. While the immune system keeps the virus at bay, later in life—usually as an adult—the virus (it is called herpes zoster, from the varicella zoster virus) can reactivate (although the reason is unknown, it is often traced to stress or a suppressed autoimmune system). This causes a different form of the viral infection called shingles—which leaves a person with an often painful skin rash.

immune response to pathogens. Along with the T memory cells, if the body is attacked in the future by the same antigen, they both reactivate rapidly to fight off the antigen. This is what doctors call "immunity" against certain diseases.

STRUCTURES INSIDE CELLS

What are the major components of the eukaryotic cell?

Many major components in eukaryotic cells exist, all of various sizes and functions. The following table lists the major structures in the cell nucleus, the cytoplasmic organelles, and the cytoskeleton (for both animals and plants):

Structure	Description
Cell Nucleus	
Nucleus	Large structure surrounded by double membrane
Nucleolus	Special body within nucleus; consists of RNA and protein
Chromosomes	Composed of a complex of DNA and protein known as chromatin; resemble rodlike structures after cell division
Cytoplasmic Organelles	
Plasma membrane	Membrane boundary of living cell
Endoplasmic	Network of internal membranes extending through reticulum (E.R.) cytoplasm
Smooth endoplasmic	Lacks ribosomes on the outer surface reticulum
Rough endoplasmic	Ribosomes stud outer surface reticulum
Ribosomes	Granules composed of RNA and protein; some attached to E.R. and some are free in cytosol
Golgi apparatus	Stacks of flattened membrane sacs
Lysosomes	Membranous sacs (in animals)
Vacuoles	Membranous sacs (mostly in plants, fungi, and algae)
Microbodies (e.g., peroxisomes)	Membranous sacs containing a variety of enzymes
Mitochondria	Sacs consisting of two membranes; inner membrane is folded to form cristae and encloses matrix
Plastids (e.g., chloroplasts)	Double membrane structure enclosing internal thylakoid membranes; chloroplasts contain chlorophyll in thylakoid membranes
The Cytoskeleton	
Microtubules	Hollow tubes made of subunits of tubulin protein
Microfilaments	Solid, rodlike structures consisting of actin protein
Centrioles	Pair of hollow cylinders located near center of cell; each centriole consists of nine microtubule triplets
Cilia	Relatively short projections extending from surface of cell; covered by plasma membrane; made of two central and nine peripheral microtubules
Flagella	Long projections made of two central and nine peripheral microtubules; extend from surface of cell; covered by plasma membrane

What are organelles?

Organelles—frequently called "little organs"—are found in all eukaryotic cells; they are specialized, membrane-bound, cellular structures that perform a specific function. For example, organelles include the nucleus, mitochondria, chloroplasts, endoplasmic reticulum, and Golgi apparatus.

What is endosymbiosis?

Endosymbiosis is the idea that cell organelles evolved from prokaryotic organisms that originally lived inside larger cells, eventually losing the ability to function as independent organisms. Because the organelles found within eukaryotic cells share a number of similarities with bacteria—such as mitochodria and chloroplasts—many scientists believe that early versions of eukaryotic cells had symbiotic relationships with certain bacteria. In particular, the eukaryote provided protection and resources, while the prokaryote specialized in converting energy (either sunlight or chemical) into forms that could be used by the eukaryotic cell (sugar or ATP, or adenosine triphosphate)—thus the term endosymbiote, meaning "shared internal life." In addition, structural similarities exist between chloroplasts, mitochondria, and free-living bacteria. (For more about ATP and energy, see the chapter "Basics of Biology.")

What are the cytoplasm and cytosol?

The many organelles of the cell are suspended in a watery medium called the cytoplasm—the location of many of the cell's chemical reactions. The cytoplasm includes everything within the plasma membrane except the nucleus. The cell membrane surrounds the cytoplasm at the cell's surface; everything within the nuclear membrane (or envelope) is called the nucleoplasm. (For more about the nucleus, see below.) Cytosol is the liquid medium of the cytoplasm. If you take out all the organelles and the other components (the insoluble nonmembrane structures), you would have the fluid inside the cell, or cytosol.

What are the largest and smallest organelles in a cell?

The largest organelle in a cell is the nucleus; the next largest would be the chloroplast—only found in plant cells—which is substantially larger than a mitochondrion. The smallest organelle in a cell is the ribosome (the site for the manufacture of proteins within the cell).

What organelles of eukaryotic cells contain DNA?

Eukaryotic cells have several DNA-containing organelles. DNA is found in nuclei (where it is called chromosomal DNA) and mitochondria (called mitochondrial DNA). And in plants—or eukaryotic cells capable of photosynthesis—chloroplasts contain DNA (called chloroplast DNA).

What is the endoplasmic reticulum?

The endoplasmic reticulum is a complex of cellular membranes—including channels and sacs—that are found in the cell's cytoplasm. It is the largest and most extensive system of the internal membranes—a collection of membrane tubes and channels in the cytoplasm—whose main function is the transport of proteins throughout both plant and animal cells.

E.R., as it is abbreviated, is divided into the sandpaper-looking rough E.R. that houses ribosomes (that manufacture proteins) and smooth E.R. that helps with such processes as the synthesis of steroid hormones and other lipids, connecting the Golgi apparatus to the rough E.R. and helping to carry out certain detoxification of drugs and other components that may be toxic or harmful to the cells.

What is a vacuole?

A cell's vacuole is a membrane-bound organelle that is primarily used for storage and contains water, enzymes, and other substances. Some vacuoles have a specific function within a cell—for example, a food vacuole stores food molecules—and in freshwater protists, contractile vacuoles maintain the cell's water balance, pumping out excess water from the cell when necessary.

What is the cytoskeleton?

The cytoskeleton is a network of protein filaments (or fibers) that extend throughout the cytoplasm. This network maintains the shape of cells, anchors the organelles to the plasma membrane, and helps with a variety of cell movements. The three types of fibers are as follows:

- *Actin filaments* (or microfilaments) are long fibers composed of two protein chains. They are responsible for cellular movements, such as contraction, crawling, "pinching" during division, and formation of cellular extensions.

- *Microtubules* are hollow tubes composed of a ring of thirteen protein filaments. They are responsible for moving materials within the cell, moving chromosomes during cell division (in particular the spindle fibers that separate the chromosomes during mitosis and meiosis), and providing the internal structure of cilia and flagella (for more about cilia and flagella, see this chapter).

- *Intermediate filaments* are tough, fibrous protein molecules in an overlapping arrangement. They are intermediate in size when compared to actin filaments and microtubules and provide structural stability to cells.

What is the major function of a cell's nucleus?

In 1831, the Scottish botanist Robert Brown (1773–1858) first named and described the small structure in a cell while studying orchids. Brown called this structure the nucleus, from the Latin word *nucula*, meaning "little nut," "kernel," or "core." Today scientists know the nucleus is the information center of the cell; it is also the storehouse of the genetic information (deoxyribonucleic acid, or DNA) that directs all of the activities of a living eukaryotic cell. It is usually the largest organelle in a eukaryotic cell and contains the chromosomes. (For more information about organelles, see this chapter; for more about chromosomes, see the chapter "DNA, RNA, Chromosomes, and Genes.")

Do all cells have a nucleus?

No, not all cells have a nucleus. Prokaryotic cells do not have an organized nucleus, but most eukaryotic cells have a single, organized nucleus. Red blood cells are the only mammalian cells that do not have a nucleus.

What are the main components of the nucleus?

A nucleus contains numerous parts. The following lists some of the main components:

- *Nuclear envelope*—The boundary surrounding the nucleus consists of two membranes—an inner one and an outer one—that form the nuclear envelope (or nuclear membrane). It separates the nucleoplasm from the cytoplasm.

- *Nuclear pores*—Nuclear pores are small openings in the nuclear envelope that permit molecules to move between the nucleus and the cytoplasm in a controlled way—for example, molecules like messenger RNA (mRNA) that are too large to dif-

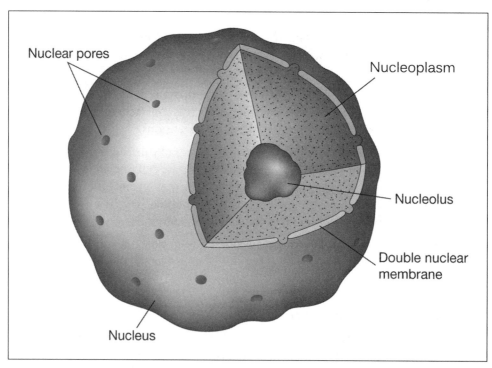

The basic parts of a cell nucleus.

fuse through the nuclear envelope. The pores are not merely perforations, but channels composed of more than one hundred different proteins. Their numbers vary, too; for example, a typical nucleus in a mammalian cell has 3,000 to 4,000 pores, which translates to around ten to twenty pores per square micrometer (mm).

- *Nucleolus*—The nucleolus is a prominent structure within the nucleus. It is a large, spherical structure present in the nucleus of a eukaryotic cell. It is the site where ribosome subunits are assembled and where both ribosomal RNA synthesis and processing occur (for more about RNA, see the chapter "DNA, RNA, and Chromosomes"). It was first accurately described by Rudolph Wagner (1806–1864) in 1835. In addition, the size of the nucleolus varies between cells. For example, cells with a high rate of protein synthesis have a large number of ribosomes. In these active cells, nucleoli tend to be large and can account for 20 to 25 percent of the nuclear volume.

- *Nucleoplasm*—The nucleoplasm is the interior space of the nucleus.

- *Chromosomes*—In addition, the DNA-bearing chromosomes of the cell are found in the nucleus. Thus, the nucleus is the repository for the cell's genetic information and the "control center" for the expression of that information.

How does DNA organize itself within the nucleus, and why is it important?

Within the nucleus, DNA (deoxyribonucleic acid)—the chemical that carries the genetic instructions for "making" living organisms—organizes with proteins into a fibrous material called chromatin. As a cell prepares to divide or reproduce, the thin chromatin fibers condense, becoming thick enough to be seen as separate structures— what we know as chromosomes. (For more about DNA and chromosomes, see the chapter "DNA, RNA, Chromosomes, and Genes.")

What is the endomembrane system of a cell?

The endomembrane system fills the cell and divides it into compartments; in other words, it includes the different membranes suspended in the cytoplasm. The structures

How much DNA is in various cells?

The amount of DNA that a cell must accommodate is significant—even for smaller organisms. For example, enough DNA is in the typical *E. coli* bacterium cell to encircle it more than 400 times. To compare, a typical human cell contains enough DNA to wrap around the cell more than 15,000 times. To look at it another way, if the DNA in a single human cell were stretched out and laid end to end, it would measure approximately 6.5 feet (2 meters). Thus, the average adult human body contains 10 to 20 billion miles (16 to 32 billion kilometers) of DNA distributed among trillions of cells. If the total DNA in all the cells from one human was unraveled, it would stretch to the Sun and back more than 600 times.

are tiny and visible only through electron microscopy. It includes lysosomes, Golgi apparatus, endoplasmic reticulum, and other related structures.

Overall, this system allows macromolecules to diffuse or be transferred from one of the components of the system to another. In particular, it processes proteins specified by the DNA within the nucleus and assembles sterols, sorting and sending them to their final destinations. Thus, in part, it is responsible for protein synthesis, modification, sorting, and transport.

What is the Golgi apparatus?

In 1898, Italian physician Camillo Golgi (1843–1926) first described an irregular network of small fibers, cavities, and granules in nerve cells. It was not until the 1940s—and the invention of the electron microscope—that the existence of the Golgi apparatus was confirmed. Today we know that the Golgi apparatus (frequently called

Italian physician Camillo Golgi first proposed the existence of what is now called the Golgi apparatus almost fifty years before electron microsopes confirmed it.

the Golgi body) is a collection of flattened stacks of membranes. It serves as the packaging center for cell products, collecting materials at one place in the cell and packaging them into vesicles, or small sacs, for use elsewhere in the cell or for transportation out of the cell. It's interesting to note that the number of Golgi bodies varies among cells. For example, protists contain one or a small number of Golgi bodies. Animal cells may contain twenty or more Golgi bodies, while plant cells may contain several hundred Golgi bodies.

What is the function of lysosomes?

Lysosomes are single, membrane-bound sacs that contain digestive enzymes. The digestive enzymes break down all the major classes of macromolecules, including proteins, carbohydrates, fats, and nucleic acids. Throughout a cell's lifetime, the lysosomal enzymes digest old organelles to make room for newly formed organelles. The lysosomes allow cells to continually renew themselves and prevent the accumulation of cellular toxins.

Who discovered lysosomes?

Lysosomes are a relatively modern discovery in cell biology; they were observed by British cytologist and biochemist Christian de Duve (1917–2013) in the early 1950s. In 1955, after six years of experiments, de Duve was convinced that he had found an organelle that had not been previously described and was involved in intracellular diges-tion (lysis)—he named the organelle a lysosome. This organelle was the first to be de-scribed entirely on biochemical criteria; the results were later verified using electron microscopy. In 1974, de Duve, Belgian biologist Albert Claude (1898–1983), and Ro-manian cell biologist George Palade (1912–2008) shared the Nobel Prize in Physiology or Medicine for their work detailing the functions of the lysosome.

What is a peroxisome, and what is its function in a cell?

Peroxisomes, discovered by British cytologist and biochemist Christian de Duve, are surrounded by a single membrane and are the most common type of microbody (an or-ganelle containing enzymes) in cells. They are especially prominent in algae, the pho-tosynthetic cells of plants, and both mammalian kidney and liver cells. Peroxisomes contain detoxifying enzymes and produce catalase, which helps to break down hydrogen peroxide into hydrogen and water.

What are ribosomes?

Ribosomes, one of the most complex aspects of the molecular machine, are the site of pro-tein synthesis in a cell. They consist of a large and small subunit composed of ribosomal RNA and protein. Ribosomes differ from most other organelles because they are not bound by a membrane; however, compared with membrane-bound organelles, ribosomes are tiny structures. The number of ribosomes differs depending on the type of cells. For example, a bacterial cell will typically have a few thousand ribosomes, while a human liver cell contains several million ribosomes. Actively growing mammalian cells contain five to ten million ribosomes—all that have to be synthesized each time the cell divides.

What are the differences between prokaryotic and eukaryotic ribosomes?

Prokaryotic and eukaryotic ribosomes resemble each other structurally, but they are not identical. They are found in the cytoplasm of both prokaryotic and eukaryotic cells as well as in the matrix of mitochondria and the stroma of chloroplasts (plant cells). In eukaryotic cytoplasm, ribosomes are found in the cystol (liquid medium of the cyto-plasm) and are bound to the endoplasmic reticulum as well as the outer membrane of the nuclear envelope. Prokaryotic ribosomes are smaller in size, contain fewer proteins, have smaller RNA molecules, and are more sensitive to different inhibitors of protein synthesis than their eukaryotic counterparts.

What are mitochondria?

In 1857, Swiss histologist and embryologist Rudolf Albert von Kölliker (1817–1905) first described "sarcosomes" (now called mitochondria) in muscle cells. The word mitochon-drion (derived from Greek for "threadlike granule") was first used in 1898; by 1948, the first functionally active mitochondria were isolated. Today we know that a mitochon-drion (singular) is a self-replicating, double-membrane body found in the cytoplasm of

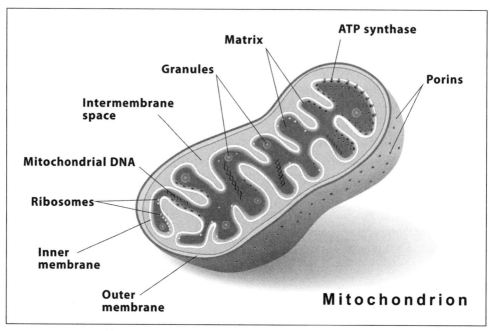

Mitochondria are essential parts of eukaryotic cells. They produce lipids and ATP and are necessary for protein synthesis.

all eukaryotic cells. The outer membrane of a mitochondrion is smooth, while the inner membrane is folded into numerous layers that are called cristae. Mitochondria are responsible for much of the metabolism necessary in protein synthesis and for the production of both ATP and lipids. (For more about ATP, see the chapter "Basics of Biology.")

How many mitochondria are in a cell?

The number of mitochondria varies according to the type of cell. The number ranges between one and 10,000, but averages about 200. For example, each cell in the human liver has over 1,000 mitochondria; cells with high energy requirements, such as muscle cells, may have many more mitochondria.

CELL WALLS AND MEMBRANES

What groups of organisms have a cell wall?

A cell wall is present in organisms in the kingdoms Archaebacteria, Eubacteria, Protista, Fungi, and Plantae (bacteria, protists, fungi, and plants). Animals are the only organisms that do not have a cell wall. (For more about the classification of organisms, see the chapter "Basics of Biology.")

In fact, the cell wall is one of the features of all the other organisms' cells that distinguish them from animal cells. For example, plant walls protect the plant cell; they also help maintain its shape and consist mainly of cellulose in a matrix of protein and sugar polymers. The cell walls in prokaryotes (for example, bacteria) also define the cell's shape and give rigidity to the cell. Though unlike plant cell walls, bacterial cell walls consist mainly of peptidoglycans—(polysaccharide chains [amino sugars] cross-linked by small peptides)—and not cellulose.

What's the difference between a primary and secondary cell wall?

A primary cell wall is laid down during cell division and is relatively thin and flexible in order to accommodate cell enlargement and elongation. As the cell matures and stops growing, the wall strengthens. A secondary cell wall is present between the plasma membrane and primary cell wall in some cells and is often deposited in several laminated layers. This type of cell wall is strong and durable and provides both cell protection and support; for example, wood consists mainly of secondary cell walls.

What scientists contributed to the study of membranes?

The study of membranes started in the late 1800s. The following lists the major developments in membrane studies:

- As early as the 1890s, British cytologist Charles Overton (1865–1933) was aware that cells seemed to be enveloped by a porous layer that allowed different substances to enter and leave cells at different rates. He recognized that lipid-soluble substances penetrated readily into cells, whereas water-soluble substances did not. He concluded that lipids were present on the cell surface as some sort of a "coat."

- American chemist and physicist Irving Langmuir (1881–1957) proposed that cells were covered by a single layer of lipids. His work became the basis for further investigation into the membrane structure. Langmuir was awarded the Nobel Prize in Chemistry in 1932.

- In 1925, Evert Gorter (1881–1954) and F. Grendel, two Dutch physiologists, suggested that a bilayer (two-layer) structure of lipids is on the cell surface. Their work was significant because it was the first attempt to understand membranes at the molecular level.

- Following the earlier work of Gorter and Grendel on cell membranes, English physiologist Hugh Davson (1909–1996) and English biologist James F. Danielli (1911–1984) proposed a sandwich model for the structure of cell membranes in 1935. This model suggested that a phospholipid bilayer was sandwiched between two layers of globular proteins.

What is the current model of the plasma membrane?

The plasma membrane is a thin membrane that surrounds and defines the boundaries of all living cells. It is only about 8 nanometers (nm) thick; in fact, it would take over 8,000 plasma membranes to equal the thickness of an average piece of paper. Since cell membranes are so fragile when observed within the living organism, scientists could only propose theoretical models for the membrane's structure. In fact, the majority of past techniques to analyze plasma membranes did not permit direct observation of their function.

But advances in the study of the membranes have been made. The current model of the plasma membrane, frequently referred to as the fluid mosaic model, is based on studies by American cell biologist Seymour J. Singer (1924–) and American biochemist and cell biologist Garth L. Nicholson (1943–). In 1972, the scientists' research revealed that the plasma membrane is a combination of proteins bobbing in a fluid bilayer of phospholipids with various proteins attached to or embedded in it. This model has been tested repeatedly and has been shown to accurately predict the properties of many kinds of cellular membranes; this structure has also been confirmed using a technique known as freeze-fracture electron microscopy.

What are phospholipids and proteins in terms of the cell membrane?

A cell membrane is mainly composed of phospholipids and proteins, which are two types of bioorganic molecules. Within the membrane, phospholipids are able to move laterally. Depending on temperature and fatty acid composition, phospholipids generally move faster than proteins.

The proteins slowly drift and bob in the fluid body of phospholipids, much like icebergs floating in an ocean. Proteins are able to change shape (also known as conformation); peripheral proteins are not embedded in the lipid bilayer, but are appendages loosely bound to the membrane surface. For example, carrier proteins are able to bind to specific molecules such as glucose in order to provide transportation for the molecule; once glucose is attached to the carrier protein, the protein changes shape and ferries the glucose inside of the cell.

What are cell junctions?

Cell junctions are the specialized connections between the plasma membranes of adjoining cells. The three general types of cell junctions are as follows: tight junctions that bind cells together, forming a barrier that is leakproof; for example, they form the lining of the digestive tract, preventing the contents of the intestine from entering the body; anchoring (or adhering) junctions link cells together, allowing them to function as a unit and form tissue, such as in the heart muscle; and communicating (or gap) junctions that allow rapid chemical and electrical communication between cells.

Are all cell membranes alike?

Although all cell membranes have the same general structure, the membrane composition is different for each species. Depending on their function, membranes vary in the amount of protein or the type of membrane receptors they contain. For example, plants that survive frigid temperatures, such as winter wheat, are able to increase their concentration of unsaturated phospholipids during the winter to prevent the membrane from freezing. Another example is muscle cells that have plasma membrane receptors for the neurotransmitter called acetylcholine—which "tells" the muscle when to contract.

PLANT CELL BASICS

What is photosynthesis?

Photosynthesis (from the Greek word *photo*, meaning "light," and synthesis, from the Greek word *syntithenai*, meaning "to put together") is the process by which plants use energy derived from light in order to make food molecules from carbon dioxide and water. It is basically a two-step process: Light energy derived from sunlight is converted to chemical energy, with oxygen (O_2) produced as a waste product of this process. The second step is carbon-fixation reactions (or the conversion of carbon dioxide [CO_2] into organic compounds) known as the Calvin cycle—a series of reactions that assemble sugar molecules from carbon dioxide (CO_2) and the energy-containing products of the light reactions. (For more about the Calvin cycle, see the chapter "Basics of Biology.")

Ultimately, photosynthesis is the process that provides food for the entire world. It is estimated that each year more than 250 billion metric tons of sugar are created through photosynthesis. Photosynthesis is a source of food not only for plants, but also all organisms that are not capable of internally producing their own food—including humans—which is why it's wise to eat your vegetables.

What scientists made significant discoveries in plant photosynthesis?

The discovery of photosynthesis—how plants make food—has a long and involved history. The following lists some of the highlights:

- Ancient Greeks and Romans believed that plants derived their food from the soil.

What light is important for photosynthesis?

Virtually all life depends on the availability of light, including the light that powers photosynthesis (the process of synthesizing energy). Light travels in waves, and its energy is contained in packets called photons. The energy of a photon is inversely proportional to the wavelength of the light—the longer the wavelength, the less energy per photon. Sunlight consists of a spectrum of colors present in light, with each color represented in a specific wavelength along the electromagnetic spectrum. In general, the most effective wavelengths of light for photosynthesis are blue (at 430 nm, or nanometers, on the electromagnetic spectrum) and red (670 nm). Curiously, green plants have the hardest time with the photosynthetic process in green light.

- The earliest experiment to test the hypothesis that plants get their food from the soil was performed by the Belgian scientist Jan Baptista van Helmont (1577–1644). He grew a willow tree in a container of soil and fed it only water; at the end of five years, the weight of the willow tree had increased by 164 pounds (74.4 kilograms), while the weight of the soil had decreased by 2 ounces (57 grams). Van Helmont concluded that the plant had received all its nourishment from the water and none from the soil.

- Chloroplasts were described and studied by Dutch scientist Antonie Phillips van Leeuwenhoek (Anton van Leeuwenhoek; 1632–1723) and English plant anatomist and physiologist Nehemiah Grew (1641–1712)—both of whom described these (and other) organelles in the latter seventeenth century—using a primitive microscope.

- English theologian and scientist Joseph Priestley (1733–1804) demonstrated that air was "restored" by plants. In 1771, Priestley conducted an experiment in which he placed a lighted candle in a glass container and allowed it to burn until extinguished by lack of oxygen. He then put a plant into the same chamber and allowed it to grow for a month. Repeating the candle experiment a month later, he found that the candle would now burn. Priestley's experiments showed that plants release oxygen (O_2) and take in carbon dioxide (CO_2) produced by combustion (the operation of burning).

- In another "oxygen-depriving" experiment, in the mid-1700s, French chemist and physicist (and often called the founder of modern chemistry) Antoine Laurent Lavoisier (1743–1794) determined that oxygen plays a role in the respiration of both plants and animals. He showed that burning fires remove oxygen from the air—and life cannot exist in such an environment. (He should have stayed with chemistry, as his eventual meddling in the taxation affairs of the French government before the Revolution would lead him to the guillotine.)

- Dutch physician Jan Ingenhousz (1730–1799) confirmed Priestley's ideas, emphasizing that air is "restored" only by green plants in the presence of sunlight.

- In 1905, evidence of the two-stage process of photosynthesis was first presented by British plant physiologist Frederick Frost Blackman (1866–1947). Blackman had identified that both a light-dependent stage and a light-independent stage occur during photosynthesis.

- In 1930, American-Dutch botanist and microbiologist C. B. van Niel (1897–1986) became the first person to propose that water, rather than carbon dioxide, was the source of the oxygen that resulted from photosynthesis.

- In 1937, British plant biochemist Robert Hill (1899–1991) discovered that chloroplasts are capable of producing oxygen in the absence of carbon dioxide when the chloroplasts are illuminated (for more about chloroplasts, see below).

Do plant cells really produce oxygen and use carbon dioxide?

Yes, plant cells produce oxygen through the process of photosynthesis. Splitting water molecules to harvest their electrons causes the release of oxygen. By submerging a small piece of an aquatic plant in a beaker containing water, one can actually see the oxygen bubbles being produced.

Plants also use carbon dioxide in order to live. A plant's cells reduce carbon dioxide to sugar by using the electrons that are produced when chlorophyll absorbs light. Simply put, six carbon dioxide (CO_2) molecules, along with six water (H_2O) molecules, can be converted into a simple sugar ($C_6H_{12}O_6$).

How do chlorophyll *a* and chlorophyll *b* play a part in photosynthesis?

During photosynthesis, light is absorbed by pigments present in organisms. Chlorophyll *a* is the primary pigment required for photosynthesis and occurs in all photosynthetic organisms except photosynthetic bacteria. Accessory pigments such as carotenoids and chlorophyll *b* absorb light that chlorophyll *a* cannot absorb; these pigments extend the range of visible light useful for photosynthesis.

What is the function of chloroplasts?

Chloroplasts are cell structures larger than any other organelle except the nucleus; in fact, their ancestors billions of years ago may have been free-living bacteria. Today, chloroplasts are the structural and functional units where photosynthesis takes place— the process whereby green plants use light energy for the synthesis of organic molecules from carbon dioxide and water, with oxygen released as a by-product. They contain the green pigment chlorophyll, which traps light energy for photosynthesis. Overall, unicellular algae may only have one large chloroplast, whereas a plant leaf cell may have between twenty and one hundred chloroplasts.

What are the main parts of chloroplasts?

Chloroplasts have outer and inner membranes, which are in close association with each other. They also have a closed compartment of stacked membranes—called grana—that lie inside the inner membrane. A chloroplast may contain one hundred or more grana, and each granum (plural for grana) may contain a few to several dozen disk-shaped structures called thylakoids; they, in turn, contain chlorophyll on their surface. The fluid that surrounds the thylakoid is called the stroma.

Do all plant cells contain chloroplasts?

No, not all plant cells contain chloroplasts; for example, onion and garlic plant cells do not have chloroplasts—and neither do the underground roots of plants. Another example is the meristem—the rapidly dividing undifferentiated plant tissue cells found in places where the plant can grow. These cells do not contain chloroplasts, but have smaller organelles called proplastids (from which a plastid develops; see below).

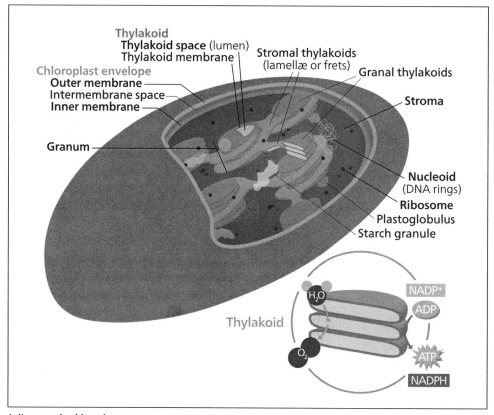

A diagram of a chloroplast.

What are plastids?

Plastids are only found in plant and algae cells and are involved in cellular storage. Depending on their location in a plant and how much light they receive, proplastids develop into one of several kinds of plastids with different functions: Chloroplasts as the site of photosynthesis; leucoplasts store starches (for example, in tubers like potatoes); amyloplasts store starches; proteinoplasts store proteins; and elaioplasts store lipids. In addition, some proplastids develop into chromoplasts that store the pigments that give some flowers, fruits, and other plants—such as carrots and tomatoes—their red, orange, and/or yellow colors. Interestingly enough, it does not include the color green. (For more about plants, see the chapter "Plant Diversity and Structure.")

What cell structures are unique to plant cells?

The chloroplast, central vacuole, tonoplast, cell wall, and plasmodesmata are only found in plant cells. The central vacuole can encompass 80 percent or more of the cell. It is usually the largest compartment in a mature plant cell. It is surrounded by the tonoplast, with functions that include storage, waste disposal, protection, and growth. Plasmodesmata are present in plant cells; they are channels or canals that occur in the cell wall, connecting the cytoplasm of adjacent cells—thus allowing molecules direct communication through those adjacent cells.

CELL DIVISION

What is the cell cycle, and how is it controlled?

The life cycle of a single eukaryotic cell is known as the cell cycle. The cycle has two major phases: interphase and mitosis. When a cell is not dividing, it is in interphase; for example, a mature neuron conducting an impulse in the brain is in interphase. Interphase is broken down into the G_1 and G_2 phases—periods of growth during which a cell increases in size, complexity, and protein content. The G_1 phase prepares the cell for DNA synthesis (known as the S phase); G_2 prepares the cell for both mitosis and the synthesis of proteins. Although many cells eventually divide, it's interesting to note that some cells remain in interphase almost indefinitely.

What type of signals control cell reproduction?

Cell reproduction is controlled by both external and internal signals. External signals are often environmental and include such things as the availability of nutrients and space for growth. An example of an internal signal is the rise and fall of protein levels at specific points in the cell cycle, which is maintained by checkpoints and feedback controls.

What are the stages of mitosis?

Mitosis involves the replication of DNA and its separation into two new daughter cells. While only four phases of mitosis are often listed, the entire process is actually comprised of six phases:

- *Interphase*—Involves extensive preparation for the division process.
- *Prophase*—The condensation of chromosomes; the nuclear membrane disappears; formation of the spindle apparatus; chromosomes attach to spindle fibers.
- *Metaphase*—Chromosomes, attached by spindle fibers, align along the midline of a cell.

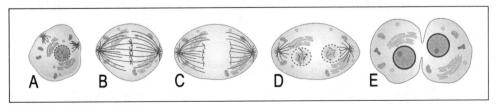

After the Interphase stage (not shown here) in which the cell prepares for division, the stages of mitosis include A) Prophase, B) Metaphase, C) Anaphase, D) Telophase, and E) Cytokinesis.

- *Anaphase*—The centromere splits, and chromatids move apart.
- *Telophase*—The nuclear membrane reforms around newly divided chromosomes.
- *Cytokinesis*—The division of cytoplasm, cell membranes, and organelles occur. In plants, a new cell wall forms.

Do all cells divide at the same rate?

No, all cells do not divide at the same rate. Cells that require frequent replenishing, such as skin or intestinal cells, may only take roughly twelve hours to complete a cell cycle. Other cells, such as liver cells, remain in a resting state (interphase) for up to a year before undergoing division. Cells such as human brain cells exist for a lifetime in a nondividing state.

How are organelles partitioned during mitosis?

Following the telophase of mitosis, cytokinesis—the physical separation of the daughter cells—occurs. Although the exact mechanisms are not clear, it appears that larger organelles, such as the endoplasmic reticulum and Golgi apparatus, undergo fragmentation into small vesicles during mitosis and later reassemble in daughter cells.

What are the major differences between cell division in plants and cell division in animals?

The major differences in plant and animal cell division are in the assembly of the spindle apparatus. The site of spindle apparatus assembly is the centrosome. In animal cells, a pair of centrioles is at the center of the centrosome. In contrast, most plants lack centrioles, but they do have a centrosome. In animal cells, a cleavage furrow forms during cytokinesis, which deepens and then pinches the parent cells in two. Plant cells, which have cell walls, do not have a cleavage furrow. Instead, a cell plate is produced in the middle of the parent cell, which grows toward the perimeter of the cell until it reaches the plasma membrane, dividing the cell in two. A new cell wall then forms from the cell plate.

What is meiosis?

Meiosis is often referred to as reduction division, meaning that the number of chromosomes present is reduced from 2N (diploid) to N (haploid). The meiotic process consists of two separate cell divisions and occurs in the gonads (female ovaries and male testes). It is important to sexual reproduction because of the genetic variation that occurs as a result of this process.

How are cell structures divided during meiosis?

Meiosis is just part of a larger process that occurs in either the ovaries or testes. During sperm formation (spermatogenesis), the cells that eventually become mature sperm go through two successive meiotic divisions. This results in four "haploid spermatids"

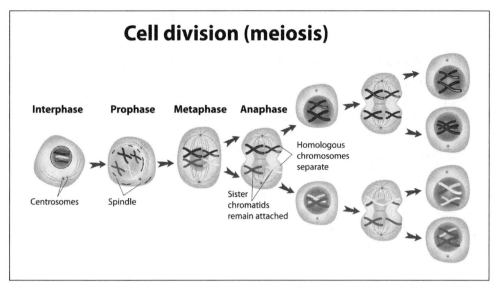

Cell division (meiosis)

Interphase **Prophase** **Metaphase** **Anaphase**

Homologous
chromosomes
separate

Centrosomes Spindle

Sister
chromatids
remain attached

In the cell division process of meiosis the resulting cells have half the number of chromosomes as the original cell, which is important in the production of sperm and egg cells.

that will develop into mature sperm cells. The sperm are then reconfigured into a cell that is specialized for one thing: fertilizing an egg. Thus, the mature sperm is basically a nucleus with one set of chromosomes, mitochondria, and flagella (for propulsion).

In humans, the cells that develop into mature eggs are present in the ovary before the oogenesis (egg formation) ever occurs. Immature eggs (oogonia) remain in a stage referred to as "meiosis I" until they mature during puberty. During the stage of "meiosis II," oocytes are ready to be released, but will not fully complete the meiotic process until after fertilization takes place. During the progression from one diploid oocyte to four haploid cells, cytoplasm is shunted unequally to only one cell. The end result is one large mature ovum and two or three very small haploid cells called polar bodies.

What happens in a cell if mitochondria are defective?

Since mitochondria are the energy producers of a cell, if a cell's mitochondria are defective, any cells that have a high metabolic rate will also be affected. Metabolic poisons can affect specific parts of mitochondrial function. These include such compounds as cyanide and dinitrophenol (an ingredient of early diet drugs). If mitochondria are damaged, they begin to leak free radicals (molecules that have a single unpaired electron in their outer shell) that can alter the DNA.

In addition, because mitochondria have their own DNA, it is also possible for them to be altered by genetic mutations. In fact, mutations in mitochondrial genes are thought to play a role in degenerative neurological diseases such as Parkinson's and Alzheimer's.   **59**

CELL RESPONSES

How do cells communicate with each other?

Cells communicate with each other via small, signaling molecules that are produced by specific cells and received by target cells. This communication system operates on both a local and long-distance level. The signaling molecules can be proteins, fatty acid derivatives, or gases. For example, nitric oxide gas is part of a locally based signaling system and is able to signal to lower a human's blood pressure.

Hormones are long-distance signaling molecules that must be transported via the circulatory system from their production site to their target cells. For example, plant cells, because of their rigid cell walls, have cytoplasmic bridges called plasmodesmata that allow cell-to-cell communication, whereas animals use gap junctions to transfer material between adjacent cells.

How do cells respond to cellular signals?

In order to respond to a signal, a cell needs a receptor molecule that recognizes the signal. A cell's response to a specific signal varies according to the signal. Some signals are local signals (for example, growth factors), while others act as distance signals (for example, hormones). The two basic types of hormone are those that bind to receptors on the cell surface and those whose receptors are found within cytoplasm. Both types can cause the cellular machinery to change its activities.

How do cells respond to steroid hormones?

Progesterone, estrogen, testosterone, and glucocorticoids are naturally occurring steroid hormones that "signal" molecules. After entering a target cell, the steroid hormone binds to a receptor protein and starts a cascade of events that ultimately activates ("turns on") or inhibits ("turns off") a specific set of genes. (For more about anabolic steroids and humans, see the chapter "Physiology: Animal Function and Reproduction.")

How do cells respond to insulin?

Protein-based hormones such as insulin bind to cell-surface receptors. While they do not enter the cell, they cause changes in the cell's metabolism. Specific cells in the pancreas secrete insulin, a hormone that regulates the concentration of glucose in the blood.

Skeletal muscles and the liver are targets for insulin. Insulin deficiency is responsible for Type I diabetes. In contrast, Type II diabetes, also known as adult-onset diabetes mellitus, is not the result of insulin deficiency but is rather due to insulin resistance. Cells with insulin resistance do not respond to increasing insulin levels by transporting glucose into cells. (For more about human diseases and biology, see the chapter "Biology and You.")

What is chemotaxis?

Chemotaxis is the process of how cells follow chemical gradients. For example, if you get a small scratch on your skin, the healing is helped by a type of white blood cell called neutrophils that removes bacteria and other foreign materials from your wound. The neutrophils' ability to leave the bloodstream and navigate to the tissue in the injury area seems uncanny, but is all because of the process of chemotaxis. This process is not only used by humans—or any other eukaryotic organisms—but is also used by prokaryotic bacterial cells that use the gradients to locate food. The tiny creatures use a kind of sampling mechanism to determine the direction of the chemical gradient, then move in that same direction. Eukaryotic cells do not move to find a chemical gradient, but rely on their size to measure the spatial differences across the cell body.

Can a cell survive in complete darkness?

Yes, cells can survive in complete darkness. A theory on the origin of life argues that living systems arose in small compartments of total darkness located within iron sulfide rocks—originally formed by hot springs on the ocean floor. Today, plant root cells live in total darkness and carry out all normal plant cell activities, with the exception of photosynthesis. And, if you think about it, cells located in the very middle of the human body also live in a dark environment!

How does a cell respond to injury?

In most tissues, injured cells die and are subsequently replaced. But sometimes, dead cells cannot be replaced. For example, if nervous tissue cells die, they are not replaced. Instead, a nerve growth factor is produced by adjoining neurons, often inducing the sprouting and growth of previously dormant neurons.

How do substances move in and out of cells?

Cells constantly transport substances across their cell membranes. Endocytosis (from the Greek *endo*, meaning "in," and *cytosis*, meaning "cell") is the process by which cells bring molecules into their structure. Exocytosis (from the Greek *exo*, meaning "out," and *cytosis*) is the process by which cells transport materials out of their structure across their own cell membrane.

How do cells "drink" and "eat"?

Cells actually do seem to "drink" and "eat" to bring molecules inside their structure (endocytosis)—but in a much different way than humans consume food and drink. The two types of endocytosis are pinocytosis (from the Greek *pino*, meaning "to drink," and *cytosis*) and phagocytosis (from the Greek *phago*, meaning "to eat," and *cytosis*). During pinocytosis, the cell membrane folds inward, forming a small pocket (vesicle) around fluid that is directly outside the cell membrane; the fluids consumed by cells may contain small molecules, such as lipids. For example, the endothelial cells that line the body's small blood vessels (capillaries) are constantly undergoing the process of pinocytosis, "drinking" from the blood within an organism's capillaries.

Phagocytosis occurs once a particle (or microorganism) is ingested; it is then wrapped within a package called a vesicle. The vesicle then fuses with a lysosome and from there, the digestive enzymes of the lysosome digest the contents of the vesicles. This process is especially important to two types of cells: amoebas (unicellular protozoa) and mammalian white blood cells (macrophages). For the amoebas, phagocytosis provides food for the protozoan; in mammals, the process plays a critical role in the immune—the mammal's primary defense—systems by getting rid of microorganisms or damaged cells.

How do cells secrete substances?

During exocytosis, material from a cell is released. First a cell gathers and packages a certain particle (this package is called a vesicle). Because the vesicle is composed of the same material that makes up the cell membrane, when it reaches the membrane, the two structures merge together much like air bubbles do in liquid. The contents of the vesicle are then expelled from the cell. For example, cells that manufacture specific proteins, such as the pancreatic cells that manufacture insulin, use the process of exocytosis to secrete insulin into the blood.

Do all cells move?

No, not all cells move; those that do move (or are motile) have distinct body features and methods for moving about. The two main features that help various types of cells move are cilia and flagella. Both have the same internal structure, but differ in their length, the number occurring per cell, and mode of beating. Cilia are about 2 to 10 microns (MGR) long, while flagella are much longer, ranging from 0.0394 microns to just under an inch (1 millimeter to several millimeters) long, although they are most commonly 10 to 200 microns (MGR) long. Both cilia and flagella are used by cells to move through watery en-

---

Why is phagocytosis so important to the human body?

Not only does phagocytosis allow us to remove minute, potentially deadly body invaders, but it also is important in the maintenance of healthy tissues. Without this mechanism, harmful materials would accumulate and interfere with the body's ability to function. A good example is the macrophages (white blood cells in the tissues) of the human spleen and liver, which dispose of more than 10,000,000,000 old (aged) blood cells a day!

vironments or to move materials across cell surfaces. Cilia move back and forth, while flagella undulate in a whiplike motion, moving in the same direction as their axis.

A cell having one or a small number of appendages can be identified as having flagella (the singular is flagellum) if they are relatively long in proportion to the size of the cell. If the cell has many short appendages, they are called cilia (the singular is cilium). Cilia can be either motile, having a rhythmic waving or beating motion, or nonmotile (primary), and instead of moving, transmit or receive signals from other cells or fluids nearby.

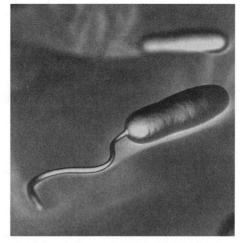

A flagella, such as the one on this sperm cell, looks like a tail and moves back and forth to propel some tiny organisms.

What is another way cells can move?

Pseudopodia, often referred to as "fake feet," are cellular extensions caused by a stretching of the cell membrane. For example, they are used by amoebas (protists) for movement and catching prey; macrophages—the white blood cells that play an important role in a human's immune system—use pseudopodia to attack and devour invading microbes.

How do some organisms use cilia and flagella?

Many examples of organisms that use cilia and flagella can be given. For example, in the human body, motile cilia in the lungs keep dust and dirt out of the bronchi (your breathing tubes) by moving a layer of sticky mucous along that keeps the area clean. Humans also have nonmotile cilia in their kidneys that bend with the urine flow and send signals to alert the cells that urine is flowing. Human (or any mammalian) sperm cells—the cells with the "long, undulating tails" that allow sperm to swim through the oviduct fluids of the female reproductive tract to reach the egg cell—are a classic example of flagella.

In other organisms, flagella and cilia are used for a multitude of processes and movements and for a variety of reasons. For example, several protists such as paramecium also have cilia and produce movement by shifting the cilia through liquid similar to how an oar propels a boat through water. Clams and mussels use cilia for obtaining food as ocean and fresh waters flow by. The ulcer-causing bacteria *Helicobacter pylori* use multiple flagella to move through mucous in order to reach the stomach's lining (epithelium).

Why do cells die?

Cells die for a variety of reasons, many of which are not deliberate. For example, cells can starve to death, asphyxiate, or die from trauma. Cells that sustain some sort of damage, such as DNA alteration or viral infection, frequently undergo a programmed cell death (see below). This process eliminates cells with a potentially lethal mutation or limits the spread of the virus. Programmed cell death can also be a normal part of embryonic development; for example, frogs undergo cell death that results in the elimination of tissues—allowing a tadpole to morph into an adult frog.

What is programmed cell death (apoptosis)?

Apoptosis, or programmed cell death, is a process by which cells deliberately destroy themselves. The process follows a natural sequence of events controlled by the genes within a nucleus: First, the chromosomal DNA breaks into fragments, followed by the breakdown of the nucleus. The cell then shrinks and breaks up into pieces (called vesicles) that are absorbed (phagocytosis) by macrophages and neighboring cells.

While programmed cell death may seem counterproductive at first glance, it plays an important part in maintaining the life and health of all living organisms. During human embryonic development, apoptosis removes the unnecessary webbing between the fingers and toes of the fetus; it is also vital to the development and organization of both the immune and nervous systems.

Can cells ever change jobs?

The more specialized of a function a cell performs, the less likely it is for the cell to change jobs within an organism. But some cells have unspecialized functions and are able to adapt to the changing needs of the body. For example, in mammals such cells include bone marrow cells, which are responsible for producing different types of cells in the blood—in all, red blood cells and five types of white blood cells (for more about blood cells, see the chapter "Anatomy: Animals Inside"). Other organisms, including slime molds of the kingdom Protista, also have cells that are capable of drastically changing their function, allowing them to change from single-celled amoebas to multicellular, reproductive spore producers.

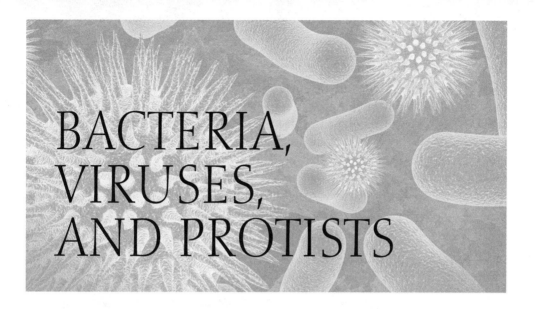

BACTERIA, VIRUSES, AND PROTISTS

HISTORICAL INTEREST IN BACTERIA

When were bacteria discovered?

Dutch scientist Antonie Phillips van Leeuwenhoek (Anton van Leeuwenhoek; 1632–1723), also a fabric merchant and civil servant, discovered bacteria and other microorganisms in 1674 when he looked at a drop of pond water through a glass lens. Early, single-lens instruments produced magnifications of fifty to 300 times real size (approximately one-third of the magnification produced by modern light microscopes). These primitive microscopes provided a perspective into the previously unknown world of small organisms, which van Leeuwenhoek called "animalcules" in a letter he wrote to the Royal Society of London. Because of these early investigations, van Leeuwenhoek is often considered to be the "father of microbiology."

How did the discovery of bacteria change the theory of spontaneous generation?

The theory of spontaneous generation proposes that life can arise spontaneously from nonliving matter. One of the first scientists to challenge this theory was the Italian physician Francesco Redi (1626–1698). In 1668, Redi performed an experiment to show that meat placed in covered containers (either glass-covered or gauze-covered) remained free of maggots, while meat left in an uncovered container eventually became infested with maggots from flies laying their eggs on the meat.

The controversy over spontaneous generation was finally solved in 1861 by French chemist Louis Pasteur (1822–1895). He showed that the microorganisms found in spoiled food were similar to those found in the air—and thus concluded that the microorganisms on spoiled food were from the air and did not spontaneously arise.

How has the classification of bacteria evolved?

Early systems for classifying bacteria were based on structural and morphological elements displayed by the organisms; for example, bacterial shape, size, and the presence or absence of several elements within the bacteria cell. They were later classified by their biochemical and physiological traits, such as the best temperatures and pH ranges for growth, respiration and fermentation, and the types of carbohydrates used as an energy source. Still later classifications were based on stains (using a dye to see an organism under the microscope), such as the Gram stain (see below). More recently, with better technology, genetic and molecular characteristics are used to understand the true evolution of bacteria and their connections to other organisms.

Who were the founders of modern bacteriology?

French chemist Louis Pasteur (1822–1895) and German bacteriologist Robert Koch (1843–1910) are considered the founders of bacteriology. In 1864, Pasteur devised a way to slowly heat foods and beverages to a high enough temperature to kill most of the microorganisms responsible for spoilage and disease. This method would not ruin or curdle the food—and is a process we now call pasteurization.

In 1882—by demonstrating that tuberculosis was an infectious disease caused by a specific bacterial species of *Bacillus*—Robert Koch set the groundwork for public-health measures that would reduce the occurrence of many diseases. His laboratory procedures, methodologies for isolating microorganisms, and four postulates (see below) for determining agents of disease gave medical investigators valuable insights into the control of bacterial infections.

What are Koch's postulates?

German bacteriologist Robert Koch (1843–1910) was the first to identify that various microorganisms are the cause of disease. His four basic criteria of bacteriology, known as Koch's postulates, are still considered fundamental principles of bacteriology.

The postulates are the four basic criteria an organism must meet in order to be identified as pathogenic (capable of causing disease). The characteristics are as follows: 1) the organism must be found in tissues of animals that have been infected with the disease, rather than in disease-free animals; 2) the organism must be isolated from the diseased animal and grown

French chemist and microbiologist Louis Pasteur was the first to realize that fermentation resulted from a biological process.

in a pure culture or *in vitro*; 3) the cultured organism must be able to be transferred to a healthy animal, which will show signs of the disease after having been exposed to the organism; and 4) the organism must be able to be isolated from the infected animal.

When was the "golden age" of microbiology?

Many scientific historians believe the "golden age" of microbiology began in 1857 with the work of French chemist Louis Pasteur (1822–1895) and lasted about sixty years. During this time, researchers identified the specific microorganisms responsible for numerous infectious diseases. For example, English surgeon Joseph Lister (1827–1912) treated surgical wounds with a

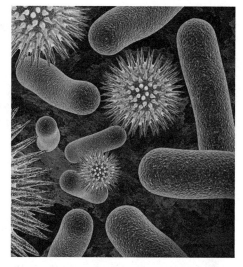

There are millions of species of bacteria on our planet, but they weren't discovered until the seventeenth century, when scientists first peered into drops of water using magnifying lenses.

phenol solution that eventually led to the advent of aseptic surgery; German physician Paul Ehrlich (1854–1915) added to the theory of immunity by synthesizing an arsenic compound that was effective in treating syphilis in humans; in 1884, Russian microbiologist Elie Metchnikoff (1845–1916), an associate of Pasteur, published a report on phagocytosis—the defensive process in which the body's white blood cells engulf and destroy microorganisms; and in 1897, Japanese physician Masaki Ogata (c. 1864–c. 1919) reported that rat fleas transmitted bubonic plague, ending the centuries-old mystery of how plague was transmitted.

What are some well-known diseases and infectious agents, and who discovered them?

The following chart identifies many diseases, their infectious agents, the scientists who discovered them, and the year in which they were discovered:

Disease	Infectious Agent	Discoverer	Year Discovered
Anthrax	*Bacillus anthracis*	Robert Koch	1876
Gonorrhea	*Neisseria gonorrhoeae*	Albert L. S. Neisser	1879
Malaria	*Plasmodium malariae*	Charles-Louis Alphonse Laveran	1880
Wound infections	*Staphylococcus aureus*	Sir Alexander Ogston	1881
Tuberculosis	*Mycobacterium tuberculosis*	Robert Koch	1882
Erysipelas	*Streptococcus pyogenes*	Friedrich Fehleisen	1882
Cholera	*Vibrio cholerae*	Robert Koch	1883
Diphtheria	*Corynebacterium diphtheriae*	Edwin Klebs and Friedrich	1883–84

Who discovered that penicillin was effective against bacterial infections?

British microbiologist Alexander Fleming (1881–1955) was the first to discover penicillin's use as an antibacterial agent. In 1928, Fleming was researching staphylococci at St. Mary's Hospital in London. As part of his investigation, he had spread staphylococci on several petri dishes before going on vacation; when he returned, he noticed a green-yellow mold contaminating one of the petri dishes—but the staphylococci had failed to grow near the mold. He identified the mold as being of the species *Penicillium notatum*. Further investigation proved that staphylococci and other organisms are killed by *P. notatum*, but it was not until the 1940s that British-Australian pathologist Howard Florey (1898–1968) and German-born British biochemist Ernst Boris Chain (1906–1979) rediscovered the benefits of penicillin and isolated it for medical use. In 1945 Fleming, Florey, and Chain shared the Nobel Prize in Physiology or Medicine for their work on penicillin.

Disease	Infectious Agent	Discoverer	Year Discovered
Typhoid fever	*Salmonella typhi*	Karl Eberth and Georg Gaffky	1884
Bladder infections	*Escherichia coli*	Theodor Escherich	1885
Salmonellosis	*Salmonella enteritidis*	August Gaertner	1888
Tetanus	*Clostridium tetani*	Shibasaburo Kitasato	1889
Gas gangrene	*Clostridium perfringens*	William Henry Welch and George Henry Falkiner	1892
Nuttall Plague	*Yersinia pestis*	Alexandre Yersin and Shibasaburo Kitasato	1894
Botulism	*Clostridium botulinum*	Emile Van Ermengem	1897
Shigellosis (bacterial dysentery)	*Shigella dysenteriae*	Kiyoshi Shiga	1898
Syphilis	*Treponema pallidum*	Fritz R. Schaudinn and P. Erich Hoffman	1905
Whooping cough	*Bordetella pertussis*	Jules Bordet and Octave Gengou	1906

How are bacteria classified on the basis of metabolic activity?

Bacteria have been classified into two types based on their metabolic activity: Heterotrophs rely on organic compounds for carbon and energy needs; most bacteria are heterotrophs and must obtain organic compounds from other organisms. The majority of heterotrophs are free-living saprobes (also known as saprophytes or saprotrophs) and obtain their nourishment from dead, organic matter. Autotrophs require inorganic nutrients and carbon dioxide as their sole source of carbon and can be photosynthetic or chemosynthetic. Photosynthetic autotrophs obtain their energy from light, while chemosynthetic autotrophs obtain their energy by oxidizing inorganic chemicals. (For more about heterotrophs, see the chapter "Cellular Basics.")

What bacterium classification is based on oxygen?

Bacteria can also be divided into four major groups based on their response to oxygen. Aerobic bacteria grow in the presence of oxygen; microaerophilic bacteria grow best at oxygen concentrations lower than those present in air—less than 20 percent oxygen. Anaerobic bacteria grow best in the absence of oxygen; and finally, facultative anaerobes can grow in the presence or absence of oxygen, with more growth when oxygen is present.

What is the Gram stain, and why is it important?

A Gram stain is produced when a scientist colors a microorganism with a dye to emphasize certain structures within the organism. The Gram stain is the most widely used stain in microbiology to identify bacteria. Developed in 1884 by Danish physician Hans Christian Gram (1853–1938), Gram stain results can be combined with other information on a bacterium's cellular structure and biochemical characteristics, generally allowing scientists to identify an unknown type of bacteria.

The Gram stain categorizes bacteria into a gram-positive or gram-negative group (often referred to as the "Gram reaction"). The most important structural difference between gram-negative and gram-positive bacteria is that gram-negative bacteria are enclosed by two layers—called a cytoplasmic membrane and an outer membrane. Between the two membranes is a thin layer that is linked to the outer membrane. In contrast, gram-positive bacteria have a thick layer, with its cell wall two to eight times as thick as the cell wall of gram-negative bacteria. (For more about stains, see the chapter "Biology in the Laboratory.")

What is a Petri dish, and who developed it?

The Petri dish—a shallow glass or plastic dish consisting of two round, overlapping halves (also called a cell culture dish)—is used to grow bacteria and other microorganisms on a certain medium, usually in the nutrient agar. The top of the dish is larger than the bottom so that when the dish is closed, a seal is created, preventing contamination of the culture; some are also loose-fitting, with either used depending on the experiment. This device was developed in 1887 by German bacteriologist Julius Richard Petri (1852–1921), a member of Robert Koch's laboratory (for more about Koch, see above). Petri dishes are very easy to use, can be stacked on each other to save space, and are one of the most common items in a microbiology laboratory.

What works are known as Bergey's Manuals?

Bergey's Manual of Determinative Bacteriology is an extensive reference manual used for bacterial classification. The first edition was published in 1923 under the sponsorship of the Society of American Bacteriologists (organized in 1899 and now known as the American Society for Microbiology [SAB]). This reference work was first conceived by bacteriologist David H. Bergey (1860–1937) with the assistance of a special committee of the SAB. By 1994, the ninth edition of the book had been published.

Bergey's book also spawned two other works under the name *Bergey's Manual of Systematic Bacteriology*. Whereas the 1994 book is used as a reference to aid in the identification of unknown bacteria, the newer manuals offer systematic information about bacteria. The first edition of *Bergey's Manual of Systematic Bacteriology* includes four volumes: Volume 1 (1984), *Gram-Negative Bacteria of General, Medical, or Industrial Importance*; Volume 2 (1986), *Gram-Positive Bacteria Other Than Actinomycetes*; Volume 3 (1989), *Archaeobacteria, Cyanobacteria, and Remaining Gram-Negative Bacteria*; and Volume 4 (1989), *Actinomycetes*. The most recent, second edition was published in five volumes beginning in 2001: Volume 1 (2001), *The Archaea and the Deeply Branching and Phototropic Bacteria*; Volume 2 (2005), *The Proteobacteria*; Volume 3 (2009), *The Firmicutes*; Volume 4 (2011), *The Bacteroidetes, Spirochaetes, Tenericutes, Acidobacteria, Fibrobacteres, Fusobacteria, Dictyoglomi, Gemmatimonadetes, Lentisphaerae, Verrucomicrobia, Chlamydiae, and Planctomycetes*; and Volume 5 (2012), *The Actinobacteria*.

BACTERIA BASICS

What are the main components of a bacterial cell?

The major components of a bacterial cell are the plasma membrane, cell wall, and a nuclear region containing a single, circular DNA molecule. Plasmids—small circular pieces of DNA that exist independently of the bacterial chromosome—are also present in a bacterial cell. In addition, some bacteria may have flagella (for movement; for more about flagella, see the chapter "Cellular Basics"); pili or fimbriae (short, hairlike appendages that allow certain bacteria to adhere to various surfaces, including the cells that they infect); or a capsule of slime around the cell wall that protects it from other microorganisms.

Do bacteria all have the same shape?

No, not all bacteria have the same shape, but vary greatly. The spherical bacteria, called cocci (such as *Staphylococcus aureus*) occur singularly in some species and as groups in other species. They have the ability to stick together and form a pair (diplococci). When they stick together in long chains, they are called streptococci; in irregularly shaped clumps or clusters of bacteria they are called staphylococci. Rod-shaped bacteria, called bacilli, occur as single rods or as long chains of rods (such as *Salmonella typhi*). Spiral- or helical-shaped bacteria are called spirilla (such as *Campylobacter jejuni*).

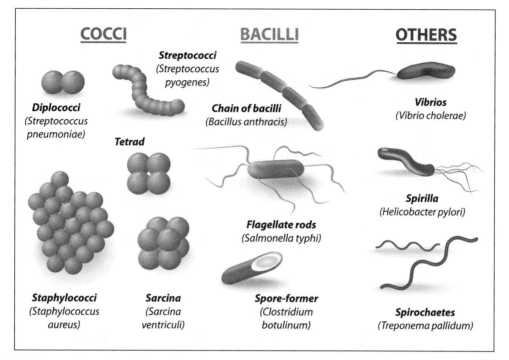

Bacteria come in a variety of shapes, the most common being varieties of cocci and bacilli.

What is a germ?

Most of us use the term "germ" to define those minute "creatures" that invade our bodies, giving us such problems as the flu, a cold, or the chicken pox. Overall, the term "germ" means any microorganism—especially those that cause disease, such as viruses, bacteria, and fungi. These are all different organisms that can cause various diseases—thus the word "germ" is a very general term.

Where are bacteria found?

Various types of bacteria inhabit every place on Earth—including places where no other organism can survive. Bacteria have been detected as high as 20 miles (32 kilometers) above the Earth and around 7 miles (11 kilometers) deep in the waters of the Pacific Ocean (in what is called the Challenger Deep in the huge Mariana Trench—a deep chasm in the seabed deep enough to swallow Mount Everest). They are found in extreme environments, such as the Arctic tundra, boiling hot springs—and all over our bodies.

Are all types of marine bacteria found in all oceans?

Although it has been said that all types of marine bacteria are distributed everywhere, this is not true. For example, in 2013, scientists found a "bipolar" species of bacteria that occur in the Arctic and Antarctic—but not anywhere else. The scientists who discovered this believe that these bacteria are more selective, with other studies showing that marine bacteria can be found in certain ocean currents on the Equator heading to other water masses. Because marine bacteria seem to have "living preferences," a warming climate like ours that affects temperature, salinity, pH levels, and circulation patterns in the oceans may also affect the distribution of these creatures. Because many bacteria are essential catalysts for chemical reactions, variations in climate could possibly change other aquatic organism populations in response.

How many groups are identified in the domain Bacteria?

Biologists recognize at least a dozen different groups of bacteria. The following lists the major groups to date (for the explanation of a Gram reaction, see above):

Major Group	Gram Reaction	Characteristics	Examples
Actinomycetes	Positive	Produce spores and antibiotics; live in soil environment	*Streptomyces*
Chemoautotrophs	Negative	Live in soil environment; important in the nitrogen cycle	*Nitrosomonas*
Cyanobacteria	Negative	Contain chlorophyll and are capable of photosynthesis; live in aquatic environment	*Anabaena*
Enterobacteria	Negative	Live in intestinal and respiratory tracts; ability to decompose form spores; pathogenic	*Escherichia, Salmonella, Vibrio*
Gram-positive cocci	Positive	Live in soil environment; inhabit the skin and mucous membranes of animals; pathogenic to humans	*Streptococcus, Staphylococcus*
Gram-positive rods	Positive	Live in soil environment or animal intestinal tracts; anaerobic; disease-causing	*Clostridia, Bacillus*
Lactic acid bacteria	Positive	Important in food production, especially dairy products; pathogenic to animals	*Lactobacillus, Listeria*
Myxobacteria	Negative	Move by secreting slime and gliding; ability to decompose materials	*Chondromyces*
Pseudomonads	Negative	Aerobic rods and cocci; live in soil environment	*Pseudomonas*
Rickettsiae and chlamydia	Negative	Very small, intracellular parasites; pathogenic to humans	*Rickettsia, Chlamydia*
Spirochetes	Negative	Spiral-shaped, live in aquatic environment	*Treponema, Borrelia*

Can bacteria be addicted to caffeine?

Yes, bacteria can be addicted to caffeine—but in this case it's because scientists have been playing with bacteria and genetics. In 2013, researchers genetically engineered *E. coli* bacteria to live on caffeine, causing the small creatures to become addicted. This isn't just to make humans feel better—many people are addicted to caffeine, too—but because scientists found that caffeine and related chemical compounds have become water pollutants due to the widespread consumption of coffee, soda pop, tea, energy drinks, some medications, and chocolate. The scientists knew that a natural soil bacterium, *Pseudomonas putida* CBB5, actually lives solely on caffeine and could be used to clean up contamination from caffeine. Thus, they took the genetic "information" from the caffeine-loving bacteria and put it into the *E. coli*—a bacteria that is much easier to handle and grow. They hope to use this new caffeine addict not only to break down caffeine in the environment, but as a possible sensor to measure caffeine levels in beverages, recover nutrient-rich by-products of coffee processing—and even help in the production of medicine.

What is the most abundant group of organisms?

The Eubacteria group is the most abundant group of organisms on Earth. To put it into perspective, more living eubacteria inhabit the human mouth than the total number of mammals living on Earth.

Are any bacteria visible to the naked eye?

Epulopiscium fishelsoni, bacteria that live in the gut of the brown surgeonfish (*Acanthurus nigrofuscus*), are visible to the naked eye. It was first identified in 1985 and mistakenly classified as a protozoan (mostly aquatic, single-celled organisms; for more about protozoa, see this chapter). Later studies analyzed the organism's genetic material that showed it to be a bacterium of unprecedented size: 0.015 inches (0.38 millimeter) in diameter—about the size of a period at the end of this sentence.

What bacteria survive inside our gut?

Bacteria known as intestinal flora survive in both the small and large intestine of the human body—thanks to the neutral pH of the intestinal environment. Because these bacteria have low requirements for oxygen and sunlight exposure, they are well suited to life in this environment. They use our digested food as a source of nutrition and even provide a benefit to us by synthesizing several of the vitamins we need, such as biotin, as well as vitamins K and B-5. In addition, if these bacteria are thriving, it is more difficult for disease-causing microbes to establish themselves in the intestines and attack our bodies.

What are Archaebacteria?

Archaebacteria (domain Archaea) are primitive bacteria that often live in extreme environments. This domain includes the following: 1) thermophiles ("heat lovers"; see below for more about thermophiles), which live in very hot environments, including the hot sulfur springs of Yellowstone National Park, which reach temperatures ranging from 140 to 176°F (60–80°C). 2) halophiles ("salt lovers"), which live in locations with high concentrations of salinity, such as the Great Salt Lake in Utah (with salinity levels from 15 to 20 percent; seawater normally has a level of salinity of 3 percent);

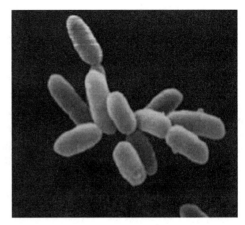

Halobacteria (shown here) are an example of Archaebacteria, a primitive form of bacteria that can survive extreme environments.

and 3) methanogens that get their energy by using hydrogen gas (H_2) to reduce carbon dioxide (CO_2) to methane gas (CH_4). (For more about organisms in extreme environments, see the chapter "Basics of Biology.")

What is the relationship between bacteria and temperature?

All microorganisms have temperature ranges that determine growth. Overall, microorganisms are unique in their ability to exist and grow at temperatures ranging from 14 to 230°F (-10–110°C). Temperature restrictions are due to limitations in cell metabolism. A microorganism's maximum temperature is the highest temperature at which growth can occur; minimum temperature is the lowest temperature at which growth can occur. A microorganism's optimum temperature is the temperature at which the growth rate is the fastest. The maximum, minimum, and optimum temperatures define the range of growth for each microorganism and are collectively referred to as the cardinal temperatures.

Bacteria are often divided into four groups on the basis of their average temperatures for growth: psychrophiles, mesophiles, thermophiles, and extreme thermophiles (sometimes called extremophiles; see the chapter "Basics of Biology"). The following chart lists the temperature ranges at which these groups can grow:

Bacteria Group	Possible Temperature	Optimum Temperature
Psychrophiles	14–77°F (-10–25°C)	50–68°F (10–20°C)
Mesophiles	50–113°F 10–45°C	68–104°F (20–40°C)
Thermophiles	86–176°F (30–80°C)	104–158°F (40–70°C)
Extreme thermophiles	176°F (80°C) and above	

What is the generation time for various bacteria?

Generation time is defined as the amount of time required for a bacterial population to double its number. For example, if a culture tube is inoculated with a cell that divides every twenty minutes (at an optimum temperature), the total cell population will grow to two cells after a period of twenty minutes, and four cells after forty minutes, and so on, with the growth continuing at this rate.

The following chart shows the generation time for selected bacteria (with corresponding genus/species, common name, and temperature range for the generation time):

Bacterium	Temperature (°F/°C)	Generation time (in minutes)
Escherichia coli (bladder infections)	98.6/37	17
Shigella dysenteriae (shigellosis)	98.6/37	23
Salmonella typhosa (typhoid fever)	98.6/37	24
Pseudomonas aeruginosa (multiple diseases)	98.6/37	31
Staphylococcus aureus (wound infections)	98.6/37	32
Bacillus subtilis (common in soil)	96.8/36	35
Clostridium botulinum (botulism)	98.6/37	35
Streptococcus lactis (used in buttermilk and cheese)	86/30	48
Lactobacillus acidophilus (ferments sugars into lactic acid)	98.6/37	66
Mycobacterium tuberculosis (tuberculosis)	98.6/37	792

What effects do pH levels have on the growth of bacteria?

The pH is the measure of the hydrogen ion activity of a solution. The pH scale ranges from 0 (very acidic) to 14 (extremely alkaline or basic). The pH, or concentration of hydrogen ions (H^+) in an environment, is critical to bacterial growth because it can affect enzyme activity. An extremely high or low pH can inactivate enzymes or disrupt cell processes. (For more about pH levels, see the chapter "Basics of Biology.")

The following table shows the pH ranges and the optimum pH that several different organisms require for growth:

Organism	pH Range for Growth	pH Optimum for Growth
Thiobacillus thiooxidans (also called *Acidithiobacillus thiooxidans*; consumes sulfur and produces sulfuric acid)	1.0–6.0	2.0–3.5
Lactobacillus acidophilus (ferments sugars into lactic acid)	4.0–6.8	5.8–6.6
Escherichia coli (associated with bladder infections)	4.4–9.0	6.0–7.0
Nitrobacter spp. (important in the nitrogen cycle)	6.6–10.0	7.6–8.6
Nitrosomonas spp. (oxidizes ammonia into nitrite)	7.0–9.4	8.0–8.8

How do bacteria reproduce?

Bacteria reproduce by binary fission. It has nothing to do with computers, but is the way certain organisms reproduce asexually. In this case, a cell divides into two similar cells. First the circular, bacterial DNA replicates, then a transverse wall is formed by an ingrowth of both the plasma membrane and the cell wall. In a favorable environment, bacteria can reproduce very rapidly. Favorable circumstances include laboratory cultures or its natural habitat. For example, under good conditions, *E. coli* can divide every twenty minutes—and a laboratory culture started with a single cell can produce a colony of 10,000,000 to 100,000,000 (or 10^7 to 10^8) bacteria in about twelve hours.

How is genetic material exchanged between bacteria during reproduction?

Although sexual reproduction involving the fusion of cells (gametes) does not occur in bacteria, genetic material is sometimes exchanged between bacteria. This is done in three different ways. The first method is a transformation in which fragments of DNA are released by a broken cell and taken in by another bacterial cell. The second possibility is transduction, in which a bacteriophage (a virus that infects and replicates within a bacteria) carries genetic material from one bacterial cell to another. The last possibility is conjugation, in which two cells of different mating types come together and exchange genetic material.

What is pasteurization?

The term pasteurization is familiar to everyone—especially those who drink milk or soy milk products. In general, it is the process of heating liquids, such as milk, to destroy microorganisms that can cause spoilage and disease. This process was developed by French chemist Louis Pasteur (1822–1895) as a method to control the microbial contamination of wine. Pasteurization is commonly used to kill pathogenic bacteria, such as Mycobacterium, Brucella, Salmonella, and Streptococcus—all common to milk and other beverages.

Three methods exist for pasteurizing milk. In the first method, low-temperature holding (LTH), the milk is heated to 145°F (62.8°C) for thirty minutes. In the second method, high-temperature short-time (HTST), the milk is exposed to a temperature of 161°F (71.7°C) for fifteen seconds—a technique also known as flash pasteurization. The most recent method allows milk to be treated at 286°F (141°C) for two seconds; this approach is referred to as ultrahigh temperature (UHT) processing. Shorter-term processing results in improved flavor and extended product shelf life—but to some people it changes the taste of the milk.

What are bioluminescent bacteria?

Bioluminescence is the production of light—with very little heat—by some organisms. The light-emitting substance (luciferin) in most species is an organic molecule that emits light when it is oxidized by molecular oxygen in the presence of an enzyme (luciferase). Bioluminescence is primarily a marine phenomenon occurring in many regions of oceans or seas. One example is the "milky sea" found in the Indian Ocean: In an area once measured to be about the size of Connecticut, the sea appears to glow an eerie blue at night—but it is actually bioluminescent marine bacteria called *Vibrio harveyi*.

Why do some bacteria cause disease and others do not?

All strains of bacteria possess genetic differences; these differences are not sufficient for them to be considered as separate species, but each strain is distinctive. For example, many different strains exist of *E. coli*. Some, such as *E. coli* 0157:H7, cause serious diseases, while others live in the intestine and can be considered beneficial because they aid digestion. In fact, although billions of bacteria exist in the world, less than 1 percent of all bacteria cause disease.

How dangerous is the bacterium *Clostridium botulinum*?

The bacterium *Clostridium botulinum* is very dangerous—resulting in a toxin called botulinum that is considered to be the most acutely toxic substance known. It can grow

---

Does a connection exist between termites and bacteria?

A definite connection—a symbiotic relationship—between termites and bacteria exists. In particular, according to a study in 2013, more than 4,500 different species of bacteria live in termite guts! These bacteria help termites to extract nutrients from woody materials—although the bacteria don't seem to have anything to do with the termite digestion (which may be genetically inherent in the termite itself). This may be of biological interest, but to the homeowner, termites often mean huge repair bills. After all, in the United States, the cost of controlling termites and repairing the damage from the creatures is about $2 billion per year.

in food products, producing botulinum (the condition is called botulism) so potent that one gram of this toxin can kill fourteen million adults! This bacterium can withstand boiling water (212°F or 100°C), but is killed in five minutes at a temperature of 248°F (120°C). This tolerance makes *Clostridium botulinum* a serious concern for people who preserve vegetables and fruits at home.

In particular, if the home canning process is not done properly, this bacterium will grow in the anaerobic conditions of the sealed container, creating an extremely poisonous food. The toxin is produced when the endospores of *Clostridium botulinum* germinate in poorly prepared canned goods—such as a leaking seal around the rim of the jar. That is why no one should ever eat food from a can that appears swollen or a can lid that is not depressed in the middle (or if the lid is easily pulled from up from the jar). More often than not, this is a sign that the can has become filled with gas released during germination of the bacteria. Consuming food from a can containing endospores that have undergone germination can lead to nerve paralysis, severe vomiting, and even death. Two major antiserums have been developed for botulism, but their effectiveness depends on how much is ingested and how long the toxin has been in the body.

Are rickettsiae and chlamydiae bacteria or viruses?

For many years, rickettsiae and chlamydiae were thought to be viruses because they are very small and are intracellular parasites. They are now known to be bacteria because they possess both DNA and RNA, have cell walls similar to those found in gram-negative bacteria, divide by binary fission, and are affected by antibiotics. They can cause several diseases, including Rocky Mountain spotted fever (carried by ticks) from the bacteria *Rickettsia rickettsii*, and Q fever from the bacteria *Coxiella bumeti*.

What is Botox?

Botox, the trade name for botulinum toxin type A, is a protein produced by the bacterium *Clostridium botulinum*—yes, the same acute toxin that causes botulism. In this case, the botulinum toxin is purified and sterilized, then converted to a form that can be injected and used in a medical setting.

Botox was not always associated with celebrities and anti-aging serums. It was first approved by the Food and Drug Administration (FDA) in December 1989 to treat two eye muscle disorders—uncontrollable blinking (called blepharospasm) and misaligned eyes (called strabismus). In 2000, the toxin was approved to treat cer-

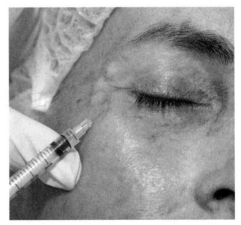

Botox injections work by introducing a bacteria under the skin that blocks muscle contractions, resulting in smoother skin.

vical dystonia, a neurological movement disorder that causes severe neck and shoulder contractions. The use of the toxin for "making you look younger" came more recently. In small doses, Botox blocks nerve cells from releasing a chemical called acetylcholine, which signals muscle contractions. By selectively interfering with a muscle's ability to contract, a person's existing frown lines and wrinkles are smoothed out, improving the appearance of the surrounding skin. But of course, injecting such a toxin into one's system raises concerns: The biggest problem is the possibility of an allergic reaction.

What are mycoplasmas?

Mycoplasmas are the smallest, free-living bacteria—and the only bacteria that exist without a cell wall. Some mycoplasmas have sterols (a type of lipid-lacking fatty acids) in their plasma membranes that provide the strength a membrane needs to maintain cellular integrity without a wall; since mycoplasmas lack cell walls that provide shape and rigidity, they have no definitive forms. For example, *Mycoplasma pneumoniae* causes a disease known as primary atypical pneumonia (PAP), a mild form of pneumonia confined to the lower respiratory tract. Because mycoplasmas do not have cell walls, penicillin is ineffective in stopping their growth. Instead, a drug called tetracycline, which inhibits protein synthesis, is recommended as the antibiotic of choice for treatment of PAP.

What is anthrax, and how does it affect humans?

Bacillus anthracis, the agent of anthrax, is a large, gram-positive, nonmoving (nonmotile), spore-forming, rod-shaped bacteria that is nasty for humans. The three major, clinical forms of human anthrax are as follows: the bacteria contracted through the skin (the most common, entering through a cut or scrape on the skin, forming lesions, blisters, then a black ulcer); lungs (breathing it into the warm, moist environment of the lungs allows the bacteria to "sprout," spreading to the lymph system—which usually takes one to six days); or gastrointestinal tract (through the ingestion of anthrax). The symptoms for each form vary; it is usually treated with antibiotics, although the effectiveness of the treatment is often dependent on how the bacteria entered the body. But no matter what, if left untreated, anthrax can spread to the bloodstream, often leading to septicemia (blood poisoning) and death.

VIRUS BASICS

What is a virus?

A virus is an infectious, protein-coated fragment of DNA or RNA. Viruses replicate by invading host cells and taking over the cell's "machinery" for DNA replication. Viral particles can then break out of the cells, spreading disease. Viruses lie dormant in any environment (land, soil, air) and on any material. They infect every type of cell from plants and animals to bacteria and fungi.

Are viruses living organisms?

All living things have around six characteristics—they adapt to their environment, have a cell makeup, have metabolic processes help them obtain and use energy, they move in response to their environment, they grow and develop, and they reproduce.

Thus, technically, viruses are not "alive." In particular, they cannot grow, they cannot reproduce (replicate) on their own—and need a host cell to become active to provide these functions. In other words, they are inert outside their living host cell. As such, they are considered to be between life and nonlife and are not considered living organisms. In fact, British biologist Sir Peter Medawar (1915–1987), a Nobel Prize recipient in Physiology or Medicine, described viruses as "...a piece of bad news wrapped in a protein." He was referring to the fact that viruses cause influenza, smallpox, infectious hepatitis, yellow fever, polio, rabies, AIDS, and many other diseases.

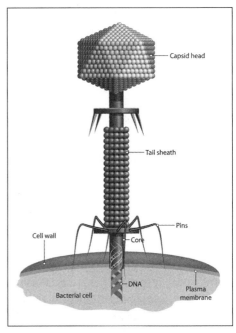

Viruses are kind of bizarre, alien creatures when you think about it—almost like robots. They are not really alive because they can't reproduce or grow without a host.

What was the first virus to be isolated in a laboratory?

In 1935, American biochemist and virologist Wendell Stanley (1904–1971) of the Rockefeller Institute (known today as Rockefeller University) prepared an extract of the tobacco mosaic virus and purified it. The purified virus precipitated in the form of crystals. During this investigation, Stanley was able to demonstrate that viruses can be regarded as chemical matter rather than as living organisms. The purified crystals retained the ability to infect healthy tobacco plants, thus characterizing them as viruses, not merely chemical compounds derived from a virus. Subsequent studies showed that the tobacco mosaic virus consisted of a protein and a nucleic acid. Further studies showed that this virus consisted of RNA (ribonucleic acid) surrounded by a protein coat. Stanley was awarded the Nobel Prize in Chemistry in 1946 for his discovery.

What is SARS?

SARS stands for Severe Acute Respiratory Syndrome—the first "new" disease of the twenty-first century that was known to jump from an animal host to humans. Once it infects a person, SARS is easily transferred from human to human, making it one of the most frightening diseases known. The first case was detected in China's Guangdong Province in November, 2002, but it was not announced until February, 2003. Around 8,000 people were infected by the disease and caused panic worldwide. To stop the spread of SARS, millions were screened at airports, schools closed, tourism took a direct hit in many countries, and hundreds in Asia and Canada were placed in quarantine. It was finally contained months later, with about 800 deaths known.

What is a coronavirus?

A coronavirus is one of a group of RNA viruses found almost everywhere; the name comes from the fact that these viruses look like a halo (corona) when viewed under an electron microscope. In humans, they are the second leading cause of the common cold (after the rhinovirus; see below for more information); two of the human coronaviruses—OC43 and 229E—cause about 30 percent of common colds. They also infect other animals, causing respiratory infections in birds and gastroenteritis (inflammation of the digestive tract) of pigs. Compared to the majority of other viruses, they are very large in size; they also have a two-step replication mechanism—also different than most viruses. When SARS (see above) was discovered in 2002, it was found to be a new type of coronavirus.

What is a norovirus?

According to the Centers for Disease Control and Prevention, a norovirus is a very contagious virus that causes gastroenteritis (inflammation of the stomach and/or intestinal lining), as a result of food poisoning or infected food, and can develop into acute nonbacterial gastroenteritis. It spreads rapidly from an infected person, contaminated water or food, or by touching contaminated surfaces. It causes your stomach or intestines (or both) to become inflamed, causing stomach pain, nausea, and diarrhea, and often vomiting. According to research, it appears that norovirus is responsible for about 50 percent of all gastroenteritis outbreaks in the United States and about 90 percent of all epidemic nonbacterial gastroenteritis outbreaks worldwide.

What's the difference between food infection and food poisoning?

The term food poisoning is actually a catch-all term and is usually considered to be synonymous with "food-borne illness." The term is broken down into food infection and food

Was a new coronavirus reported in 2013?

Yes, a new coronavirus, MERS-CoV (CoV is for "coronavirus"), was announced by the Centers for Disease Control and Prevention in 2013—one that seemed to have started around early 2012. The new virus was different from any other known human coronavirus, including the one that caused Severe Acute Respiratory Syndrome (SARS). Considered a worldwide disease—thought to have originated from the Arabian Peninsula or neighboring countries (to date, no cases have been reported in the United States)—it is known to pass from human to human, causing severe respiratory illness.

intoxication and with either one, you will have similar symptoms. With a food infection, a microorganism (most often a virus or bacteria, such as *Salmonella*) grows inside your body and becomes the source of your symptoms. With food intoxication, a chemical or natural toxin (often a by-product of the microorganism present in the food, such as *Staphylococcus aureus*; it's known as an exotoxin) causes you to be sick—and most bacterial food poisonings are actually food infections. Both can cause many identical symptoms—headaches, vomiting, abdominal pain, cramps, diarrhea, dehydration—which is why it's so hard for a doctor to determine just which one you are experiencing.

Why do noroviruses have so many names?

Noroviruses have been around for a long time—thus they have been referred to by many names, usually based on the region in which an outbreak occurs. The first norovirus was probably noticed in 1929 by American pediatrician John Zahorsky (1871–1963); it was called the "winter vomiting disease." Another bad gastrointestinal outbreak occurred in Norwalk, Ohio in 1968, and the virus was named the Norwalk agent (also the Norwalk-like virus, or NLV; small, round-structured virus, or SRSV; and even Snow Mountain virus). And after genetic studies, it was classified as a member of the viral family *Caliciviridae* (a single strand of RNA for its genome); by 2002, the official name of the genus became *Norovirus*. One of the more recent norovirus outbreaks occurred in 2012 called GII.4 Sydney—it began in Australia and has since spread to the United States and other countries.

What is a rhinovirus?

A rhinovirus is one that most of us are very familiar with: the viruses that are responsible for the common cold. They are also the most common viral infective agents that attack the human body, with around ninety-nine recognized types. They spread mostly by airborne particles (someone sneezes next to you) or contaminated surfaces (you open a door an infected person has touched—and in particular, you then rub your eyes or nose, where the virus is easily picked up by your mucus membranes). And this virus is fast—sticking to surface receptors within fifteen minutes after it enters your respiratory tract.

What is the structure of a virus?

Viruses consist of strands of the genetic material nucleic acid—either as RNA or DNA, but not both—surrounded by a protein coat called a capsid. The capsid protects the genome (collection of genes); it is often subdivided into individual protein particles called capsomeres—features that create the shape of the virus. The capsid protein coat of a virus can come in three main shapes: Helical, resembling a wound spring (such as the tobacco mosaic virus); isosahedral, a multifaceted virus (such as the herpes simplex); and complex, as the name implies, can be combination of shapes (such as T-4 bacteriophage).

How do viruses compare to bacteria?

A virus and a bacterium have numerous differences. The following lists some of the general differences:

Characteristic	Bacteria	Viruses
Able to pass through bacteriological filters	No	Yes
Contains a plasma membrane	Yes	No
Contains ribosomes	Yes	No
Possesses genetic material	Yes	Yes
Requires a living host to multiply	No	Yes
Sensitive to antibiotics	Yes	No
Sensitive to interferon	No	Yes

What is the average size of a virus?

Viruses are much smaller than bacteria. The smallest viruses are about 17 nanometers in diameter, and the largest viruses are up to 1,000 nanometers in length. By comparison, the bacterium *E. coli* is 2,000 nanometers in length, a cell nucleus is 2,800 nanometers in diameter, and an average eukaryotic cell is 10,000 nanometers in length. The following are the average sizes of some specific common viruses (note: 1 nanometer is equal to 0.001 micrometers; the head of a pin is about 1,000 micrometers in diameter):

Virus	Size (in nanometers)
Smallpox	250
Tobacco mosaic (seen in plants)	240
Rabies	150
Influenza	100
Bacteriophage	95
Common cold	70
Polio	27
Parvovirus (often seen in domesticated animals)	20

How do viruses enter their host cells to reproduce?

A virus is able to enter a host cell by either tricking the host cell to pull it inside, as the cell would do to a nutrient particle, or by fusing its viral coat with either the host cell's wall or membrane and then releasing its genes into the host. Some viruses inject their genetic material into the host cell, leaving their empty viral coats outside of the host cell.

Where did viruses originate?

The most widely accepted hypothesis is that viruses are bits of nucleic acid that "escaped" from cells. According to this view, some viruses trace their origin to animal cells, some to plant cells, and others to bacterial cells. The variety of origins may explain why viruses are species-specific—that is, why some viruses only infect species that they are closely related to or the organisms from which they originated. This hypothesis is supported by the genetic similarity between a virus and its host cell.

Do viruses contain both DNA and RNA?

Viruses have either DNA or RNA as their genomic material, whereas cells—including bacteria—have both. In DNA viruses, the synthesis of viral DNA is similar to how the host cell would normally carry out DNA synthesis; in other words, the virus inserts its genetic material into the host's DNA. In RNA viruses, transcription (the first stage of protein synthesis, in which messenger RNA is produced from the DNA) takes place with the help of RNA polymerase.

What is influenza?

It is an infectious, acute respiratory disease commonly called the "flu," caused by RNA viruses of the family Orthomyxoviridae (the actual influenza viruses). It is transmitted most often through the air by coughs or sneezing, which is why it can spread so fast. The flu is characterized by fever, chills, headache, generalized muscular aches, and a frequent cough; it can also lead to viral or bacterial pneumonia or secondary bacterial pneumonia—even for healthy individuals. In fact, influenza affects people of all ages, but can be particularly severe for the very young, the very old, and people with complications due to other diseases.

Why is influenza considered to be one of the most lethal viruses?

The influenza virus is probably the most lethal virus in human history. In fact, the symptoms of human influenza were described by Greek physician Hippocrates about 2,400 years ago. Since that time, many outbreaks have occurred: Some say Christopher Columbus brought the virus to the Antilles in 1493, causing close to the entire population to die; it was also responsible for fifty million deaths during a worldwide outbreak in 1918 (see below); and the 1957 Asian flu and 1968 Hong Kong flu killed millions of people. In fact, according to statistics, seasonal outbreaks around the world often average between 250,000 to 500,000 deaths from the virus.

What is the difference between an epidemic and a pandemic?

An epidemic and a pandemic are different when it comes to viral outbreaks. The epidemics are usually confined to certain areas, such as a yearly statewide epidemic of flu during the winter months. A pandemic is much more widespread and includes affected people worldwide. For example, one of the most well-known pandemics was the worldwide influenza outbreak of 1918 in which 200 million cases of influenza were discovered, causing fifty million deaths.

How did the historic viral outbreak of 1918 help vaccine supplies in 2009?

In 2012, scientists were able to genetically sequence the 1918 virus, confirming that the influenza pandemics of 1957, 1968, and 2009 were direct descendants of the 1918 virus. In fact, because scientists were able to understand the structural similarities of the 2009 virus versus the 1918 virus, they were able to pinpoint what age groups would be most

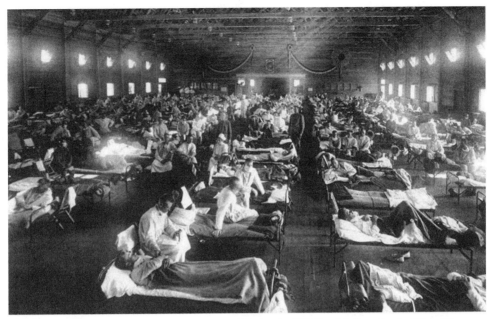

This photo was taken in a Camp Funston, Kansas, ward during the 1918 Spanish flu epidemic. Fifty million people died during that horrible outbreak.

vunerable to infection by the 2009 influenza virus. This allowed them to concentrate the vaccine to those who truly needed the supplies—in this case, to younger people.

How does a host body fight off an invading a virus?

A host body uses its "immune system" to fight off a virus. Once any foreign substance is introduced, the body responds through what is called an "immune response"—the way the body produces antibodies. These antibodies are specific to each invader and can be thought of as natural "antibiotics"—destroying an invader and preventing the host from contracting the same disease in the future.

The antibodies specific to the infection take about seven days to work. In the meantime, the body's cells that are exposed and infected with the virus produce naturally occurring interferons, or proteins (called cytokines, proteins that carry signals between cells). They stop a disease when they are released into the bloodstream and then bind to receptors on the surface of cells. They are usually released within three to five days after exposure to the virus, preventing infection in neighboring cells until the antibodies are made. (Note: in contrast to antibodies, interferons are not virus specific, but host specific; for example, viral infections of human cells are inhibited only by human interferons.)

Which animal viruses are DNA viruses and which are RNA viruses?

DNA and RNA viruses in animals (yes, humans included) have several differences. The following lists some of the more familiar diseases caused by these two types of viruses:

DNA Viruses

Type of Virus	Examples
Adenoviruses	Approximately forty types of viruses infecting human respiratory and intestinal tracts, causing sore throats, tonsillitis, and conjunctivitis
Herpesviruses	Herpes simplex type 1 (cold sores), herpes simplex type 2 (genital herpes), varicella-zoster (causing chicken pox and shingles)
Papovaviruses	Human warts, degenerative brain diseases, polyomas (tumors)
Parvoviruses	Infections in dogs, swine, arthropods, and rodents; causes gastroenteritis in humans after eating infected shellfish
Poxviruses	Smallpox, cowpox

RNA Viruses

Type of Virus	Examples
Paramyxoviruses	Rubeola (measles), mumps, distemper (in dogs)
Orthomyxoviruses	Influenza in humans and other animals
Picornaviruses	Polioviruses, Hepatitis A, human colds; coxsackievirus and echovirus cause aseptic meningitis; enteroviruses infect the intestine; rhinoviruses infect the respiratory tract
Reoviruses	Vomiting and diarrhea
Retroviruses	AIDS, some types of cancer
Rhabdoviruses	Rabies
Togaviruses	Rubella, yellow fever, encephalitis

Do interferons come in different types?

Yes, different types of interferons in the body help protect against viruses. In 1957, British virologist Alick Isaacs (1921–1967) and Croatian virologist Jean Lindenmann (1924–) identified a group of over substances that were later designated as alpha, beta, and gamma interferons depending on their molecular structure; more recently, two other classes have been discovered—omega and tau. So far, more than twenty kinds of interferon-alpha have been found, along with only one interferon-beta and one interferon-gamma identified.

Interferons can be further broken down into type 1 (including interferon-alpha, -beta, -omega, and -tau) and type 2 (interferon-gamma), depending on their amino acid sequence. In general, type 1 interferons make cells resistant to the viral infection; type 2 is responsible for regulating the overall immune system function.

Why is it difficult to treat viral infections with medications?

Antibiotics are ineffective against viral infections because viruses lack the structures (for example, a cell wall) with which antibiotics interfere. In general, it is difficult to treat viral infections with medications without affecting the host cell as viruses use the host cell's machinery during replication. Thus, even though some antiviral medications exist, it is truly the body's immune system that fights off viral infections.

But still, several antiviral drugs have been developed that are effective against certain viruses—the following chart lists some of them:

Disease	Viral pathogen	Antiviral drug
AIDS	Human immunodeficiency virus	Azidathymidine (AZT), didanosine, dideoxycytosine
Chronic hepatitis	Hepatitis B or C	interferon alfa-2B
Genital herpes, shingles, chicken pox	Herpesvirus	Acyclovir, idoxuridine, trifluridine, vidarabine
Influenza A	Influenza	Amatadine

Can a virus infect bacteria?

Yes, in particular, a bacteriophage (also called a phage) is a virus that can infect bacteria. The term "bacteriophage" means "bacteria eater" (from the Greek word *phagein*, meaning "to devour"). Phages consist of a long nucleic acid molecule (usually DNA) coiled within a protein head, with many of the phages having a tail attached to the head. It is the fibers extending from the tail that are often used to attach the virus to the bacterium.

Bacteriophages are classified in two ways: lytic phages destroy the host cell. When a lytic virus infects a susceptible host cell, it uses the host's metabolic machinery to replicate viral nucleic acid and produce viral proteins. Temperate phages do not always destroy their host cell. After attachment and penetration, the DNA from a temperate phage becomes incorporated into the host bacterial DNA; it is then referred to as a prophage. The prophage replicates at the same time as the bacterial DNA—but the viral genes may be repressed indefinitely.

When were bacteriophages first discovered?

Bacteriophages were discovered in the early 1900s by British scientist Frederick W. Twort (1877–1950) and French scientist Felix d'Hérelle (1873–1949). In 1915, Twort reported observing a filterable agent that destroyed bacteria growing on solid media; d'Hérelle independently confirmed the discovery in 1917. It was actually d'Hérelle who named the agent "bacteriophage." However, very few scientists accepted these findings and the work on the growth and infectious nature of bacteriophages. It was not until the 1930s that Martin Schlesinger, a German biochemist, characterized bacteriophages, establishing their own unique place in the microbial world.

When did scientists first observe a virus infecting a bacterial cell?

It may seem like something out of a science fiction movie, but in 2013, scientists observed, for the first time, the detailed changes in a virus's structure—called T-7—as it infected a bacterium, *E. coli*. The researchers observed that while the virus searched for prey, it extended one or two of six ultrathin fibers that are normally folded at the base of the virus' "head." When the prey is spotted, the T-7 virus extends the feeler-fibers, essentially walking around the cell until it finds the perfect spot to infect the prey's cell. Like a minitransformer, the T-7 virus changes its structure, injecting some of its proteins through a protein path in the bacterium's cell membrane—thus allowing the virus to easily send its genetic material into the prey. Once the transfer of the viral DNA is in the prey's cell, the pathway in the membrane seals up—and the infection is complete. Although this experiment only included the T-7 virus and *E. coli*, scientists believe this may be typical of how many viruses attack other organisms.

Biochemist and neurologist Stanley Prusiner coined the word "prion" to describe abnormal forms of proteins.

What is a prion?

Prions are abnormal forms of natural proteins. In 1982, American neurologist and biochemist Stanley Prusiner (1942–) used the term "prion" in place of the expression "proteinaceous infectious particle" when describing an infectious agent. Scientists still have not discovered exactly how prions work. Current research shows that a prion is composed of about 250 amino acids, but no nucleic acid component has been found. They appear to accumulate in lysosomes. In the brain, it is possible that the filled lysosomes burst and damage cells. As diseased cells die, the prions contained in the cells are released and are able to attack other cells—thus, like viruses, prions are considered infectious agents. It is thought that prions are responsible for the group of brain diseases known as transmissible spongiform encephalopathies (TSEs)—including bovine spongiform encephalopathy (mad cow disease) when it occurs in cattle and Creutzfeldt-Jakob disease when it occurs in humans.

PROTISTS

What are protists?

Protists (or protoctista) are an extremely diverse group of predominantly microscopic organisms. All protists are eukaryotic, and while many are unicellular and have only one nucleus, they may be multicellular, have multiple nuclei, or form colonies of identical, unspecialized cells. Although most are microscopic, some are much larger, reaching lengths of over an inch (just over 2 millimeters). In early, traditional classifications, they were listed on the basis of being neither plant nor animal nor fungus; further evidence suggested that protists exhibit characteristics of the plant, animal, and fungi kingdoms—but their classification is still highly debated.

> ## What can deep-sea protists tell scientists about ancient protists?
>
> In 2008, scientists studying modern deep-sea protists used their observations to extrapolate information about ancient protists. In particular, they studied the profusion of macroscopic groovelike traces in rock called trace fossils—the remnants of trails made by ancient organisms. Comparing 1.8 billion-year-old protist trace fossils to the trails of living, grape-sized protists called *Gromia sphaerica*, the scientists found a great similarity, concluding that many of these organisms were around in profusion much earlier than previously thought.

Why is the study of ancient protists important?

The study of ancient protists is important to scientists trying to understand life on early Earth. This is because the early protists—especially the protozoa (single-celled organisms that live mostly in water)—are thought to have been the first complex organisms on Earth. It is thought that they evolved from primitive cells 2 to 1.5 billion years ago, and those early organisms may have been a cross between plants and animals. They may have also had symbiotic relationships with bacterialike organisms, eventually leading to multicelled organisms. All of this is speculation, mainly because the fossil record of such creatures is so sparse—thanks to the organism's size (they are difficult to find in rock) and the fact that it probably decays quickly, leaving no trace.

What are some major groups of Protista?

Little agreement exists among scientists on how to classify the protists, but these small creatures are often divided into several general groups that share certain characteristics of locomotion, nutrition, and reproduction. The following chart is just one of many general groupings of protists:

Group	Characteristics
Sarcodinas	Amoebas and related organisms that have no permanent locomotive structure
Algae	Single-celled and multicellular organisms that are photosynthetic
Diatoms	Photosynthetic organisms with hard shells made of silica
Flagellates	Organisms that propel themselves through water with flagella
Sporozoans	Nonmotile parasites that spread by forming spores
Ciliates	Organisms that have many short, hairlike structures on their cell surface associated with locomotion
Molds	Heterotrophs with restricted mobility that have cell walls composed of carbohydrates

What human diseases are caused by certain protists?

Many diseases are caused by protists. Some familiar ones from various regions around the world are: *Entamoeba histolytica*, a parasitic amoeba which causes amoebic dysentery, an intestinal disorder. It's estimated that up to ten million individuals in the United States have parasitic amoebas, but only two million exhibit symptoms of the disease. In tropical areas, up to half the population may be infected. Species of *Trypanosoma* (called trypanosomes) are the cause of the "African sleeping sickness" (both *T. gambiense* and *T. Rhodesiense*, transmitted by biting flies of the genus *Glossina*, or tsetse fly); and from the American tropics, Chagas's disease (caused by *T. cruzi*, spread by blood-sucking bugs of the genera *Triatoma* and *Rhodnius*). And the protist *Plasmodium* causes malaria, in which an individual of the genus *Plasmodium* enters the human body through a bite of a mosquito of the genus *Anopheles* that has been infected by the protist. (For more diseases and other effects of protists, see below.)

How many species are in protist phyla, and what are some of their effects?

Biologists estimate that millions of species of protists could exist. The following lists, to date, the protist phyla, the estimated number of species, and some examples of the protists, such as in human disease (note: as more protists—and numbers of species in the various phyla—are found each year, these numbers will no doubt change):

Phylum	Number of Species	Some Examples
Rhizopoda	Hundreds	Amoebas (intestinal parasite *Entamoeba histolytica*; can cause amoebic dysentery)
Foraminifera	>4,000	Mostly oceans, for example, plankton
Actinopoda	Hundreds	Radiolarians
Chlorophyta	7,000	Green algae
Rhodophyta	2,500–6,000	Red algae
Phaeophyta *Macrocystis*	1,500–2,000	Brown algae; for example, a kelp called
Chrysophyta	11,500	Diatoms
Pyrrhophyta	2,100	Dinoflagellates
Euglenophyta	1,000	Mostly single-celled aquatic algae
Zoomastigophora (also called Zooflagellates)	Thousands	
Apicomplexa	>4,500	*Plasmodium* (malaria); *Toxoplasma gondii* (toxoplasmosis)
Ciliophora	8,000	Fresh or saltwater protozoa; paramecium
Oomycota	580	Water molds; downy mildew (for example, of grapes, *Plasmopara viticola*)
Slime molds:		
Dictyosteliomycota (or in part, Acrasiomycota)	>70	Cellular slime molds
Myxogastria	500	Plasmodial slime molds

What are some differences in characteristics between some protists?

The great diversity among protists has made their grouping and classification difficult. Characteristics used to classify protists include mode of locomotion, presence or absence of flagella and cilia, body form and coverings, pigmentation, the ability to conduct photosynthesis, type of nutrition, and whether the organism is unicellular or multicellular. The following list includes one of the many classifications, along with some characteristics of certain protists:

Phylum	Morphology	Body form/ covering	Locomotion	Pigment/ photosynthesis
Rhizopoda	Single cell	No definite shape; shells in some	Pseudopodia (means "false foot")	None
Foraminifera	Single cell	Shells	Cytoplasmic projections	None
Actinopoda	Single cell	Skeletons	Pseudopods reinforced with microtubes	None
Chlorophyta	Single cell and colonial multicellular	Cellulose in cell wall	Flagella; some species are nonmotile	Chlorophyll
Rhodophyta	Mostly multicellular	Cellulose	Nonmotile	Chlorophyll and phycoerythrin
Phaeophyta	Multicellular	Cellulose	Flagella	Chlorophyll
Chrysophyta	Single cell; some colonial	Shells with silica	No flagella; move by gliding over secreted slime	Chlorophyll
Pyrrhophyta	Single cell; some colonial	Mostly cellulose plates	Flagella	Chlorophyll; carotenoids
Euglenophyta	Single cell	allowing	Flexible pellicle carotenoids euglenoids to change their shape	Flagella; Chlorophyll
Zoomastigophora	Single cell	None	Flagella	None
Apicomplexa	Single cell	Spores	Nonmotile	None
Ciliophora	Single cell	None	Cilia	None
Dictyosteliomycota (or, in part, Acrasiomycota)	Single cell during most of the life; multicellular during reproductive stages	Cellulose (spores)	Pseudopods for single cells	None

91

Phylum	Morphology	Body form/ covering	Locomotion	Pigment/ photosynthesis
Myxogastria	Multinucleate mass of cytoplasm	None	May have flagella	None
Oomycota	Coenocytic mycelium	Cellulose	Flagella	None

How have scientists recently reclassified protists?

The classification of protists has traditionally included species such as protozoa and algae, some fungal-like organisms, and other organisms that didn't fit into the idea of a plant or animal. By 2005, the first scientific community-wide effort to categorize all the protists was made, but it still was limited by the technology available at the time.

Thus, in 2012, with more DNA and RNA sequencing on many organisms, another classification was presented. The most significant change was the introduction and recognition of new super groups—a way of understanding the relationships between ancient and modern protists and even the protists' connections to animals and plants. One such change is what scientists call *Amorphea*, a super group that links animals, fungi, and protists to a diverse group of protists largely dominated by various amoeboid cells (such as macroscopic slime molds); another super group is SAR, which includes the most common algae, microbial predators, and parasites. It will be a few more years until all the classifications are modified to accept the newest data.

How are protists divided based on their way of gathering nutrients?

Protists gather nutrients in several ways, depending on the organism.

- The autotrophs synthesize organic compounds from carbon dioxide in the atmosphere.

- Heterotrophs obtain carbon from organic compounds in the environment, or, like amoebas, are active predators. (For more about heterotrophs, see the chapter "Cellular Basics.")

- The mixotrophs can switch between the autotrophic and heterotrophic ways of obtaining nutrients. For example, some euglenoids contain chloroplasts, using photosynthesis to synthesize organic particles in the presence of light;

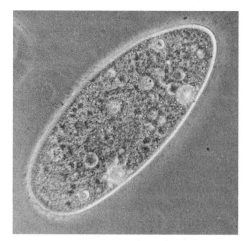

A paramecium is a unicellular animal of the Ciliate protozoa and is commonly found in freshwater environments.

How were the white cliffs of Dover, England, formed?

The white cliffs of Dover are composed of a variety of protist fossil shells, including coccolithophores (a type of algae) and foraminiferans. Their process of formation took millions of years: After these protists died, their shells were deposited on the bottom of the ocean in a fine gray mud; after time, layer upon layer of sediment deposited above compressed the mud. After even more time, the mud hardened, forming a type of limestone we call chalk. Eventually, geologic processes—such as the uplift of the land and erosion by water or ice—exposed the outcrop of white limestone.

when it is dark, they switch to a heterotrophic mode of eating food—easily switching back and forth depending on conditions.

How do protists move?

Protists move in several different ways. Because they are exclusively found in aquatic or very moist environments, they need certain appendages to remain motile. Two of the most common ways are with flagella and cilia, both used by cells to move through watery environments. Cilia move back and forth, while flagella undulate in a whiplike motion, moving in the same direction as the cell's axis. (For more about flagella and cilia, see the chapter "Cellular Basics.") Other protists use pseudopods (or "false feet") to move, and are large, lobe-shaped extensions of the organism.

But note: not all protists move. Some are sessile—attached using certain structures (usually stalks) that adhere to a substrate. And some protists are both sessile and mobile; for example, many of the brown algae have free-floating sperm, whereas mature algae are attached to rock or other substrate—and are not mobile.

How do protists reproduce?

Protists reproduce in many ways depending on their environment, life cycle, and type, either asexually—by budding, binary fission, or mitosis—or sexually. During binary fission, the organism's DNA replicates and the cells divide; during budding, the organism produces a smaller bud of itself that will grow into an individual protist identical to the original.

Some protists use a variety of reproductive methods; for example, paramecium reproduce using binary fission, but after so many hundreds of times essentially splitting apart, the paramecium use sexual reproduction to exchange their genetic material. Scientists do not know what triggers this sudden urge on the protist's part to sexually reproduce—but it may have to do with surrounding environmental stresses. (For more about the environment see the chapter "Environment and Ecology.")

93

Where are amoebas found?

Amoebas are found in soil, freshwater, and saltwater. They have no definite body shape and continually change form (amoeba is derived from the Greek word meaning "change") as they move using their pseudopods, which are lobe-shaped projections of the amoebas' cytoplasm. As the cytoplasm extends and fills, the amoebas move. These same pseudopods also surround and capture food in a vacuole, ingesting the particle or organism by phagocytosis, or the digestion and absorption of the organism by phagocytes.

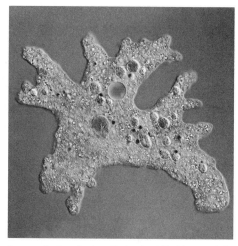

An amoeba is a single-celled animal that has no definite body shape. They move and capture food by extending pseudopods out into their surrounding environment.

What protists are indicators of polluted water?

Of the many protists in aquatic environments, the euglenoids seem to be the best bio-indicators. These small creatures—they are typically among the largest algal cells, but are often thought to act more like animals than other algae—are considered to be unicellular flagellates and live in water habitats that can vary in pH and light. Some euglenoids are capable of photosynthesis; others do not carry on photosynthesis (they apparently lost their chloroplasts over time) and are predatory. But in general, because their population thrives under high nutrient levels—and human pollution often carries the decomposing organics needed by euglenoids—they are useful bio-indicators of polluted water.

What is a slime mold?

A slime mold is often thought of as an organism that uses spores in order to reproduce, often resembling a gelatinous slime. They were once thought to be fungi, but are now known to be a separate group, with about 900 known species. Currently, they are divided into two groups—the true slime molds (Myxogastria) and the cellular slime molds (Dictyosteliomycota or, in part, Acrasiomycota). Of the latter, a cellular slime mold called *Dictyostelium discoideum* has been studied as a model for the developmental biology of complex organisms. Under the best conditions, this organism lives as individual, amoeboid cells. When food is scarce, the cells stream together into a moving mass resembling a slug that can change into a stalk with a spore-bearing body at its top. This structure then releases spores that can grow into a new amoeboid cell.

What are diatoms?

Diatoms are microscopic algae of the phylum Chrysophyta in the kingdom Protista. Almost all diatoms are single-celled algae and dwell in both fresh- and saltwater; they are abundant in the cold waters of the northern Pacific and the Antarctic Oceans. Diatoms are yellow or brown in color and are an important food source for marine plankton and many small animals. Diatoms have hard cell walls; these "shells" are made from silica that has been extracted from the water. It is unclear how the extraction of silica from water is accomplished. When they die, their glassy shells—called frustules—sink to the bottom of the sea and harden into rock called diatomite. One of the most famous and accessible diatomites is the Monterey Formation along the coast of central and southern California. Diatoms are familiar to gardeners, too—since diatomaceous earth is often used to control garden insect pests.

What is a water mold?

A water mold is one of a group of protists called oomycetes that also includes downy mildews and white rusts. Like many protists, because they look similar and are also decomposers and/or parasites, they were once classified as fungi. Plus, many of the

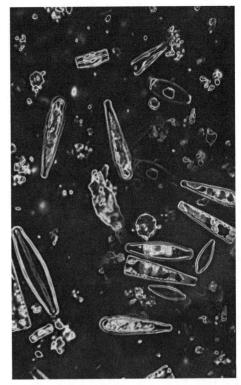

Algae of the phylum Chrysophyta are called diatoms. Common in cold ocean waters, they have a tough outer shell made of silica and are an important part of the food chain.

oomycetes have long filament structures similar to the hyphae found in fungi. But their cell walls are made of cellulose and they have flagellated cells—two reasons why they cannot be classified as fungi.

How did the protist *Phytophthora infestans* influence Irish history?

Once thought to be a fungus, the protist *Phytophthora infestans*—an oomycete—is of the potato plant's most lethal pathogens. It causes what is known to many farmers and gardeners as the "late blight" of potato—a disease that has popped up in numerous places throughout the centuries. It was first noticed in the 1500s in Europe; a few centuries later, it was responsible for the Irish potato famine of 1845 to 1849 (it's estimated that 400,000 people perished during the famine due to malnutrition).

P. infestans thrives and produces spores under humid and moist environments; thus, each year, the temperature and rainfall affect the timing and severity of the late blight disease. It decays the leaves and stem of the potato plant, eventually causing the tuber to stop growing; from there, the tubers rot after they are attacked by the pathogen.

FUNGI

HISTORICAL INTEREST IN FUNGI

What is the scientific study of fungi called?

The scientific study of fungi (plural for fungus) is called mycology, from the Greek word *mycote*, meaning "fungus." The first scientific mention of a fungus was in Europe during the mid-1500s—the potato "late blight" (*Phytophthora infestans*) that was introduced from South America, devastating the potato crops; this blight was first recorded in the United States in 1830. Interestingly enough, the late blight was eventually found to be *not* caused by a fungus, but by a protist. (For more about late blight, see the chapter "Bacteria, Viruses, and Protists.")

Who wrote some of the early studies of fungi?

In 1801, South African-born Dutch mycologist Christiaan Hendrik Persoon (1761–1836) wrote one of the first treatises on modern mycology. In 1821, the first extensive taxonomical study of fungi, *Systemia Mycologicum*, was published by Swedish mycologist Elias Fries (1794–1878)—a book that is still considered an authoritative resource on fungal taxonomy. He is also considered the "father of mycology" and devoted his career to the study of botany, concentrating specifically on fungi and lichens.

CLASSIFYING FUNGI

What were the early steps taken to classifying fungi?

In the earliest classification systems, fungi were classified as plants. The first classification system to recognize fungi as a separate kingdom was proposed in 1784. Since that

What famous children's author studied and drew illustrations of fungi?

English author Beatrix Potter (1866–1943), perhaps best known for having written *The Tale of Peter Rabbit* in 1902, began drawing and painting fungi in 1888. She eventually completed a collection of almost 300 detailed watercolors, which are now in the Armitt Library in Ambleside, England. In 1897, she prepared a scientific paper on the germination of Agaricineae spores for a meeting of the Linnean Society of London. Although Potter's findings were originally rejected, experts now consider her ideas correct.

time, researchers identified four characteristics shared by all fungi: they lack chlorophyll; the cell walls of fungi contain the carbohydrate chitin (a tough, nitrogenous substance that can be found in the outer shells of some crabs or mollusks); they are not truly multicellular since the cytoplasm of one fungal cell mingles with the cytoplasm of adjacent cells; and they are heterotrophic eukaryotes (unable to produce their own food from inorganic matter, absorbing their food from whatever they are growing on), while plants are autotrophic eukaryotes.

What are some of the most recent classifications of fungi?

Classifying fungi is not an easy task—especially because of the number of different species, findings thanks to DNA sequencing, and new methods used by scientists to study the fungi. The following lists two of the most recent classifications, both based on (as most fungi classifications) the reproductive spore a fungus produces. And although both these lists will no doubt change as more studies are made, these are the terms you would see when exploring the world of fungi (for example, in 2011, a potentially new phyla—Cryptomycota—was suggested).

One classification system developed in 2007 has four main groups (phyla): Ascomycota (sac fungi, with spores called ascospores), Basidiomycota (club fungi, with spores called basiodiospores), Chytridiomycota, and Zygomycota (bread molds, with spores called zygospores); a few years later, a group was taken out of the Zygomycota and named the phyla Glomeromycota, and the phyla Microspordia was also named. In yet another classification, fungi are divided into about ten phyla—the same names, but new phyla added: Chytridiomycota, Monoblepharidiomycota, Neocallimastigomycota, Blastocladiomycota, Microsporida, Zygomycota 1, Zygomycota 2, Endomophthorales, Glomeromycota, Ascomycota, and Basidiomycota. Of all the phyla in this classification, the Ascomycota have the most species (around 64,163 species); next in line are the Basidiomycota (with 31,515 species), then Micorspordia (with 1,300 species). The lowest number of species is in the phyla Neocallimastigomycota (with twenty species).

What are deuteromycetes—"imperfect fungi"—mentioned in many fungi classifications?

In many classifications, if the reproductive, sexual spore of a fungus species has not been identified or examined, that fungus is placed in the phyla Deuteromycota. Imperfect fungi are also called deuteromycetes or conidial fungi. They are a collection of distinct fungal species known to reproduce only asexually; the reason for this special listing is that, to date, the sexual reproductive features—or even sexual reproduction—has not been seen in any of these fungi. Most imperfect fungi are thought to be ascomycetes that have lost the ability to reproduce sexually.

This aspergillus is one example of an imperfect fungus, or deuteromycete, a fungus that reproduces only asexually.

Deuteromycetes are mostly free-living and terrestrial, but some are pathogenic—those fungi that affect humans and other animals. For example, pathogenic deuteromycetes include athlete's foot (*Epidermophyton floccosum*) and ringworm (*Microsporum canis*). On the positive side, other famous deuteromycetes include *Penicillium roquefortii* and *Penicillium camemberti*—used to make Roquefort and Camembert cheeses, respectively.

How can fungi be defined based on their life cycles?

Many ways to separate and define fungi exist. One way is based on their life cycles, the presence of a fruiting body, and the types of spores produced. Based on these criteria, fungi can be divided into three groups:

- *Multicellular filamentous molds*—These fungi are made up of very fine threads called hyphae that grow at the tips and divide repeatedly to create branching chains. They usually break down organic matter in the soil; they also can grow spores on the branches above ground. Environmental conditions (rain, wind, etc.) help to spread the spores that can eventually result in new fungi.

- *Macroscopic filamentous fungi*—These fungi produce a mycelium (a mass of hyphae that is not a reproductive structure) below ground, but also produce large fruiting bodies above the ground—or what most of us call mushrooms or toadstools—that hold their spores. This fruiting body is actually hyphae that are tightly packed together; the lines (gills) under the mushroom cap hold the spores, with a cap 10 centimeters in diameter capable of producing up to one hundred million spores per hour.

- *Single-celled microscopic yeasts*—Yeasts are single-celled organisms that are about the same size as a human red blood cell. They only reproduce by budding—a lobe of the yeast breaks off from the original (parent) cell, producing another smaller (daughter) cell.

Fungal reproductive spores vary greatly in size, shape, color, and surface texture. They range from 1 to 100 microns (1 micron = $3.93700787 \times 10^{-5}$ inch) in diameter—the largest approximately one-tenth the thickness of a dime—with the majority about 2 to 20 microns in size. Yet another comparison is that twenty million 5-micron spores could fit into the space of a postage stamp.

FUNGI BASICS

How old are fungi?

Although not as old as bacteria (fossil evidence suggests they may be at least 3.5 billion years old), fungi have been on the Earth for hundreds of millions of years. The earliest fungi fossil evidence is from the Ordovician period, about 460 to 455 million years ago. Some scientists suggest that fungi may have played an important part in allowing vascular plants to colonize land about 425 million years ago. Even when the dinosaurs roamed the planet over sixty-five million years ago, fungi were there, with some ninety-four-million-year-old fossils displaying mushroom-forming fungi similar to those that exist today.

How many species of fungi have been identified?

Scientists have identified between 70,000 and 80,000 species of fungi; almost 2,000 new species are discovered each year. For example, in 2013, two new species of mushrooms were discovered on the Iberian Peninsula—both belonging to the *Hydnum* genus (known as ox tongue mushrooms), a type of fungus commonly used in cooking. Some mycologists estimate that about 1.5 million species exist worldwide—placing fungi second only to insects in the number of different species.

What organisms are often misinterpreted as fungi?

As scientists find out more about various fungi, they become better at dividing the organisms into their true, respective groups. Thus, several types of organisms are often mistakenly called by an incorrect name. For example, some organisms called mold or fungus are not part of the kingdom Fungi—in particular, slime molds and water molds. These organisms are called Oomycota; slime molds are now known to be a mix of three or four unrelated groups, while all the oomycetes of this group are classified in the Cromista, along with diatoms and brown algae.

Where are fungi found?

Fungi seem to grow everywhere—but they grow best in dark, moist habitats wherever organic material is available. Moisture is necessary for their growth, and some fungi obtain water from the atmosphere and/or from the medium upon which they live. When the environment becomes very dry, fungi survive by going into a resting stage or by producing spores that are resistant to drying. The optimum pH for most species is 5.6, but some fungi can tolerate and grow in pH levels ranging from 2 to 9 (for more about pH, see the chapter "Basics of Biology"). Certain fungi can grow in concentrated salt solutions or sugar solutions such as jelly or jam. Fungi also thrive in a wide range of temperatures—even refrigerated food may be susceptible to fungal invasion.

What are hyphae?

Most fungi are made up of many cells (or multicellular)—with the exception of yeasts. The many cells make up a mass of intertwined filaments—also called branching tubes—known as hyphae (a mass of hyphae is called a mycelium). Some hypha contain internal crosswalls (septa) that divide the hyphae into separate cells, but some do not; for example, coenocytic hyphae lack septa. Cytoplasm in many species flows freely throughout the hyphae, passing through major pores present in the septa. Because of this streaming, proteins and other materials that are synthesized in the hyphae can be carried to their tips, which are actively growing. As a result, fungal hyphae may grow very rapidly when food and water are abundant and the temperature is optimum.

What material makes up the cell walls of fungi?

Unlike plant cell walls, a fungus's cell wall is composed of chitin—called a polysaccharide or nitrogenous substance—and one that is most often found in the outer skeleton (exoskeleton) of certain land animals, such as grasshoppers, and outer shells of some marine animals, such as crabs and mollusks. (To compare, the cell walls of plants and some protists are composed of cellulose.) The rigidity of the chitin is for protection and also helps to slow down dehydration of the fungi, especially in times of drought.

What structures, or organelles, are found in a typical fungus?

A typical fungus has a nucleus—sometimes more than one per cell—that holds its genetic material (chromosomes). Other organelles that help a fungus cell to function include the cytoplasm (for movement of materials through the hydrae), mitochondria (converts energy), rough and smooth endoplasmic reticulum (makes the more complex proteins), and Golgi apparatus (forms many types of proteins and enzymes). Fungi also have, like bacteria, ribosomes, but those in bacteria are smaller and have a different way of reproducing.

What organelle is absent in fungi?

Centrioles that divide during cell division processes, mitosis and meiosis, are lacking in all fungi. (They are found only in animal cells and are self-replicating organelles that appear to help organize cell division, but aren't essential to the process; for more about mitosis and meiosis, see the chapter "Cellular Basics.")

What main carbohydrate is stored by fungi?

The main carbohydrate stored by fungi is glycogen—also the main storage carbohydrate of animals. This fact suggests that fungi are more closely related to animals than plants, which store starch as their main carbohydrate.

How do fungi obtain food?

Fungi are heterotropic—they cannot photosynthesize. In general, instead of taking food inside its body and then digesting it as an animal would, most fungi digest food outside their bodies by secreting special strong enzymes onto the food. In this way, complex organic compounds are broken down into simpler compounds that a fungus can absorb (which means fungi are osmotrophic) through their cell walls and cell membranes. In addition, fungi do not swim, walk, or run to get their nutritional needs—but move to the food by growing toward it.

What is a fruiting body?

Macrofungi, or macroscopic fungi—meaning large fungi such as mushrooms and toadstools—produce fruiting bodies. Also called sporophores, the fruiting body is a structure that enables the organism to disperse spores for reproduction. The fruiting body of a mushroom is easy to see—it's the structure that is visible above the ground. They are found in a variety of shapes, ranging from the common cap-and-stem mushrooms to the more exotic, antlerlike, coral-like, cagelike, trumpet-shaped, or club-shaped mushrooms. The method of spore dispersal for the various types of macrofungi is related to the shape of the fruiting body.

What are puffballs?

Puffballs are round or pear-shaped fruiting bodies that sit directly on the ground or on rotted wood. They come in several different shapes; their relatives include the Earth star and the giant puffball. In the eastern and central United States and Canada, they are frequently found sitting on the ground in the summer and fall—most looking like a medium-brown ping-pong ball (although some can reach the size of a small watermelon). If you squeeze the puffball, the thin skin will crack open and the insides will fall out like a puff of smoke. In reality, the "smoke" particles are spores—each about 3.5 to 5.5 micrometers (MGRm; 0.00014–0.00019 inch) in diameter. Thousands of spores are in one puffball; in fact, a giant puffball (*Calvatia gigantean*) can contain more than seven trillion spores!

What is the size of most fungi?

The sizes of fungi vary greatly depending on the type of fungus. In general, most microscopic—or smaller—fungi are 2 to 10 micrometers (MGRm) in diameter and several tenths of an inch in length. The average size of fungi hyphae are 5 to 50 micrometers (MGRm) in length.

On the other hand, macroscopic mushrooms are much larger—especially the part that is visible, or the fruiting body, also called sporophores (see above about fruiting

Can fungi be categorized based on how they obtain nutrients?

Yes, another way to define fungi is to categorize them by how they obtain their nutrients. Most fungi species are saprotophic, or they absorb nutrients from waste and decomposing dead matter and organisms. Other species are parasitic, or a type of symbiosis in which one organism benefits and the "host" is harmed; these fungi are also called necrotrophs, fungi that kill the host cells in order to obtain their nutrients. One example is athlete's foot fungus on a human being. Still other fungi are mutualistic, a type of symbiosis in which both organisms in the relationship benefit; these fungi are called biotrophs, obtaining their nutrients from a living host (plant or animal). For example, a fungus and an alga in lichens—the fungus provides the moist surface for the alga, and the alga manufactures food for the fungus.

bodies). These range in size from 8 to 10 inches (20–25 centimeters) and heights of 10 to 12 inches (25–30 centimeters). Some of the larger fungi, such as the bracket fungi, can reach over 16 inches (40 centimeters) in diameter; even larger are some giant puffballs, the largest measuring 5 feet (150 centimeters) in diameter.

What are dimorphic fungi?

Many fungi, particularly those that cause disease in humans, are dimorphic—that is, they have two forms. For example, in response to changes in temperature, nutrients, or other environmental factors, some fungi can change from yeast to a mold form.

Does a relationship exist between plants and fungi?

Yes, a relationship does exist between some plants and fungi—something many gardeners know about when it comes to certain plants. And it has to do with symbiosis—the close association of two or more different organisms. One type of symbiosis is known as mutualism, defined as an association that is advantageous to both parties. The most common—and possibly the most important since it represents almost four-fifths of mature plants—mutualistic relationship in the plant kingdom is known as mycorrhiza.

The word mycorrhiza is derived from the Greek *mykes*, meaning "fungus," and *rhiza*, meaning "root." Mycorrhiza is a specialized, symbiotic association between the roots of plants and fungi that occurs in the vast majority of plants—both wild and cultivated. In a mycorrhizal relationship, the fungi assist their host plants by increasing the plants' ability to capture water and essential elements such as phosphorus, zinc, manganese, and copper from the soil and transferring them into the plant's roots. The fungi also provide protection against attack by pathogens and nematodes. In return for these benefits, the fungal partner receives energy from carbohydrates, amino acids, and vitamins essential for its growth directly from the host plant—mostly during the process of the plant photosynthesis.

What specific plants are helped through mycorrhiza fungi?

Many plants benefit from mycorrhiza fungi. For example, the basidiomycetes (mushrooms, bracket fungi, etc.) are the fungal, mycorrhizal partners of trees and other woody plants. Zygomycetes (molds, etc.) are the fungal partners of nonwoody plants. In fact, it's estimated that mycorrhizal fungi amount to 15 percent of the total weight of the world's plant roots.

What are the two types of mycorrhiza fungi?

Plants have a relationship with fungi in two ways: In endomycorrhiza, the hyphae of the fungus penetrate the outer cells of the plant root and extend into the surrounding soil; in ectomycorrhiza, the hyphae surround but do not penetrate the roots. Endomycorrhiza are much more common than ectomycorrhiza. The most common plants associated with ectomycorrhiza are trees and shrubs growing in temperate regions, including pines, firs, oaks, beeches, and willows. These plants tend to be more resistant to extreme temperatures, drought, and other harsh environmental conditions. In addition, some ectomycorrhizal fungi may provide protection from acid precipitation (for more about acid pollution, see the chapter "Environment and Ecology").

Do fungi affect living matter?

Yes, fungi affect more than just dead or decaying organic matter—they may also do harm to the environment by attacking and giving diseases to living plants and animals. In fact, fungi are some of the most harmful pests to living plants and are responsible for billions of dollars in agricultural losses each year. In addition, food products that have been harvested and stored are not immune to fungal decay. Fungi often secrete substances into the foods they attack, making the foods unpalatable or even poisonous.

What organisms are known as "humongous fungus"?

The "humongous fungus" are two different fungi—both enormous, underground fungus found in the United States in two different states. The first, the fungus *Armillaria gallica*, was found growing in 1992 in northern Michigan. It encompasses 35 acres (0.056 square miles, or 0.145 square kilometers) and is thought to be at least 1,500 years old. Also, in the Malheur National Forest in Oregon is a giant fungus of the same genus (different species), *Armillaria ostoyae*, or honey mushroom. This organism spans 2,200 acres (8.9 square kilometers)—making it, to date, the largest organism by area.

Scientists can prove that these huge organisms are a single fungus by taking samples of each fungus and performing DNA analyses on each sample. In both cases, it turns out that the genetic material of the respective samples were identical.

How do bacterial and fungal spores differ?

The main purpose of bacterial spores—known as endospores—is to protect bacterial cells so they can survive extreme, harsh conditions. In comparison, fungi reproduce both sex-

Are fungi disappearing?

No, fungi are not disappearing, because every year more are discovered. Plus, thanks to genetic technologies, scientists are learning more about the connections between fungi and their environment. But in some areas, certain types of fungi are disappearing—depending on where you look. For example, the average number of fungal species in the Netherlands has declined significantly— thought to be because of an increase in air pollution. Fungi are more sensitive to air pollution than plants because fungi have no protective covering, whereas the upper parts of plants are protected by cuticles and bark. In addition, while plants extract water from soil using their roots, some fungi absorb water directly from the atmosphere—along with the pollution that may be present in the air. Thus, poor air quality definitely contributes to the decline of fungi.

ually and asexually through the formation of spores. Asexual spores are formed by the hyphae of one organism; the organisms that form from these spores are identical to their parents. Sexual spores result from the fusion of nuclei from two strains of the same species of fungus; organisms from sexual spores derive characteristics from each parent.

What are the main types of asexual fungi spores?

The main types of asexual spores among the fungi are arthrospores, chlamydospores, sporangiospores, and conidia. Conidia and sporangiospores are produced from a fruiting body, but neither arthrospores nor chlamydospores involve a fruiting body. Arthrospores are formed by fragmentation of the hyphae, whereas chlamydospores are thick-walled spores formed along the margin of the hyphae.

What are the haploid and diploid phases of fungal reproduction?

Fungi are organisms with life cycles that exhibit both a haploid phase and a diploid phase. During the haploid phase, reproduction is conducted through gametes, or a mature reproductive cell (male or female) that units with the opposite sex and fuse together to form a zygote. During the diploid phase, reproduction is conducted by spores, which develop individually and divide through mitosis, producing the gametophytes (sexual form of a plant) of the next generation. (For more information about mitosis, see the chapter "Cellular Basics.")

FUNGI IN THE ENVIRONMENT

Why are fungi important in recycling?

Fungi are the ultimate recyclers on Earth. As the primary decomposers in the biosphere, they break down organic matter, including dead plants and other vegetation. As fungi actively decompose materials, carbon, nitrogen, and the mineral components present in organic compounds are released—all of which can also be recycled, with carbon dioxide released into the atmosphere and minerals returned to the soil. It is estimated that, on average, the top 8 inches (20 centimeters) of fertile soil contain nearly 11,023 pounds (5 metric tons) of fungi and bacteria per 2.47 acres (1 hectare)! Without fungi acting as decomposers, dead, organic matter would overpower the world, and life on Earth would eventually become impossible.

Does fungi have an economic impact?

Yes, fungi have had a direct economic impact over time. For example, fungi produce gallic acid, which is used in photographic developers, dyes, and indelible ink, as well as in the production of artificial flavoring, perfumes, chlorine, alcohols, and several acids. Fungi are also used to make plastics, toothpaste, soap, and in the silvering of mirrors. In Japan, almost 1.10231131×10^9 pounds (500,000 metric tons) of fungus-fermented soybean curd (tofu and miso) are consumed annually. But on the negative side, different strains of the rust fungus *Puccinia graminis* cause billions of dollars of damage annually to food and timber crops throughout the world. (For more about rust fungi, see below.)

How does a plant disease called ergot affect humans and cattle?

A fungus known as *Claviceps purpurea* is known as the plant disease ergot. Ergot does have good uses, as it is used pharmaceutically to produce drugs used to induce labor in pregnant women and to control bleeding after childbirth; ergotamine, an ergot alkaloid, is used to treat migraine headaches.

But it also has a bad side: Eating bread and other grain products contaminated with ergot causes a disease called St. Anthony's fire. Common during the Middle Ages, this

What fungus may have played a role in the Salem witch trials?

Scientists now believe that the Salem witch hunts of 1692 may have initially been caused by an infestation of a microbiological poison: The fungus *Claviceps purpurea*, commonly known as rye smut that produces the poison ergot. When ingested, this poison produces symptoms similar to those exhibited by the girls who accused others of being witches in Salem. Historians and biologists have reviewed environmental conditions in New England from 1690 to 1692 and have found that conditions were perfect for an occurrence of rye smut overgrowth, as the weather was particularly wet and cool. In addition, rye grass had replaced wheat as the principal grain because wheat had become seriously infected with wheat rust during the cold and damp weather. The symptoms of ergot poisoning include convulsions, pinching or biting sensations, and stomach ailments, as well as temporary blindness, deafness, and muteness.

disease caused sensations of intense heat followed by a complete loss of sensation in an infected person's limbs; if cattle that graze on grains infected with ergot ingest enough, it can cause death or the spontaneous abortion of fetuses. But ergot is now much rarer thanks to improved techniques in grain production and milling.

What fungus uses a "shotgun" approach to survive?

The fungus *Pilobolus* spends much of its life around a cow pasture—specifically in cow dung. They are decomposers, breaking down organic matter as it feeds on the dung. But in order to survive, the fungus needs to get *into* the cow's dung: First, the cow eats the spores of the *Pilobolus* while it grazes; the tough spores pass through the animal's digestive tract, and when the animal excretes, the fungus is there, ready to grow.

But in order to spread, the fungus has evolved a way to shoot its spores into the grass to make it easier for the cows to grab and ingest. The fungus's stalk contains spore at the top, and as water pressure in the stalk increases, the tip eventually explodes— "shooting" the spores away at around 35 feet (10.8 meters) per second—at a height that can jump over cows and land up to 8 feet (2.5 meters) away. From there, the cows eat the spores that land in the grass, starting the cycle all over again.

Does a "good" relationship between fungi and ants exist anywhere?

The leaf-cutting ants, found in Central and South America as well as in the southern United States, have a mutually beneficial, or symbiotic, relationship with certain fungi of the genus *Septobasidium*. Because the ants are not able to digest the cellulose found in leaves, the fungus breaks down the cellulose—a food source for the fungus—and converts it into carbohydrates and proteins, which is perfect for the ants to digest. The ants then eat the fungus. Thus, the ants provide the fungus with a guaranteed food supply and eliminate competing fungi. Interestingly enough, the ants and fungus are not known to occur independently from each other.

Does a "bad" relationship between fungi and ants exist anywhere?

Of course, organisms don't always have a good relationship, including other species of ants and fungi. One bad relationship is the fault of an organism called the "zombie-ant fungus." This parasitic fungus changes the behavior of a tropical carpenter ant, *Camponotus leonardi*, found in the rainforest canopy of Thailand, causing the ants to become zombielike and die in a place that's advantageous for the fungi's reproduction. It's a process straight from a science fiction story: The fungus actually fills the ant's head and body, causing the muscles to atrophy; it also attacks the ant's central nervous system. The zombie-ant walks randomly, eventually falling to the leafy understory area—a cooler, more moist spot perfect for the fungus to thrive. The fungus continues its assault until the ant is dead, eventually growing a fruiting body through the ant's head—releasing more spores to be picked up by another ant.

Sometimes ants can be infected by a fungus that eventually kills them and uses the insect's body for nutrients.

It took until 2012 for researchers to figure out why the entire ant colony is not destroyed by the zombie-ant fungus. It turns out that a parasite fights the fungus—in fact, it's yet another fungus!

What are the special challenges of producing effective antifungal drugs?

Since fungi are eukaryotic, their cellular structure is similar to—if not the same as—that of animals and humans. Drugs that affect fungi often affect corresponding host cells, which can result in the host experiencing drug toxicity. Many antifungal drugs can only be used topically, but very few drugs have been found to be selectively toxic—that is, toxic to fungi and not their human hosts. What scientists try to do to make antifungal drugs work is interfere with the function or synthesis of ergosterol. Ergosterol is found in the cytoplasmic membrane of fungal cells but is not found in human cells. Thus, some antifungal drugs interfere with fungus-specific structures and functions, such as the cell wall.

What fungus plays an important role in human organ transplantation?

The fungus *Tolypocladium inflatum* is the source of cyclosporin, a medication that suppresses the immune reactions that cause organ transplant rejections. Cyclosporin does not cause the undesirable side effects that other immune-suppressing medications do. This remarkable drug became available in 1979, making it possible to resume organ transplants, which had essentially been abandoned. As a result of cyclosporin, successful organ transplants are almost commonplace today.

What is the difference between mold and mildew?

Molds and mildew are both types of fungi—but that is often where the similarities end. They differ in color: the molds are usually black, red, blue, or green; mildew is usually gray or white. Molds are most often slimy and fuzzy; mildew is typ-

If you find something dark and fuzzy growing on food, it is probably mold, not mildew.

ically powdery. Mold grows as multifilament hyphae, while mildew grows flat. Molds are usually found on rotted food, ceilings, and walls. Mildew is most often found in showers, bathtubs, and sinks—and even plants; it comes as powdery and downy. But both grow in any place in which damp, humid conditions exist.

Why are species of the genus *Neurospora* important?

Pink bread molds of the genus *Neurospora* have long served as powerful laboratory models used to study genetics, biochemistry, and molecular biology. Scientists first demonstrated the concept that one gene produces a corresponding protein by studying *Neurospora*. Its ease of growth and the extensive genetic information available for this organism make it a convenient model for the study of many processes found in higher plants and animals. Among the fungi, it is second only to yeast as a basic model organism.

How many species of fungi are plant pathogens?

More than 8,000 species of fungi cause disease in plants. In fact, most diseases found in both cultivated and wild plants are caused by fungi. Some pathogenic fungi grow and multiply in their host plants; others grow and multiply on dead organic matter and host plants. Fungi that are pathogenic to plants can occur below the soil surface, at the soil surface, and throughout the body of a plant. Fungi are responsible for leaf spots, blights, rusts, smuts, mildews, cankers, scabs, fruit rots, galls, wilts, and tree diebacks and declines, as well as root, stem, and seed rots.

What are rusts and smuts, and what effect do they produce in crops?

Rusts and smuts are very small fungi responsible for many serious plant diseases. Cereals and other grains are highly susceptible to attack by rusts and smuts. Many rusts and smuts have complicated life cycles, as they are known to use more than one plant species as a host during their lifetime. For example, wheat rust spends a portion of its life in barberry plants and a portion in wheat plants.

How were fungi involved in World War I?

During World War I the Germans needed glycerol to make nitroglycerin, which is used in the production of explosives such as dynamite. Before the war, the Germans had imported their glycerol, but the British naval blockade during the war prevented such imports. The German scientist Carl Neuberg (1877–1956) knew that trace levels of glycerol are produced when the yeast *Saccharomyces cerevisiae* was used during the alcoholic fermentation of sugar. He sought and developed a modified fermentation process in which the yeast would produce significant quantities of glycerol and less ethanol. The production of glycerol was improved—and Neuberg's procedure was implemented with the conversion of German beer breweries to glycerol plants, which produced 1,000 tons of glycerol per month. After the war ended, the production of glycerol was not in demand, so it was suspended.

What are some plant diseases caused by fungus?

Most of the discoveries concerning fungi were tied to the organisms' effects on plants and food. For example, in 1873, the fungus *Armillaria mellae* caused havoc to citrus plants by invading the root system of citrus trees. Since that time, over a hundred plant diseases have been discovered that are caused by fungi.

What are the ideal conditions for fungi to attack wood?

Fungi tend to attack wood when temperatures range between 50 and 90°F (10–32°C). Wood needs to be moist for fungi to grow: The most serious decay occurs when the moisture content of the wood is approximately 30 percent; wood with a moisture content of less than 20 percent will usually not decay, and any infection will not progress. Wood that is too wet will not decay because the excess moisture does not allow fungi sufficient access to air, thus impeding their growth.

What trees are highly resistant or highly susceptible to fungal decay?

In general, black locust, walnut, white oak, cedar, and black cherry trees are highly resistant to fungal decay. Species that are highly susceptible to fungal decay include aspen, willow, silver maple, and American beech trees.

What fungi rot different parts of trees?

Most fungi that rot wood in standing trees are basidiomycetes, the group that includes mushrooms and fleshy shelves. Most wood-rot fungi attack only one or two related species of trees, with conifers and deciduous trees more prone to fungi damage. In addition, many wood-rot fungi damage only specific parts of a tree. For example, *Ganoderma lucidum* and *Heterobasidion annosum* specifically rot roots and are rarely found in the higher portion of trees, whereas *Cerenna unicolor* and *Climacodon septentrionale* are more common in high sections of trees and are rarely found in the roots. *Laetiporus sulfurous* and *Fomitopsis pinicola* are found in all sections of a tree except in the smallest branches.

What two major United States tree species have been adversely affected by a fungus?

Two major tree species have been devastated by fungus in the United States—the American chestnut and the elm. The two species have been devastated by the chestnut blight and the Dutch elm disease. The American chestnut once made up almost half of the population of hardwood forests in central and southern Pennsylvania, New Jersey, and southern New England. In its entire range, the species dominated deciduous forests, making up almost one-quarter of the trees. By the mid-1900s, the fungus *Cryphonectria parasitica* caused a disease commonly known as chestnut blight and destroyed nearly every mature specimen of the American chestnut tree in the United States. The blight most often attacks older trees, infecting a tree's layers of living bark and the adjacent layers of wood; this often creates cankers that cause the bark on the trunk to crack. The fungus continues to kill the cells that carry the food made in the leaves to other parts of a tree, and nutrients are not able to reach various parts of the tree. Although the fungus leaves the roots unaffected, allowing a tree to send up new sprouts and saplings, within a number of years, the "older" bark and wood of those trees can also become infected.

The other major fungal disease attacked elms—trees that are susceptible to the fungus *Ophiostoma novo-ulmi* and *Ophiostoma ulmi*, which causes Dutch elm disease. The fungus spores (carried by bark beetles) enter the cells in the outermost wood of trees. As the fungus grows, the tree's cells become plugged, and water and nutrients are not able to move from the roots to the top of a tree—eventually causing the tree to die. Dutch elm disease is believed to have originated in the Himalayas and traveled to Europe from the Dutch East Indies in the late 1800s. It emerged in Holland just after World War I and was first identified in 1930 in Cincinnati, Ohio—thought to have been carried on elm logs imported from Europe. By 1940, the disease had spread to nine states; by 1950, it was found in seventeen states and had spread into southern Canada. Today it is found wherever elm trees grow throughout North America.

Have fungi been effective in biocontrol?

Biocontrol is defined as the use of one living organism to kill or control another organism. Fungi that parasitize insects are a valuable weapon for biocontrol. The spores of a parasitic fungus are sprayed on pest insects, attacking and controlling its host. Another fungus was identified as killing populations of silkworms as early as 1834; the spores of the same fungus are now used as a mycoinsecticide—a parasitic fungus used to kill insects—to control Colorado potato beetles. The spores of other fungi are used to control spittlebugs, leaf hoppers, citrus rust mites, and other insect pests.

MUSHROOMS AND EDIBLE FUNGI

What mushrooms have been identified as fossils?

Mushrooms are rarely found in fossil records since little in the structure of a soft mushroom can be fossilized—most eventually decay, leaving no imprint. But in 1990, two researchers in the Dominican Republic discovered a fossil of a fleshy, gilled mushroom. The mushroom, *Coprinites dominicana*, is about thirty-five to forty million years old (although this is debated) and is now one of five known fossil mushrooms from the tropics—together they are called coprinites. Microscopic examination suggests they are related to modern-day inky-cap mushrooms.

Are all large fungi shaped like common, store-bought mushrooms?

No, not all large fungi are shaped like the type of mushrooms we usually buy in the grocery store. In fact, the fruiting bodies of fungi come in a seemingly endless array of forms and colors. Many are variations on the familiar stalk-and-cap pattern of the common mushrooms sold in stores, although some have minute spore-bearing pores instead of gills on the undersides of their caps. And many fungi do not resemble mushrooms at all; for example, puffballs are solid, fleshy spheres, and bird's nest fungi form little cups containing "eggs" packed with spores. One kind of fungus looks like a head of cauliflower, and others resemble upright, branching clumps of coral. Some protrude from tree trunks like shelves, while others look like glistening blobs of jelly.

How many kinds of mushrooms are edible?

Among the basidiomycetes, approximately 200 varieties of edible mushrooms and about seventy species of poisonous ones exist. Some edible mushrooms are cultivated commercially; more than 844 million pounds (382,832 metric tons) are produced in the United States each year.

What are some of the more common, edible mushrooms?

Humans—and other animals—have consumed many mushrooms and other edible fungi over time. The following lists some of the more popular ones:

Common Name	Scientific Name
American matsutake	*Tricholoma magnivelare*
Beech mushroom	*Hypsizygus tessulatus*
Blewit	*Lepista nuda*
Black truffle*	*Tuber melanosporum*
Chanterelle*	*Cantharellus cibarius*
Chicken mushroom	*Laetiporus sulphureus*
Golden mushroom	*Flammulina velutipes*
Honey mushroom	*Armillaria mellea*
Meadow mushroom	*Agaricus campestris* (and others)

What mushrooms were considered sacred by the Aztecs?

Mushrooms of the genus *Conocybe* and *Psilocybin*—both with hallucinogenic properties—were considered sacred by the Aztecs, an empire located in Mexico before the 1500s. *Psilocybin*, which is chemically related to lysengic acid diethylamide (or LSD), is a component of both genera and responsible for the trancelike state and colorful visions (or psychedelic properties) experienced by those people who ingest these mushrooms. In fact, they are still used in certain religious ceremonies by the descendants of the Aztecs.

Common Name	Scientific Name
Morels*	*Morchella esculenta* (and others)
Oyster mushroom	*Pleurotus ostreatus*
Porcino*	*Boletus edulis*
Shiitake	*Lentinula edodes*
Straw mushroom	*Volvariella volvacea*
Trompette des morts*	*Craterellus cornucopioides*

*Wild, uncultivated mushroom.

What purpose do the gills of mushrooms serve?

Gills—the usually linear structures present on the underside of a mushroom's cap—serve two main purposes: The first is to maximize the surface area where the spores are produced, thus allowing for an increased number of spores; the second purpose is to help hold up the cap of the mushroom. Spores are produced in the basidia—specialized cells that line the surface of the gills. It has been estimated that a mushroom with a cap that is 3 inches (7.5 centimeters) in diameter can produce as many as forty million spores per hour.

What is a spore print?

A spore print is an important tool in the identification of mushrooms. To make a spore print, the stalk is removed and the cap placed gill side down on a piece of paper. The mushroom cap is covered with a glass and left undisturbed for several hours or overnight. The spores that are present in the mushroom will drop, and based on the color of the spore print, the sizes and shapes of the spores, and the overall pattern of the mushroom's gills, it can usually be identified. It is more difficult to get a spore print from a younger mushroom—the process works best with mature mushrooms.

How are mushrooms grown commercially?

The most common, commercially grown mushroom is the white button mushroom, *Agaricus bisporus*. Mushroom farms consist of special planting beds in buildings with controlled temperature and humidity. The beds are filled with soil mixed with material rich in organic matter. They are then inoculated with mushroom spawn—a pure culture of the mushroom fungus grown in large bottles on an organic-rich medium. The mycelium grows and spreads throughout the soil mixture for several weeks. Mushroom formation is induced by adding a layer of casing soil to the surface of the bed. Mushrooms appear on the surface of the bed through a process known as a "flash" with the mushrooms collected immediately after flashing, while they are still fresh.

Can poisonous mushrooms be identified reliably?

No general, reliable rules exist to identify poisonous mushrooms. Some of the edible varieties are quite easily recognized, but some edible varieties closely resemble poisonous mushrooms and can only be distinguished by an expert. The common lore that poisonous mushrooms make silver spoons turn black while mushrooms that can be peeled are edible is not true. Some of the deadliest mushrooms, amanitas, do not turn silver spoons black and can be peeled! The only rule to follow is that one must be able to identify a mushroom with certainty prior to eating it.

What toxic substances are produced by poisonous mushrooms?

The most toxic substances produced by poisonous mushrooms are amatoxins and phallotoxins. These toxins act by interfering with RNA and DNA transcription, inhibiting the formation of new cells. Besides the DNA and RNA damage, the toxins also collect in the liver, ultimately leading to liver failure.

Does an antidote for mushroom poisoning exist?

At this time, no effective antidote for mushroom poisoning in humans exists. The toxins produced by mushrooms accumulate in the liver and lead to irreversible liver damage. Unfortunately, it's possible that no indication of poisoning will happen for several hours after ingesting a toxic mushroom. When the symptoms do present, they often resemble typical food poisoning. Liver failure becomes apparent three to six days after ingesting a poisonous mushroom, with the only solution usually being an eventual liver transplant.

Can mushrooms grow up overnight?

A mushroom is only the fruiting body—that is, reproductive structure—of a much-larger fungus body that grows unseen in rotting logs, rich humus, and dark, damp places. Many familiar mushrooms have fruiting bodies that are fleshy and umbrella shaped. Warm, damp weather triggers their sudden appearance. Usually first to be noticed are small, round "button caps" composed of densely packed hyphae. Soon after the outer covering ruptures, the stem elongates, and the cap enlarges to its full size. This entire process can indeed happen overnight!

One such fast-growing fungus is the stinkhorn type called *Dictyophora indusiata*—considered one of the world's fastest-growing organisms. It pushes out of the ground at a rate of about 0.2 inches (0.5 centimeters) per minute. The growth rate is so fast that a crackling sound can be heard as the tissues of the fungus swell and stretch. During growth, a delicate, netlike veil forms around the fungus, giving this fungus its common name, "the lady of the veil." The fungus then decomposes and in the process produces a strong odor that is similar to the smell of decaying flesh. This odor attracts flies that crawl over the fungus and collect its spores on their feet. This process ensures that the spores are carried to new areas. Although the odor produced by species of *Dictyophora* is quite unpleasant, members of this genus are considered delicacies in China, where they are marketed as aphrodisiacs.

How is a fairy ring formed?

Long ago it was believed that the circles of mushrooms that sometimes form in meadows marked the locations where fairies gathered at night to dance. Fairy rings, or fungus rings, are frequently found in grassy areas. The three types of rings are those that do not affect their surrounding vegetation, those that cause increased vegetation growth, and those that damage their surrounding environment. The rings are started from the underground, food-absorbing part of a fungus (mycelium). The fungus growths are circular because a round, inner band of decaying mycelium forms underground. This band uses up the resources present in the soil that is directly above it. When the fungus forms caps that present above ground, the mushrooms grow around the mycelium, creating a ring effect. Each succeeding generation grows further from the center.

What are truffles, and where do they come from?

Truffles, a delight of gourmets, are arguably the most prized edible fungi. Found mainly in western Europe, they grow near the roots of trees (particularly oak, but also chestnut, hazel, and beech) in open woodlands. Unlike typical mushrooms, truffles develop 3 to 12 inches (7.6–30.5 centimeters) underground, making them difficult to find. Truffle hunters use dogs and pigs that have been specially trained to find the flavorful morsels. Both animals have a keen sense of smell and are attracted to the strong, nutlike aroma of truffles. In fact, trained pigs are able to pick up the scent of a truffle from 20 feet (6.1 meters) away. After catching a whiff of a truffle's scent, the animals rush to the origin of the aroma and quickly root out the precious prize. Once the truffle is found, the truffle hunter (referred to in French as a *trufficulteur*) carefully scrapes back the dirt to reveal the fungus. Truffles should not be touched by human skin, as doing so can cause the fungus to rot.

These black truffles are highly prized as a gastronomical delicacy. They are not only tasty, but hard to find and harvest, which makes them expensive.

What do truffles look like?

Truffles have a rather unappealing appearance—they are somewhat round, but irregularly shaped, and have thick, rough, wrinkled skin that varies from off-white to almost black in color. The fruiting bodies present on truffles are fragrant, fleshy structures that usually grow to about the size of a golf ball; they range from white, gray, or brown to nearly black in color. Nearly seventy varieties of truffles are known, but the most desirable is the black truffle—also known as black diamond—that grows in France's Perigord and Quercy regions, as well as Italy's Umbria region. The flesh of the black diamond appears to be black, but it is actually dark brown and contains white striations. The flesh has an aroma that is extremely pungent. The second most popular is the white truffle (actually off-white or beige) of Italy's Piedmont region. Both the aroma and flavor of this truffle are earthy and garlicky. Fresh truffles are available from late fall to midwinter and can be stored in the refrigerator for up to three days.

LICHENS

What are lichens?

Lichens are organisms that grow on rocks, tree branches, and bare ground. They are a combination of two different organisms living together in a symbiotic relationship: a population of either algal (green) or cyanobacterial cells that are single or have filaments and fungi. Lichens do not have roots, stems, flowers, or leaves. The fungal component of a lichen is called the mycobiont (from the Greek terms *mykes*, which means "fungus," and *bios*, meaning "life"), and the photosynthetic component is called the photobiont (from the Greek terms *photo*, meaning "light," and *bios*, meaning "life"). The scientific name (genus/species) given to the lichens is the name of the fungus, which is most often an ascomycete (for more about this type of fungus, see this chapter).

Who is considered the "father of lichenology"?

Swedish botanist Erik Acharius (1757–1819) is considered the father of lichenology. He was the founder of modern lichen taxonomy, having described and arranged lichens into forty distinct genera. His four major works of research—*Lichenographi suecic prodromus* (1798), *Methodus lichenum* (1803), *Lichenographia universalis* (1810), and *Synopsis methodica Lichenum* (1814)—formed the foundations of modern lichenology.

Do lichens have a symbiotic relationship with fungi?

Yes, lichens have a symbiotic relationship with fungi and were one of the first organisms recognized as having such a connection. Because the fungus has no chlorophyll, it cannot manufacture its own food, but it can absorb food from algae. The algae is a layer of usually single-celled plants near the surface, just below a jellylike layer of fungal hyphae (for more about hyphae, see this chapter).

In this mutualistic relationship, the algae captures nutrients that land on its surface and provides energy through the process of photosynthesis. In return, the fungus provides the algae with protection from the Sun, thus decreasing the loss of moisture and giving structure to the algae. Recent evidence, however, indicates that the photosynthetic cells may grow faster when they are separated from the fungus, and that may mean that the fungus is acting more like a parasite. A unique feature of this relationship is that it is so perfectly developed and balanced that the two organisms behave as a single organism, but it is interesting to note that these organisms are otherwise independent and can survive without each other.

Lichens such as these cup lichens are organisms that can grow well on bare ground, rocks, and bark. They are formed from two organisms living in a symbiotic relationship: a fungus and a photobiont (also called a phycobiont).

What is unusual about the natural distribution of lichens?

Lichens grow everywhere around the world. They are widespread because they are able to live and grow in some of the harshest environments on Earth, from arid desert regions to the Arctic; in fact, in Antarctica more than 350 species of lichens grow but only around two species of plants. Lichens grow on bare soil, tree trunks, rocks, fence posts, buildings, and on alpine peaks. Some lichens are so tiny that they are almost invisible to the naked eye; others like reindeer "mosses"—actually a lichen—can cover acres of land with ankle-deep growth. One species of the genus *Verrucaria* even grows underwater as a submerged marine lichen.

How many species of lichens exist—and are any edible?

Many species of lichens are found around the world. It is estimated that at least 20,000 different species exist, and more are found every year. And yes, some lichens are edible—but like hunting for wild, edible mushrooms, you really have to be an expert, as some lichens contain toxic acids. Historically, for example, the Black Tree lichen (*Alectoria*) was used by Montana Indians who would wash and soak the lichens, then cook them for one to two days in a steam pit. It was eaten or dried and powdered, then used as a mush or thickener later. (The Flathead Indian families reportedly ate 25 pounds [11.34 kilograms] of lichens each year.) Another edible lichen is the Reindeer moss (*Cetraria*)—often used in Switzerland for its benefits to the digestive system—and added to meats and pastries to retard spoilage. In fact, a lichen called Old Man's Beard (*Usnea*), once used for dye, is thought to contain antibiotic substances, and also used as an antimicrobial compress and dressing for open wounds. And Iceland moss *(Cetraria is-*

landica), found in the mountains of northern regions, is considered nutritious—but it was used mainly in earlier times for soups, stews, and even in breads.

How do lichens anchor themselves to a rock or other material?

The fungi produce a potent acid, which is used to etch holes in rock, wood, buildings, and other material to stabilize the lichens. Threadlike appendages are then inserted into the holes to anchor the organism, which is why we often see the lichens seemingly defying gravity on its side, or even upside-down, on a tree or rock ledge.

What creates the often bright colors of lichens?

Lichens come in varied and often bright colors, from bright greens to almost fluorescent-looking grays. The fungi's natural acid, when combined with an alcohol from the algae, produce acid crystals that help produce the colors of the lichens. The colorful pigments also play a role in protecting the photosynthetic partner—the algae—from the Sun's destructive rays.

What pigments produced by lichens are used in manufacturing?

Lichens are often strikingly colored, and their pigments are often extracted and used as natural dyes. The traditional method of manufacturing Scotland's famous tweeds made use of fungal dyes; synthetic dyes are now commonly used. Orchil, used to dye woolens, is a pigment produced specifically by lichens. In addition, several species of the lichen *Rocella* (from Africa) were used to produce litmus, a widely used acid-base (pH) indicator.

What lichens are often used by the perfume and cosmetic industries?

The musklike fragrance and fixative properties of lichens known as *Evernia prunastri* and *Pseudevernia furfurnacea* make them popular components of perfumes and cosmetics. The essential oils of these lichens are extracted with solvents; both lichens are common to southern France, Morocco, and the Serbo-Croatian peninsula.

What is the difference between a moss and a lichen?

Most of the "mosses" people see on trees—especially thanks to the saying "moss grows mainly on the north side of trees"—are really lichens. True mosses are definitely and distinctly green like other plants, whereas lichens are not as vividly green. In general, their shapes also differ, and of course, so do their reproductive habits.

What is the ecological role of lichens?

Lichens account for approximately 8 percent of the vegetation covering Earth's surface and are extremely important ecologically. In certain environments, such as regions of tundra, they cover vast areas of land. When they cover the ground, they prevent soil

> ## How were lichens and the food chain affected following the Chernobyl nuclear disaster?
>
> In 1986, following the Chernobyl nuclear power station disaster in Ukraine, arctic lichens as far away as Lapland were tested and showed levels of radioactive dust that were as much as 165 times higher than had been previously recorded. In addition, a human connection was found between the nuclear disaster and lichens: The lichens—being efficient absorbers of air-borne particles—are a primary winter food source for reindeer (*Rangifer tarandus*) in Scandinavia, and reindeer meat is commonly consumed by humans who live in those regions of tundra. The accumulated level of radioactive dust containing radiocesium became so high in the lichens that the reindeer meat became unsuitable for human consumption—and tragically, hundreds of tons of reindeer carcasses were disposed of as toxic waste.

from drying out; in desert areas they are able to capture and conserve the moisture present in fog and dew. Lichens also release nutrients, such as nitrogen and phosphorus, which are important for tree growth in many regions. Lichens are also an important food source for many species of animals, including wild turkeys and reindeer of the Arctic tundra. Birds such as the olive-headed weaver of Madagascar and the goldfinch of Europe use lichens to build their nests. It is thought that lichens have played a part in delaying global warming by consuming significant amounts of carbon dioxide (CO_2) during photosynthesis, but other factors may negate their effect.

What is the relationship between lichens and air pollution?

Because they absorb minerals from the air, rainwater, and directly from their substrate, lichens are extremely sensitive to pollutants in the atmosphere—especially sulfur dioxide—and thus can be used as bioindicators of air quality. In particular, lichens can absorb airborne pollutants in toxic concentrations. Thus, a high level of pollutants can cause the destruction of the lichens' (in other words, algae's) chlorophyll, which leads to a decrease in the occurrence of photosynthesis. Pollutants also upset the physiological balance between the fungus and the algae (or cyanobacterium), causing the lichens to degrade or completely die off.

YEASTS

How do yeasts differ from other fungi?

Yeasts are unicellular—single cells—throughout their entire life. About 600 different species of yeast have been discovered. Only a few of them cause disease in plants, animals, and humans; a few are used commercially; and the majority live in the soil, on plant surfaces, and decaying organic matter. The average yeast cell is 4 to 12 micrometers (MGRm) long, but with all the different species, this is merely an estimate.

How do yeasts reproduce?

Most yeast species' reproductive growth is very rapid, especially in certain substances like sugar. Certain yeast cells reproduce by fission, or splitting in two. But most yeasts reproduce by budding, in which the cell wall of the yeast swells, then forms a growth called a bud. The bud then breaks off, forming a new independent yeast cell, with each bud that separates from its mother yeast cell having the ability to grow into a new yeast cell.

How is yeast utilized in food and beverage manufacturing?

Yeast has often been called the first "domesticated animal," as they have been used in winemaking, beer making, and bread making for centuries. Yeast converts food into alcohol and carbon dioxide (CO_2) during fermentation. In the manufacture of wine and beer, the yeast's manufacture of alcohol is a desired and necessary component of the final product, with the CO_2 giving beer and champagne their bubbly attributes. Bread making requires the production of CO_2 by yeast, thus allowing certain doughs to rise. Yeasts used in brewing and baking are cultivated strains carefully kept to prevent contamination and used over and over; for example, certain bakers and bakeries make sourdough bread with yeasts that have been cultivated for tens of years. (For more about fermentation, see the chapters "Basics of Biology" and "Biology and You.")

What is the difference between baker's yeast and brewer's yeast?

Baker's yeast is used as a leavening agent to increase the volume of baked goods. It comes as active dry and compressed fresh yeast, and both are used as leavening agents to help a foodstuff like bread to rise. Active dry yeast is comprised of tiny, dehydrated granules of yeast. Although the granules are alive, the yeast cells are dormant due to

Is yeast used in making soy sauce?

Yes, yeast is used in the making of soy sauce, the dark-brown, salty liquid that was first produced in Japan to make soy beans more palatable. The old way of fermenting soy sauce is in two stages (note: not all soy sauce is made in this way): The soy beans are soaked, cooked, and mixed with roasted wheat. Then the fungus *Aspergillus oryzae* is added and kept aerobically active for up to forty hours. A paste forms and is put in a deep vat; the yeast *Saccharomyces rouxii* and lactobacilli are added—both preventing further growth of the *A. oryzae*. A month later, a liquid forms—with large concentrations of amino acids, simple sugars, and some vitamins—to produce what we call soy sauce.

their lack of moisture; because the cells are dormant, dry yeast has a long shelf life. Active dry yeast becomes active when mixed with a warm liquid. Compressed fresh yeast is moist and extremely perishable. It must be stored under refrigeration and used within one to two weeks.

Brewer's yeast is a much different foodstuff. As the name implies, it is a special non-leavening agent used in beer making—it converts the sugars in malted barley into alcohol. Dozens of specialized strains of yeast produce by-products that give each type of beer its unique taste, from all types of lagers to ales. And because it is a rich source of vitamin B, it is also used as a food supplement.

Why is the yeast *Saccharomyces cerevisiae* important in genetic research?

Biologists have studied *Saccharomyces cerevisiae*, a yeast used by bakers and brewers, for many decades because it offers valuable clues to aid in the understanding of how more advanced organisms work. For example, humans and yeast share a number of similarities in their genetic makeup. The DNA present in certain regions of yeast contain stretches of DNA subunits that are nearly identical to those in human DNA. These similarities indicate that humans and yeast both have similar genes that play a critical role in cell function.

In 1996, an international consortium of scientists from the United States, Canada, Europe, and Japan completed the genome of *S. cerevisiae*—the first eukaryotic organism to be completely sequenced. They found that the genome is composed of about 12,156,677 base pairs and 6,275 genes, organized on sixteen chromosomes; only about 5,800 are believed to be true functional genes. With their rapid generation time, yeasts continue to be the organism of choice to provide significant insights into the functioning of eukaryotic systems (plus, it's now known that yeast shares about 31 percent of its genome with that of humans). Thus, since the first sequencing, regular updates have been made to two main databases: the Saccharomyces Genome Database—a highly annotated and cross-referenced database for yeast researchers—and the Munich Information Center for Protein Sequences (MIPS).

PLANT DIVERSITY

EARLY PLANTS

What are some (approximate) major dates in plant evolution?

Although these dates change periodically—depending on new studies or fossils discovered—the following dates are important in plant evolution (remember, these dates are merely approximations):

- 3.6–2.8 billion years ago—The first blue-green algae to produce oxygen appeared.

- 472 million years ago—The first plants—probably liverworts, simple plants that lack stems or roots—adapt to land-water regions, such as tidal pools, marking the first steps aquatic plants took to move on land.

- 425 million years ago—The first vascular plants evolved on land.

- 350 million years ago—The first ferns and gymnosperms appeared on land.

- 300 million years ago—The first coal deposits formed from decaying plants.

- 190–140 million years ago—The first flowering plants evolved (although this is highly debated; another estimate is 215 million years ago).

- 65 million years ago—Around this time, most dinosaurs became extinct; it was also when the first flowering plants appeared.

What organisms represent the kingdom Plantae?

The kingdom Plantae (plants) now includes all eukaryotic organisms that have chlorophylls *a* and *b* as part of their photosynthetic package—they obtain most of their energy from sunlight through photosynthesis using chlorophyll contained in chloroplasts (chlorophyll is the pigment found in the chloroplasts that give plants their green color). The kingdom in some classifications may also include plants that are parasitic and may not

123

produce normal amounts of chlorophyll or photosynthesize. This kingdom includes green algae (such as *Ulva* and *Chara*), flowering plants, conifers, ferns, and mosses.

What conditions led to the origin of land plants?

Plants moving from the oceans to land probably occurred as the result of two main conditions: oxygen and changes in some photosynthetic organisms. Initially, the Earth's atmosphere had very low concentrations of oxygen; about one billion years ago, oxygen-generating organisms—bacteria and algae—eventually produced enough oxygen not only for plants to evolve on land, but also allowed the development of the ozone layer. This layer absorbed much of the Sun's ultraviolet rays—responsible for disrupting DNA—and thus, protected the organisms.

The changes in some of the photosynthetic organisms also contributed to the origin of land plants. In particular—and this took millions of years—some algae developed the ability to withstand short periods in which no water was available (desiccation); this probably included the development of a thicker layer to protect the cells. Eventually, these organisms would have to develop ways of moving nutrients from the ground (no longer the water) to the cells and special cells to help the plant support itself on land.

From what organisms did land plants probably originate?

Many scientists believe land plants evolved from green algae, especially the charaophytes (from *Charaphyceae*). Not only are the chloroplasts in the green algae the same as those of land plants, but the two share a number of biochemical and metabolic traits. In addition, both contain the same photosynthetic pigments: carotenes, xanthophylls, and chlorophylls *a* and *b*; cellulose is a major component of the cell walls of plants and algae, and both store their excess food (carbohydrates) as starch.

Why is an extinct plant called *Cooksonia* important?

To date, an extinct plant called *Cooksonia*—named for paleobotanist Isabel Cookson (1893–1973)—is important to understanding how plants eventually evolved on land. The first fossils of these macroscopic land plants were found in the Silurian period in Ireland and are about 425 million years old. The plant was very small (not even an inch high) and simple. It had a bifurcated (branched) stem topped with small spheres in which the spores formed; it also had water-conducting cells in the stems, but no roots or leaves. It is thought that this tiny organism spread in humid places on land and may be the ancestor of the flowers in your gardens.

How does a plant become a fossil?

Fossilization is dependent upon where organisms grow and how quickly they are covered by sediment. Because the organics in plants readily decay or are chewed apart by other organisms, paleobotanists rarely find the fossil remains of whole plants. Instead, only fossilized parts of plants are usually found. Because of this, scientists estimate that very few of the billions of flora that have lived on the Earth since plant life evolved have become fossils.

Fossilization of plants can occur in several ways. Compression fossils are often formed in water, in which heavy sediment flattens leaves or other plant parts. Over millions of years, as more layers of sediment accumulate, the weight literally squeezes out water present in the plant tissue, leaving only a thin film of tissue in the resulting rock. An impression fossil is left behind when the organism's remains have been completely destroyed, leaving only the imprint of the plant. Fossil molds and casts are formed when animal or plant tissues become surrounded by hardened sediment; the tissue then decays away. The hollow created by the tissue is called a mold; if a fossil mold becomes filled with sediment (or even minerals) over time, it results in a fossil called a cast.

Petrified wood is actually a fossil that is formed when the wood of ancient trees is chemically mineralized over thousands of years.

How does petrified wood form?

Certain plants do not become fossils but actually turn to stone. Petrified wood is one such fossil, as the plant remains are chemically altered by mineralization, thus eventually turning to stone. In a process that takes thousands of years, when water containing dissolved minerals—such as calcium carbonate or silicate—infiltrates wood (or plants), the minerals either replace or enclose the organism so the structural details of the plant remain. Botanists find these types of fossils to be very important since they allow for the study of the internal details of extinct plants.

How old are ginkgo trees?

The genus *Ginkgo*, commonly known as maidenhair trees, comprises one of the oldest living trees on Earth. This genus is native to China, where it has been cultivated for centuries. It has not been found in the wild; it is likely that it would have become extinct if it had not been cultivated. Fossils of 200-million-year-old ginkgoes show that the modern-day ginkgo is nearly identical to its ancestors, but only one living species of ginkgo remains, *Ginkgo biloba*, with its characteristic, fan-shaped leaves that turn a brilliant yellow and drop all at once in the autumn. The fleshy coverings of the seeds produced by females of the species *G. biloba* have a distinctly foul odor as they age; the male plants do not have seeds, of course, and thus, many horticulturalists and gardeners tend to cultivate the male trees.

Did land plants cause ice ages in the past?

In 2012, scientists were puzzled by evidence of ice sheets advancing in the Ordovician period between 488 and 444 million years ago. At that time, the continents were all clustered over what is now the South Pole and stretched as far as the Equator. Some scientists now believe that plants were the culprits: As the plants spread across and took root over dry land, they extracted minerals from the rocks they lived on. The scientists further offered several scenarios involving what happens to these minerals—all of which could eventually cause carbon levels to lower, along with the temperatures (in other words, almost the opposite of what we call "global warming"). The researchers suggest that the spread of terrestrial plants could have brought about a series of ice ages.

What is the relationship between ancient plants and coal formation?

Most of the coal mined today was formed from prehistoric remains of primitive land plants, particularly those of the Carboniferous period (approximately 300 million years ago). Five main groups of plants contributed to the formation of coal: three groups were all seedless, vascular plants (ferns, club mosses, and horsetails) and two groups which are now extinct (seed ferns and the primitive gymnosperms). Forests of these plants were located in low-lying, swampy areas that periodically flooded. When these plants died, they decomposed, but as they were covered by flood water, they did not decompose completely. As this material accumulated, layers of sediment covered the decaying plants, with the resulting heat and pressure of the layers eventually converting the plant material to coal. Various types of coal (lignite, bituminous, and anthracite) were formed, depending on the layers' exposure to varying temperatures and pressures.

HISTORICAL INTEREST IN PLANTS

What was the significance of *De Materia Medica*, written by the Greek physician Dioscorides?

De Materia Medica (*About Medicinal Materials*) was written by Greek physician Dioscorides (ca. 40–90 C.E.) in the first century C.E. The manuscript included the names and uses of the 600 plants that, at the time, were known to have medicinal properties. The purpose of the publication was to improve medical service in the Roman Empire. In addition to its medical use, it became the book most often used for plant classification in the Western world for nearly 1,500 years. During the fifteenth and sixteenth centuries, European botanists and physicians used *De Materia Medica* to formulate their "herbals"—illustrated books on the medicinal uses of plants.

Who is known as the "founder of botany"?

The ancient Greek scientist Theophrastus (ca. 372–287 B.C.E.) is known as the "founder of botany." His two works on botany, *On the History of Plants* and *On the Causes of Plants*, were so comprehensive that 1,800 years passed before any new significant botanical information was discovered. He integrated the practice of agriculture into botany and established theories regarding plant growth and the analysis of plant structure. He also related plants to their natural environment and identified, classified, and described 550 different plants.

What is *Gray's Manual of Botany*?

Gray's Manual of Botany, first published in 1848 by American botanist Asa Gray (1810–1888) under the title *Manual of the Botany of Northern United States, from New England to Wisconsin and South to Ohio and Pennsylvania Inclusive*, was the first accurate and modern guide to the plants of eastern North America. The publication contained keys and thorough descriptions of plants. The eighth, and centennial, edition was largely rewritten and expanded by Merritt Lyndon Fernald (1873–1950) and published in 1950. This edition was corrected and updated by R. C. Rollins and reprinted in 1987 by Dioscorides Press.

What contributions did John and William Bartram make to botany?

The first American botanist was John Bartram (1699–1777); along with his son, William Bartram (1739–1823), he traveled throughout the American colonies observing the flora and fauna. Although John Bartram never published his observations, he was considered the authority on American plants. In 1791, his son William published his notes on American plants and animals as *Bartram's Travels*.

What are some plant hormones and those responsible for their identification?

Many important plant hormones have been discovered over the years. The following lists some of the more interesting ones and their discoverers (for more about plant hormones, see the chapter "Plant Structure, Function, and Use"):

Auxins—English naturalist Charles Robert Darwin (1809–1882) and his son, English botanist Francis "Frank" Darwin (1848–1925), performed some of the first experiments on growth-regulating substances. They published their results in 1881 in *The Power of Movement in Plants*. In 1926, Dutch biologist Frits Warmolt Went (1903–1990) isolated the chemical substance responsible for elongating cells in the tips of oat (genus *Avena*) seedlings. He named this substance auxin, from the Greek term *auxein*, meaning "to increase."

Gibberellins—In 1926, the Japanese scientist Eiichi Kurosawa discovered a substance produced by a fungus, *Gibberella fujikuroi*, that caused a disease in rice called "foolish seedling disease"; the seedlings would grow rapidly but appear sickly and then fall over. In 1938, Japanese chemists Teijiro Yabuta (d. 1977) and Yasuke Sumiki (1901–1974) isolated the compound and named it gibberellin.

Cytokinins—In 1941, Johannes van Overbeek discovered a potent growth factor in coconut (*Cocos nucifera*) milk. In the 1950s, Swedish-born American plant physiologist Folke Skoog (1908–2001) was able to produce a thousandfold purification of the growth factor but was unable to isolate it. American biologist Carlos O. Miller (1923–2012), Skoog, and their colleagues succeeded in isolating and identifying the chemical nature of the growth factor, naming the substance kinetin—and the group of growth regulators to which it belonged cytokinins—because of their involvement in cytokinesis (or cell division).

Ethylene—Even before the discovery of auxin in 1926, ethylene was known to have effects on plants. In ancient times the Egyptians would use ethylene gas to ripen fruit. During the 1800s, trees near streetlamps (they burned ethylene gas for illumination) would become defoliated from leaking gas. In 1901, Russian botanist Dimitry Neljubov demonstrated that ethylene was the active component of illuminating gas.

Absicisic acid—Welch botanist Philip F. Wareing (1914–1996) discovered large amounts of a growth inhibitor in the dormant buds of ash and potatoes that he called dormin. Several years later in the 1960s, American botanist Frederick T. Addicott (1912–2008) reported the discovery in leaves and fruits of a substance capable of accelerating abscission (or the shedding of various parts of an organism, in this case, when a plant drops its leaves, fruit, flower, or seed) that he called abscisin. It was soon discovered that dormin and abscisin were chemically identical.

What were some early ideas about plant classification?

The earliest classifications of plants were based on whether the plant was considered medicinal or was shown to have other uses. *De re Rustica* by Roman statesman Marcus Porcius Cato (also known as Cato the Censor or Cato the Elder, 234–149 B.C.E.) lists 125 plants and was one of the earliest catalogs of Roman plants. Roman author and naturalist Gaius Plinius Secundus (23–79 C.E.), known as Pliny the Elder, wrote *Historia Naturalis*; published in the first century, it was one of the earliest significant plant catalogs in the ancient world, describing more than 1,000 plants. Plant classification became more complicated as more and more plants were discovered. One of the earliest plant taxonomists was the Italian botanist Caesalpinus (1519–1603). In 1583, he classified more than 1,500 plants according to various attributes, including leaf formation and the presence of seeds or fruit.

What early botanists published the first plant classifications?

English naturalist and botanist John Ray (1627–1705) was the first scientist to base plant classification on the presence of multiple similarities and features. His *Historia Plantarum Generalis*, published between 1686 and 1704, was a detailed classification of more than 18,000 plants. The book included a distinction between monocotyledon and dicotyledon flowering plants.

How have tree rings been used to date historical events?

The study of tree rings is known as dendrochronology. Every year, trees produce an annular ring composed of one wide, light ring and one narrow, dark ring. During spring and early summer, tree stem cells grow rapidly and larger, thus producing the wide, light ring. In winter, growth is greatly reduced and cells are much smaller, producing the narrow, dark ring. In the coldest part of winter or the dry heat of summer, no cells are produced. Comparing pieces of dead trees of unknown age with the rings of living trees allows scientists to establish the date when the fragment was part of a living tree. This technique has been used to date the ancient pueblos throughout the southwestern United States. A subfield is called dendroclimatology, or the study of the tree rings of very old trees to determine climatic conditions of the past. The effects of droughts, pollution, insect infestations, fires, volcanoes, and earthquakes are all visible in tree rings.

Other scientists added to the lists of early plant classifications. The French botanist J. P. de Tournefort (1656–1708) was the first to characterize genus as a taxonomic rank that falls between the ranks of family and species and included 9,000 species in 700 genera. The French botanist Antoine Laurent de Jussieu (1686–1758) published *Genera Plantarum*. The tome *Prodromus Systematis Naturalis Regni Vegetabilis* was started in 1824 by the Swiss botanist Augustin Pyrame de Candolle (1778–1841) and completed fifty years later. Yet another classification work, *Genera Plantarum*, was published between 1862 and 1883 by the English botanists George Bentham (1800–1884) and Sir Joseph Dalton Hooker (1817–1911).

Why was Carolus Linnaeus important to plant classification?

The Swedish naturalist Carolus Linnaeus (born Carl von Linné, 1707–1778) published *Species Plantarum* in 1753—a work that organized plants into twenty-four classes based on their reproductive features. Even today, the Linnaean system of binomial nomenclature remains the most widely used system for classifying all Earth organisms. It is considered an artificial system since it often does not reflect natural relationships.

What was the "flower clock" of Linnaeus?

Linnaeus also invented a floral clock to tell the time of day. He had observed over a number of years that certain plants constantly opened and closed their flowers at particular times of the day—with the times varying from species to species. Thus, he could deduce the approximate time of day according to which species opened or closed its flowers. Linnaeus planted a garden displaying local flowers, arranged in sequence of flowering throughout the day, that would flower even on cloudy or cold days. He called it a *horologium florae* or "flower clock."

129

Can a plant be patented?

New kitchen gadgets aren't the only things that need a patent; living plants are also patented. The first patent in the United States was received by Henry F. Bosenberg, a landscape gardener: U.S. Plant Patent no. 1 was given on August 18, 1931 for a climbing or trailing rose. And plenty more have been patented since: Between 1977 and 2011, just over 8,500 plant patents were granted to inventors (California and Florida with the most patents); worldwide, including the U.S., 18,376 plant patents were granted.

The plant patent inventors—growing plants from cucumbers to hydrangeas—have to follow certain criteria. According to the U.S. Patent and Trademark Office, the inventor "has to have invented or discovered and asexually reproduced a distinct and new variety of plant (not including a tuber-propagated plant or a plant found in an uncultivated state)." The grant lasts for twenty years and protects the inventor's rights by stopping others from asexually reproducing, selling, or using the plant so reproduced. Interestingly enough, according to the patent office, algae and macro fungi are considered plants, but bacteria are not!

Who developed plant breeding into a modern science?

American botanist Luther Burbank (1849–1926) developed plant breeding as a modern science. His breeding techniques included crosses of plant strains native to North America and foreign strains. He obtained seedlings that were then grafted onto fully developed plants for an appraisal of hybrid characteristics. His keen sense of observation allowed him to recognize desirable characteristics, enabling him to select only varieties that would be useful. One of his earliest hybridization successes was the Burbank potato, from which more than 800 new strains and varieties of plants—including 113 varieties of plums and prunes—were developed. More than twenty of these plums and prunes are still commercially important today. (For more about breeding and plants, see the chapter "DNA, RNA, Chromosomes, and Genetics.")

What were some of the steps to modern plant classifications?

Like most organism classification systems, plants have had a varied history. English naturalist Charles Robert Darwin's (1809–1882) ideas on evolution began to influence systems of classification during the late nineteenth century. The first major phylogenetic (relating to the evolutionary development of organisms) system of plant classification was proposed around the close of the nineteenth century. *Die natürlichen Pflanzenfamilien* (*The Natural Plant Families*), one of the most complete phylogenetic systems of classification and still in use through the twentieth century, was published between 1887 and 1915 by the German botanists Adolf Engler (1844–1930) and Karl Prantl (1849–1893). Their system recognizes about 100,000 species of plants, organized by their presumed evolutionary sequence. Systems of classifications in the twentieth century often focused on groups of plants, especially flowering plants, rather than all plants. American taxonomist Charles Bessey (1845–1915) was the first scientist to publish a system of classification in the early twentieth century.

What does the term cladistics mean?

Cladistics is one of the newest approaches to classification. It is often defined as a set of concepts and methods for determining cladograms, which portray branching patterns of evolution. Overall, it is a method of classification of animals and plants according to the proportion of measurable characteristics that they have in common. Thus, the organisms are grouped together based on whether or not they have one (or more) shared characteristics that are unique and that come from the group's last common ancestor; in addition, the characteristics cannot be present in more distant ancestors.

BOTANY BASICS

What is botany?

Botany is the study of plants, and although this is a simplistic explanation, today's plants are descendants of some of the oldest known organisms on Earth.

What are some subdisciplines in botany?

Botany is divided into a multitude of subdisciplines. The following lists some of the studies within the study of botany:

Agronomy—Using the plant sciences and applying it to crop production.

Bryology—The study of mosses and liverworts.

Economic botany—The study of how humans use plants—and how it affects the overall economy of countries.

Ethnobotany—The study of how indigenous peoples use plants, such as the native peoples of the Amazon rain forests.

Forestry—The study of forest management; it also includes people who study the use of forest products.

Horticulture—The study of ornamental plants, vegetables, and fruit trees.

Paleobotany—The study of fossil plants.

Palynology—The study of pollen and spores.

Phytochemistry—The study of plant chemistry, including the chemical processes that take place in plants.

Plant anatomy—The study of plant cells and tissues throughout their life cycles.

Plant ecology—Studying the role plants play in the environment.

Plant genetics—The study of genetic inheritance in plants.

Plant morphology—The study of plant forms and their life cycles.

Plant pathology—The study of plant diseases—their spread and how they affect humans and other animals.

Plant physiology—The study of plant function and development.

Plant systematics—The study of the classification and naming of plants.

How many species of plants live on Earth?

The exact number of plant species on Earth is hard to determine—mainly because so many plants have yet to be found and identified, such as in the deep forests of the Amazon River Basin. But overall, many scientists believe about 300,000 to 315,000 species exist, with the great majority (estimated at about 260,000–290,000) being seed plants.

What is the alternation of generations in plants?

All plants' sexual life cycles are characterized by something called the Alternation of Generations (capitalized or not), which means the generations essentially alternate between what are called diploid sporophytes and haploid gametophytes. Sporophytes (often represented as 2n) produce haploid spores as a result of meiosis; the spores grow into multicellular, haploid individuals known as gametophytes (often represented as n). Spores are the first cells of the gametophyte generation, and gametophytes produce ga-

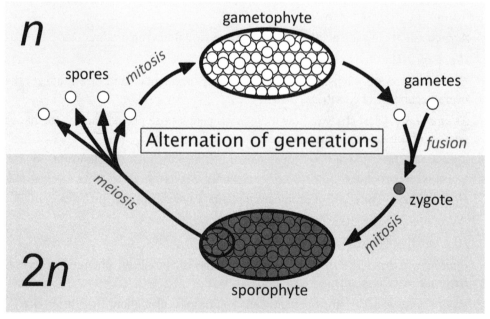

This diagram illustrates the Alternation of Generations in plants in which a a generation of diploid sporophytes is followed by one of haploid gametophytes.

metes as a result of mitosis. Male and female gametes fuse to form a zygote, which grows into a sporophyte, with the zygote being the first cell of the following sporophyte generation. Thus, the life cycle of the plant is complete. (For more about meiosis and mitosis, see the chapter "Cellular Basics.")

What are homosporous and heterosporous plants?

Certain plants are considered to be homosporous and heterosporous plants. Homosporous plants produce only one type of spore; heterosporous plants produce or have two types of spores (microspores and megaspores) that develop into the male gametophyte and female gametophyte respectively. (The gametophyte is the phase that produces the sexual organs of a plant; for more about gametophytes, see above.)

What are the phyla of plants?

The phyla—the major species of an organism—of plants include a multitude of species, all varying in characteristics, function, and number. The following lists the many phyla of plants (note: this is only one plant classification system—many others exist):

Phyla	Number of Species	Characteristics	Example
Bryophyta	12,000	Nonvascular	Mosses
Hepaticophyta	6,500	Nonvascular	Liverworts
Anthocerotophyta	100	Nonvascular	Hornworts
Psilophyta	6	Vascular, homosporous, no differentiation between root and shoot	Whisk ferns
Lycophyta	1,000	Vascular, homosporous or heterosporous	Club mosses
Arthrophyta	15	Vascular, homosporous	Horsetails
Pterophyta	12,000	Vascular, homosporous	Ferns
Cycadophyta	100	Vascular, heterosporous, seed-forming	Cycads (commonly known as "sago palms")
Ginkgophyta	1	Vascular, heterosporous, seed-forming, deciduous tree	Ginkgo
Gnetophyta	70	Vascular, heterosporous, seed-forming	Ephedra, shrubs, vines
Coniferophyta	550	Vascular, heterosporous, seed-forming	Conifers (pines, spruces, firs, yews, and redwoods)
Anthophyta	240,000	Vascular, heterosporous, seed-forming	Flowering plants

What are the major groups of plants?

Dividing plants is an interesting study, as so many types and varieties exist. But scientists have managed to divide plants into groups based on whether the plants have transport vessels or not. The following is the way most scientists organize the two groups (for more about these groups, see this chapter):

Bryophytes—These are the nonvascular plants. Because bryophytes lack a system for conducting water and nutrients, they are restricted in size and live in moist areas close to the ground. Examples of bryophytes are mosses, liverworts, and hornworts and are most often found in moist environments—mainly because they require water to reproduce sexually. (However, species inhabit almost every environment, from hot, dry deserts to the coldest regions of the Antarctica continent.) Overall, they are the second largest group of land plants after flowering plants. They are generally small, compact plants that rarely grow to more than 8 inches (20 centimeters) tall. They have parts that appear leaflike, stemlike, and rootlike and lack vascular tissue (xylem and phloem). Most species have certain structures that help them to retain moisture around their sperm-producing and egg-producing structures and large gametophytes that hold on to sporophytes.

Tracheophytes—These are the vascular plants, which are further divided into seedless plants and those that contain seeds. Plants with seeds are divided into flowering and nonflowering groups. Examples of seedless, vascular plants are ferns, horsetails, and club mosses. The cone-bearing conifers are seed-bearing, nonflowering vascular plants. It's interesting to note that the majority of plants on Earth are seed-bearing, flowering, vascular plants—known as angiosperms.

BRYOPHYTES

Why are bryophytes important to the study of early plants?

Some scientists believe bryophytes called liverworts are some of the closest living relatives of early land plants; they are thought to have evolved from freshwater, multicellular, green algae. Fossils of liverwort plants were found in 2010 in the Central Andean Basin of northwest Argentina. The scientists who made the discovery believe that this bryophyte plant—which lacked stems or roots—may be evidence that plants evolved on land ten million years earlier than previously thought. They found spores from the liverwort fossil that dated from between 473 and 471 million years ago, making these very simple plants the oldest land plant remains found to date.

What phase is dominant in the life cycles of bryophytes?

Like all plants, bryophytes—mosses, liverworts, and hornworts—exhibit alternation of generations. In this case, the haploid, or gametophyte, generation is the most conspicuous, dominant phase and persists for most of the plant's life. For example, a mat of moss consists of haploid gametophytes. The diploid, or sporophyte, generation lives only for a short time, growing out of the top of the gametophyte and depending on it for its food. (For more about the alternation of generations, see this chapter.)

How are bryophytes used as bio-indicators?

Bio-indicators are physiological, chemical, or behavioral changes that occur in organisms as a result of changes in the environment. Bryophytes of the genus *Hypnum* are particularly sensitive to pollutants, especially sulfur dioxide. As a result, most bryophytes are not found in cities and industrial areas. Mosses and liverworts, especially *Hypnum cupressiforme* and *Homalotecium serieceum*, have also been used as bio-indicators—one example was to monitor radioactive fallout from the Chernobyl nuclear reactor accident in 1986.

What bryophytes are closely related to green algae?

Hornworts are more closely related to green algae than to any other group of plants. Hornwort cells usually have a single, large chloroplast with a granular, starch-containing body (pyrenoid) similar to those in green algae. Mosses and liverworts are like all other plants because they have many dish-shaped chloroplasts per cell.

What are rhizoids?

Rhizoids are a characteristic feature of mosses, liverworts, and hornworts. Rhizoids are slender, usually colorless projections that consist of a single cell or a few cells. They serve to anchor mosses, liverworts, and hornworts to a substrate—rock, building, or most solid surfaces—and absorb water.

What feature of liverworts hints at their name?

Liverworts were named during the Middle Ages, when herbalists followed the theoretical approach known as the "Doctrine of Signatures." The core philosophy of this work was that if a plant part resembled a human organ or body part, it would be useful in treating ailments of that organ or part. For example, the thallus of thalloid liverworts resembles a lobed liver; therefore, in line with this philosophy, the plant was used to treat liver ailments. The word "liver" was combined with "wort" (meaning "herb"); thus, the name "liverwort." Today, liverworts provide food for foraging animals. Due to their ability to retain moisture, liverworts also assist in the decay of logs and aid in the disintegration of rocks into soil.

What plants are erroneously called mosses?

Not all plants called mosses are bryophytes. Irish moss (*Chondrus crispus*) and related species are actually red algae. Iceland moss (*Cetraria islandica*) and reindeer moss (*Cladonia rangiferina*) are lichens. Club mosses (genus *Lycopodium*) are seedless, vascular plants, and Spanish moss (*Tillandsia usneoides*) is a flowering plant in the pineapple family. (For more about lichens, see the chapter "Fungi.")

What is Elfin Gold?

Elfin Gold is another name for cave moss or luminous moss (*Schistostega pennata*)—a small plant with reflective, somewhat spherical cells at its tips. Native to the Northern Hemisphere, this moss gives off an eerie gold and greenish glow. The luminosity is caused by the reflection of light from the chlorophyll in the filament growing from the plant's germinating spore. Most often it forms green mats in caves, or cavities in wood, between rocks, or tree roots. In Japan, the plant has been the subject of numerous books, television shows, newspaper and magazine articles, and even an opera. A national monument to this species is located near the coast of Hokkaido, where the moss grows near a small cave.

Why are mosses important?

Some mosses are decomposers that break down the substrata and release nutrients used by complex plants. Mosses also play an important role in controlling soil erosion, providing ground cover and absorbing water. Mosses can also be indicators of air pollution, as under conditions of poor air quality, few mosses can exist. These organisms are also among the first organisms to invade areas that have been destroyed by a fire or volcanic eruption.

What are the uses of peat moss?

Peat moss (genus *Sphagnum*) grows mostly in bogs and is favored by gardeners for its ability to increase the water-holding capacity of soils. Due to large, dead cells in the leaflike parts, it is able to absorb five times as much water as cotton plants. Peat moss is also used as semimoist cushions by florists to keep other plants and flowers damp.

Species of *Sphagnum* also have medicinal purposes. Certain aboriginal people use peat moss as a disinfectant and, due to its absorbency, as diapers. Peat moss is acidic and an ideal dressing for wounds. Native North Americans used species of the genera *Mnium* and *Bryum* to treat burns. During World War I, the British used more than one million wound dressings made of peat moss.

TRACHEOPHYTES–FERNS

What are tracheophytes—vascular plants?

Scientists classify plants as bryophytes—nonvascular plants (see above)—or tracheophytes, which are vascular plants. They are represented by the seedless plants (such as ferns that reproduce by spores) and seed plants (separated into gymnosperms, or conifers, and angiosperms, or flowering plants). The word "vascular" comes from the Latin word *vasculum*, meaning "vessel" or "duct." Members of the extinct genus *Cooksonia*—named for the paleobotanist Isabel Cookson (1893–1973)—were the first ancient vascular plants to be identified. (For more about early plants, see this chapter.)

What is the oldest group of vascular plants?

The oldest group of vascular plants is still a hotly debated topic among botanists—especially as new plant fossils are found. It is believed that the first vascular plants were members of the division *Rhyniophyta*, which flourished around 400 million years ago but are now extinct. Still other scientists think the fossil record points to the club mosses (lycopods) as the most ancient—and in particular a genus of early lycopsid plants called Baragwanathia, found in Australian rocks from about 420 million years ago.

What are some types of ferns?

Ferns are seedless, vascular plants that reproduce by spores instead of seeds; they produce only one type of spore that develops into a bisexual gametophyte—giving these seedless plants the name homosporous. Their sporophytes are large, dominant, and nutritionally independent. They include the ferns of the genus *Pterophyta*, the largest group; the whisk ferns of the genus *Psilophyta*; the club mosses of the genus *Lycophyta*; and the horsetails of the genus *Arthrophyta*. These plants have leaves, roots, cuticles, stomata, specialized stems, conducting tissues, and, in most cases, seeds.

Tree ferns like these in Southland, New Zealand, evoke a sense of a primitive forest from the Cretaceous Period.

137

What vascular plant contains silica and was once used to clean and polish?

The outer tissues of a vascular plant called horsetail (a division of ferns) contains abrasive particles of silica—a mineral absorbed from the soil to give strength to the plant structure. In fact, because of their roughness, these plants were once called scouring rushes and were used by Native Americans to polish bows and arrows. Early North American settlers also used horsetails to clean their pots and pans along stream banks—these plants are found in abundance in such areas. You can still see horsetails growing today along stream and river banks, and moist areas, such as ditches. They may not be used as scouring pads anymore, but in herbal medicine, they have been used as a mild diuretic and for the healing of broken bones.

How does light affect the growth of ferns?

Light (and water, of course) controls spore germination in ferns. Wavelengths in the red range of the spectrum induce spore germination, while wavelengths in the blue range of the spectrum prevent spore germination. The term for the effects of wavelength on fern spore germination is called photomorphogenesis (this is not to be confused with photosynthesis, how light is used to produce energy for a plant to use, or phototropism, how the plant grows in the direction of light.)

What is a fiddlehead fern?

A fiddlehead is a type of fern composed of a rhizome (an underground stem that grows horizontally) that produces roots and leaves called fronds. As each young frond emerges from the ground, it is tightly coiled and resembles the top of a violin, hence the name fiddlehead. Cooks have also been known to enjoy fiddleheads—cooking the fern by steaming, simmering, or sautéing, and even raw in salads. The fern has a chewy texture and is said to taste like a cross between okra, asparagus, and green beans.

What is a *Lycopodium*?

A *Lycopodium* is a type of club moss—a division of ferns. They have horizontal branching stems, either above or below ground. Depending on the species, they send up tall shoots, ranging from around a half inch to a foot tall. They are an old group: About 200 to 250 million years ago, swampy forests of the northern hemisphere included a now extinct species of the club moss family—one that grew up to 100 feet (30 meters) in height. They were once used as flash powder for stage performances. Herbally, Native Americans used the tea as an analgesic to relieve pain after childbirth; more recently, the spores have been used for rashes.

TRACHEOPHYTES–GYMNOSPERMS

What are gymnosperms?

Gymnosperms (from the Greek terms *gymnos*, meaning "naked," and *sperma*, meaning "seed") produce seeds that are totally exposed (thus "naked," or not enclosed inside a fruit like angiosperms; for more about angiosperms, see below) or borne on the scales of cones. These "naked-seed" plants—truly a misnomer because the seeds are really highly protected under the scales of developing cones—first included the cycad and ginkgo; later, the sequoia, cypress, and pines evolved.

What are the major gymnosperm phyla?

The major phyla of gymnosperms are as follows: Coniferophyta (or Pinophyta) are conifers that include pine, spruce, hemlock, and fir; Ginkgophyta, that consists of one species, the ginkgo or maidenhair tree; Cycladophyta, the cycads or ornamental plants; and Gnetophyta, a collection of very unusual vines and trees, such as the *Ephedra* and *Gnetum*.

Why are gymnosperms important to the wood and paper industry?

Gymnosperms account for approximately 75 percent of the world's timber and a large amount of the wood pulp used to make paper. In North America the white spruce, *Picea glauca*, is the main source of pulp wood used for newsprint and other paper. Other spruce wood is used to manufacture violins and similar string instruments because the wood produces a desired resonance. The Douglas fir, *Pseudotsuga menziesii*, provides more timber than any other North American tree species and produces some of the most desirable lumber in the world. The wood is strong and relatively free of knots. Uses for the wood include house framing, plywood production, structural beams, pulp wood, railroad ties, boxes, and crates. Since most naturally occurring areas of growth have been harvested, the Douglas fir is being grown in managed forests. The wood from the redwood *Sequoia sempervirens* is used for furniture, fences, posts, some construction, and has various garden uses.

How else are gymnosperms important to other industries?

In addition to the wood and paper industry, gymnosperms are important in making resin and turpentine. Resin, the sticky substance in the resin canals of conifers, is a combination of turpentine, a solvent, and a waxy substance called rosin. Turpentine is an excellent paint and varnish solvent but is also used to make deodorants, shaving lotions, medications, and limonene—a lemon flavoring used in the food industry. Resin has many uses; it is used by baseball pitchers to improve their grip on the ball and by batters to improve their grip on the bat; violinists apply resin to their bows to increase friction with the strings; dancers apply resin to their shoes to improve their grip on the stage.

Do cycads produce the largest seed cones?

Yes, the largest seed cones are produced by cycads—short palmlike plants. They may be up to around 1 yard (1 meter) in length and weigh more that 3.3 pounds (15 kilograms).

Cycads are also considered one of the most ancient forms of naked-seed plants, covering forests by about 220 to 130 million years ago. An increase in the abundance of cycads also occurred during the Jurassic period—a time that is often referred to as the "Age of Cycads."

How do you tell fir, pine, and spruce trees apart?

The best way to tell the difference between these trees is by their cones and needles:

Species	Needles	Cones
Balsam fir	Needles are 1–1.5 in (2.54–3.81 cm) long, flat, and arranged in pairs opposite each other	Upright, cylindrical, and 2–4 in (5–10 cm) long
Blue spruce	Needles are roughly 1 in (2.54 cm) long, grow from all sides of the branch, are silvery blue in color, and are very stiff and prickly	3.5 in (8.89 cm) long
Douglas fir	Needles are 1–1.5 in (2.54–3.81 cm) long, occur singularly, and are very soft	Cone scales have bristles that stick out
Fraser fir	Similar to Balsam fir, but needles are smaller and more rounded	Upright, 1.6—2.4 inches (4–6 cm) long
Scotch pine	Two needles in each bundle; needles are stiff, yellow green, and 1.5–3 in (3.81–7.62 cm) long	2–5 in (5–12.7 cm) long
White pine	Five needles in each bundle; needles are soft and 3–5 in (7.62–12.7 cm) long	4–8 in (10–20.3 cm) long
White spruce	Dark-green needles are rigid but not prickly; needles grow from all sides of the twig and are less than an inch (2.54 cm) long	1–2.5 in (2.54–6.35 cm) long and hang downward

Do pine trees keep their needles forever?

Pine needles occur in groups, called fascicles, of two to five needles. A few species have only one needle per fascicle, while others have as many as eight. Regardless of the number of needles, a fascicle forms a cylinder of short shoots that are surrounded at their base by small, scalelike leaves that usually fall off after one year of growth. The needle-bearing fascicles are also shed a few at a time, usually every two to ten years, so that any pine tree, while appearing evergreen, has a complete change of needles every five years or less. Only a few native conifers shed all of their leaves in the fall—the bald cypress (*Taxodium distichum*) and the Larch (*Larix larcina*).

How long does it take to produce a mature pinecone?

From the time young cones appear on the tree, it takes nearly three years for them to mature. The spores of a pine tree are located on scalelike features (sporophylls) that are

Is a hemlock poisonous?

Two species are commonly known as hemlock: *Conium maculatum* and *Tsuga canadensis*. The first is a weedy plant, and all parts of it are poisonous. In ancient times, minimal doses of the plant were used to relieve pain, although with a great risk of poisoning from this form of treatment. This hemlock was also used to carry out death sentences in ancient times; for example, the Greek Athenian philosopher Socrates (469 B.C.E.–399 B.C.E.) was condemned to death and sentenced to drink a potion made from hemlock. The poisonous species should not be confused with *Tsuga canadensis*, a member of the evergreen family. Its leaves are not poisonous and are often used to make tea.

densely packed in structures called cones. Conifers, like all seed plants, are heterosporous, meaning that male and female gametophytes develop from spores produced by separate cones. Small pollen cones produce microspores that develop into the male gametophytes or pollen grains, and larger cones make megaspores that develop into female gametophytes. Each tree usually has both types of cones. This three-year process culminates in the production of male and female gametophytes, brought together through pollination, and the formation of mature seeds from the fertilized ovules. The scales of ovulate cones then separate, and the seeds are scattered by wind or wildlife. A seed that lands on a habitable place germinates, its embryo emerging as a pine seedling. (For more about gametophytes, see this chapter.)

What tree was once known only in fossil form?

Deciduous trees called dawn redwoods (*Metasequoia*) have bright green leaves in the summer that turn coppery red in the fall before they drop. Previously known only as a fossil, a living tree was found in China in 1941 and has been grown in the United States since the 1940s. The U.S. Department of Agriculture eventually distributed seeds to experimental growers in the United States—and the dawn redwood tree now grows all over the country.

Are giant redwood trees found only in California?

No, but although redwoods extend somewhat into southern Oregon, the vast majority of giant redwoods are found in California. The closest relative to this form of redwood is the Japanese cedar found in regions of Asia. This tree grows to a height of 150 feet (45.7 meters) with a circumference of 25 feet (7.6 meters). The genus *Sequoia* has two species, which are commonly known as the redwood and big tree. Both can be seen in either Redwood National Park or Sequoia National Park. At the latter park, the most impressive tree is known as the General Sherman Tree at 272 feet (83 meters) tall (see below); other trees found in Sequoia National Park exceed 300 feet (91.4 meters) in height, but are more slender.

What tree species from the United States have lived the longest?

Of the 850 different species of trees in the United States, the oldest species is the bristlecone pine, *Pinus longaeva*. This species grows in the deserts of Nevada and southern California, particularly in the White Mountains. Some of these trees are believed to be over 4,600 years old—with the potential life span estimated to be 5,500 years.

How do hardwoods differ from softwoods?

"Hardwood" and "softwood" are terms used commercially to distinguish woods. Hardwoods are the woods of dicots, regardless of how hard or soft they are, while softwoods are the woods of conifers. Many hardwoods come from the tropics, while almost all softwoods come from the forests of the northern temperate zone.

The "General Sherman" redwood tree in California's Sequoia National Park is one of the largest in the world, standing at height of 272 feet (83 meters).

Why do tree leaves turn color in the fall?

The carotenoids (pigments in the photosynthesizing cells)—responsible for the fall colors—are present in the leaves during the entire growing season. However, the colors are eclipsed by the green chlorophyll. Toward the end of summer, when chlorophyll production ceases, the other colors of the carotenoids (such as yellow, orange, red, or purple) become visible. Two factors are necessary in the production of red autumn leaves: Warm, bright, sunny days, during which the leaves manufacture sugar, which must be followed by cool nights with temperatures below 45°F (7°C). This weather combination traps the sugar and other materials in the leaves, thus resulting in the manufacture of red (anthocyanin). But a warm cloudy day restricts the formation of bright colors. With decreased sunlight, sugar production is decreased, and this small amount of sugar is transported back to the trunk and roots, where it has no color effect. (For more about plant pigments, see the chapter "Plant Structure, Function, and Use.")

What are some other long-living tree species in the United States?

Besides the bristlecone pine, the longest-living tree species in the United States are the giant sequoia (*Sequoiadendron giganteum*) with an average life span of 2,500 years, the redwood (*Sequoia sempervirens*) that can live an average life span of 1,000 to 3,500 years, the Douglas fir (*Pseudotsuga menziesii*) that can live to a ripe-old average age of 750 years, and the bald cypress (*Taxodium distichum*), with an average life span of 600 years.

What is, to date, the oldest living tree(s) in the world?

The potential age of the bristlecone pine is very young when compared to the world's oldest recorded tree—a spruce tree in the Dalarna province of Sweden that is thought to be 9,550 years old. This discovery, made around 2008, included several more old spruce trees, including another that was 9,000 years old (the ages were determined using carbon-14 dating methods), and others that were over 8,000 years old.

What is the oldest surviving species of tree in the world?

The oldest surviving species in the world is thought to be the maidenhair tree (*Ginkgo biloba*) of China. This species of tree first appeared during the Jurassic period, some 160 million years ago. Also called *icho*, or the *ginkyo* (meaning "silver apricot"), this species has been cultivated in Japan since 1100 B.C.E. Ginkgos are also grown in the United States and are known, of course, for their hardiness. (For more about ginkgos, see this chapter.)

What is the second oldest living tree?

The General Sherman Tree at Sequoia National Park, California, is about 4,000 years old—making it the oldest living tree next to the bristlecone pine. It is 272 feet (83 meters) tall and has a diameter of 32 feet (9.75 meters) and a circumference of 101 feet (30.8 meters). The weight of the tree is estimated to be more than 6,000 tons. Approximately 150 million years ago, these giant trees were widespread across the Northern Hemisphere. While the size of these giant trees implies that they are composed of very strong wood, the opposite is true. The wood is useless as timber because it is brittle and shatters into splintery, irregular pieces when struck. Perhaps the weakness of the wood is why so many of these giant trees still survive and have not been harvested by the logging industry.

TRACHEOPHYTES–FLOWERING PLANTS (ANGIOSPERMS)

What is one of the oldest-known fossil flowers found?

One of the oldest-known fossil flowers was discovered in 2002—a flowering plant estimated to be 125 million years old. Found in China by local farmers, the plant (*Archaefructus sinensis*) was about 20 inches (51 centimeters) high with thick stems stretching up in the water; from there, the stem would expose its pollen and seed organs above the water, with the seeds no doubt falling into the water and floating along the shore—germinating in the more shallow waters. The flower never bloomed—and does not look like modern angiosperms—but it has the qualifications of a flower. Besides being the oldest-known fossil flower, it is also thought to be an indication that modern flowers had their origins underwater.

143

What factors have contributed to the success of seed plants?

Seed plants became prolific over time for many reasons. For example, seed plants do not require water for sperm to swim to an egg during reproduction; pollen and seeds have allowed them to grow in almost all terrestrial habitats; the sperm of seed plants is carried to eggs in pollen grains by the wind or animal pollinators such as insects; the seeds themselves are fertilized eggs that are protected by a seed coat until conditions are proper for germination and growth; and finally, the smallness of most seeds make them easily dispersed by wind or animals.

What are the major characteristics of angiosperms?

Angiosperms are seed plants whose reproductive structures are fruits and flowers. They are the most diverse plant species in the world and include about 90 percent of all plants—more than 240,000 species. When angiosperms reproduce, their sperm does not have to "swim" with flagella (for more about flagella, see the chapter "Cellular Basics") in order for fertilization to occur, but can use numerous other ways to get the sperm to fertilize the plant, such as by wind, insects, and wildlife.

What define the two major groups of angiosperms—monocots and dicots?

Angiosperms are classified into two major groups, monocots and dicots. The description of monocots and dicots is based on the first leaves that appear on the plant embryo: monocots have one seed leaf, while dicots have two seed leaves. Approximately 65,000 species of monocots and 175,000 species of dicots exist. Orchids, bamboo, palms, lilies, grains, and many grasses are examples of monocots. Dicots include most trees that are nonconiferous, shrubs, ornamental plants, and many food crops.

Several more differences exist between the two groups. The seed leaves, also called cotyledons, differ: monocots have one cotyledon, while dicots have two cotyledons. Other differences include the floral parts—monocots have them usually in threes, dicots usually in fours or fives; roots—monocots have fibrous roots, dicots have taproots; and vacular bundles in the plant's stem—monocots are parallel, dicots are in a ring.

What is the most important angiosperm family and why?

Flowering plants (angiosperms) include the grass family, considered more important than any other family of flowering plants. The reason has to do with what humans around the world have eaten for centuries: the edible grains of these cultivated grasses, known as cereals, are (and have been) a basic food of most civilizations. Wheat, rice, and corn are the most extensively grown of all food crops, with other important cereals being barley, sorghum, oats, millet, and rye.

What is the most widely cultivated cereal in the world?

Wheat is the most widely cultivated cereal in the world; the grain supplies a major percentage of the nutrients needed by the world's population. Wheat is one of the oldest domesticated plants—with origins in the Near East at least 9,000 years ago—and it has been argued that it laid the foundation for Western civilization. Wheat grows best in temperate grassland biomes (for more about biomes, see the chapter "Environment and Ecology") that receive 12 to 36 inches (30 to 90 centimeters) of rain per year and have relatively cool temperatures. Some of the top wheat-producing countries are Argentina, Canada, China, India, the Ukraine, and the United States.

What are some economically important angiosperms?

Angiosperms produce lumber, ornamental plants, and a variety of foods. Just a very few of the economically important angiosperms are listed as follows:

Common Name	Genus Name	Economic Importance
Gourd	*Cucurbitaceae*	Food (melons and squashes)
Grass	*Poaceae*	Cereals, forage, ornamentals
Lily	*Liliaceae*	Ornamentals and food (onions)
Maple	*Aceraceae*	Lumber and maple sugar
Mustard	*Brassicaceae*	Food (cabbage and broccoli)
Olive	*Oleaceae*	Lumber, oil and food
Palm	*Arecaceae*	Food (coconut), fiber, oils, waxes
Rose	*Rosaceae*	Fruits (apple), ornamentals (roses)

What are the fastest-growing land and ocean plants?

Bamboo (*Bambusa spp.*), native to tropical and subtropical regions of Southeast Asia and islands of the Pacific and Indian Oceans, is the plant that gains height most quickly. Bamboo can grow almost 3 feet (1 meter) in 24 hours. This rapid growth is produced partly by cell division and partly by cell enlargement. The ocean plant that seems to set all records is a species of marine algae called the Pacific giant kelp (*Macrocystis pyrifera*), which can grow up to 150 feet (46 meters) or more in length—and has even been measured growing about 18 inches (45.72 centimeters) in a day.

What are carnivorous plants, and how are they categorized?

Carnivorous plants are plants that attract, catch, and digest animal prey, absorbing the bodily juices of prey for the nutrient content. Carnivorous plants can be divided into more than 400 species, with each one

The Venus Fly Trap, which is native to the U.S. east coast, adapted to nutrient-poor soil by becoming carnivorous.

classified according to the nature of their trapping mechanism. All carnivorous plants have traps made of modified leaves with various incentives or attractants—such as nectar or an enticing color—that can lure prey.

The three major types of traps are as follows: The Venus flytrap, *Dionaea muscipula*, and the bladderwort, *Utricularia vulgaris*, have active traps that rapidly imprison victims. Each leaf is a two-sided trap with trigger hairs on each side. When the trigger hairs are touched, the trap shuts tightly around the prey. Semi-active traps—such as those of the sundew (*Drosera capensis*) and butterwort (*Pinguicula vulgaris*)—employ a two-stage trap in which the prey is caught in adhesive fluid. As prey struggles, the plant is triggered to slowly tighten its grip. Finally, passive traps—with shapes resembling a vase or pitcher, such as the pitcher plant (*Sarracenia purpurea*)—entice insects using nectar. Once lured to the leaf, the prey falls into a reservoir of accumulated rainwater and drowns.

What is the difference between poison ivy, oak, and sumac?

These North American woody plants grow in almost any habitat and are quite similar in appearance. Each variety of plant has three-leaf compounds that alternate berrylike fruits and rusty brown stems. Poison ivy (*Rhus radicans*) grows like a vine rather than a shrub and can grow very high, covering tall, stationary items such as trees. The fruit of *R. radicans* is gray in color and is without "hair," and the leaves of the plant are slightly lobed. *Rhus toxicodendron*, commonly known as poison oak, usually grows as a shrub, but it can also climb. Its leaflets are lobed and resemble the leaves of oak trees, and its fruit is hairy. Poison sumac (*Rhus vernix*) grows only in acidic, wet swamps of North America. This shrub can grow as high as 12 feet (3.6 meters). The fruit it produces hangs in a cluster and ranges from gray to brown in color. Poison sumac has dark-green leaves that are sharply pointed, compound, and alternating; it also has inconspicuous flowers that are yellowish green. All parts of poison ivy, poison oak, and poison sumac can cause serious dermatitis.

What are succulents?

A group of more than thirty plant families, including the amaryllis, lily, and cactus families form what is known as the succulents (from the Latin term *succulentis*, meaning "fleshy" or "juicy"). Most members of the group are resistant to droughts, as they are dry-weather plants. Even when they live in moist, rainy environments, these plants need very little water.

PLANT STRUCTURE, FUNCTION, AND USE

PLANT STRUCTURES

How are plants defined?

Plants are defined in many ways due to their great number. In general, a plant is a multicellular, eukaryotic organism with cellulose-rich cell walls and chloroplasts that has starch as its primary carbohydrate food reserve. Most are photosynthetic and are primarily terrestrial, autotrophic (capable of making their own food) organisms.

What are the major characteristics of vascular (tracheophyte) plants?

The majority of vascular plants consist of roots, shoots, and leaves. The root system penetrates the soil below ground and anchors the plant; there, the roots absorb water and various materials necessary for plant nutrition. The shoot system consists of the stem and the leaves and is the part of the plant above ground level. The stem provides the framework for the positioning of the leaves; the leaves are the sites of photosynthesis. Growing plants maintain a balance between the size of the root system and the shoot system. The total water- and mineral-absorbing surface area in young seedlings usually far exceeds the photosynthesizing surface area. As a plant ages, the root-to-shoot ratio decreases. Additionally, if the root system is damaged, shoot growth is reduced by lack of water, minerals, and root-produced hormones.

What are some specialized cells in vascular plants?

All plant cells have several common features, such as chloroplasts, a cell wall, and a large vacuole. In addition, a number of specialized cells are found only in vascular plants. The following lists the main plant cells (for more information about cells, see the chapter "Cellular Basics")—also called ground tissue (that functions mainly in support, storage, and photosynthesis):

149

Parenchyma cells—Parenchyma (from the Greek *para*, meaning "beside," and *en* + *chein*, meaning "to pour in") are the most common cells found in leaves, stems, and roots. They are often spherical in shape with only primary cell walls and play a role in food storage, photosynthesis, and aerobic respiration. For example, most nutrients in plants such as corn and potatoes are contained in starch-laden parenchyma cells; they are also in the photosynthetic tissue of a leaf, the flesh of fruit, and the storage tissue of roots and seeds.

Collenchyma cells—Collenchyma (from the Greek term *kola*, meaning "glue") cells have thickened primary cell walls and lack secondary cell walls. They provide support for parts of the plant that are still growing, such as the stem.

Sclerenchyma cells—Sclerenchyma (from the Greek term *skleros*, meaning "hard") cells have tough, rigid, thick secondary cell walls hardened with lignin (the main chemical component of wood) that makes the cell walls more rigid. The two types are fiber cells, which are long, slender cells that usually form strands or bundles, and sclereid cells, sometimes called stone cells, which occur singly or in groups and have various forms with a thick, very hard secondary cell wall.

What are the types of plant tissue?

Plants have three types of tissue, each with its own function. They include the dermal tissue, which includes the epidermis, and the vascular tissue, which includes the xylem and phloem. The following gives more detail about these tissues:

Epidermis—Several types of specialized cells occur in the epidermis (outside), including guard cells and root hairs. Flattened epidermal cells—one layer thick and coated by a thick layer of cuticle—cover all parts of the primary plant body.

Xylem—Xylem (from the Greek term *xylos*, meaning "wood") is the main water- and mineral-conducting tissue of plants and consists of dead, hollow, tubular cells arranged end to end. The water transported in xylem replaces water that is lost through evaporation (through the leaf's stomata). Water-conducting cells come in two types. Water flows from a plant's roots up through the shoot via pits in the secondary walls of cells called tracheids; vessel element cells have perforations in their end walls that allow the water to flow between cells.

Phloem—The phloem conducts foods for the plant—including carbohydrates (mainly sucrose), hormones, amino acids, and other substances for the plant's growth and nutrition. The two kinds of cells in the phloem (from the Greek term *phloios*, meaning "bark") are sieve cells (in the seedless vascular plants and gymnosperms) and sieve-tube members (in angiosperms). Both are elongated, slender, tubelike cells arranged end to end with clusters of pores at each cell junction. Sugars (especially sucrose), other compounds, and some mineral ions move between adjacent food-conducting cells.

What are some of the major plant pigments?

Four major plant pigments give a plant its green color (for example, stems and leaves) and other colors (for example, certain colorful flowers). The following lists the pigments and where they are found (for more about plant pigments, see this chapter):

Betalains—The common types of these pigments are betacyanins and betaxanthins. They are found in flowers and fungi, and they represent the colors red to violet and yellow to orange.

Chlorophyll—Chlorophyll is what makes a plant's foliage and leaves green. It is also what enables plants to produce oxygen during photosynthesis.

Flavonoids—So far, scientists have identified more than 3,000 naturally occurring flavonoids. They are placed into twelve different classes, including the anthocyanins (the most prevalent), aurones, chalcones, flavonols, and proanthocyanidins. They are the reason for the many colors of certain flowers and are common in such plants as berries, eggplant, and citrus fruits, such as lemons and oranges. The colors are usually yellow, red, blue, and purple.

Carotenoids—The carotenoids consist of pigments such as carotene. They are most often found in green plants (often masked by the chlorophylls), mangoes, yams, carrots, and other brightly colored plants. They are typically orange, red, yellow, and pink. (Carotene is also important to humans, as our bodies break down that molecule to produce vitamin A molecules.)

Are flavonoids good for humans?

Yes, humans benefit from flavonoids, such as those found in dark chocolate, walnuts, strawberries, cinnamon, and blueberries. It is thought that the flavonoids all may lower cholesterol levels; in addition, some flavonoids even have health-benefitting antioxidant properties. And it's not only humans: The flavonoids in fruit attract animals that eat the fruit and thereby disperse the plant's seeds, while still other flavonoids influence flower coloration that attracts insect pollinators.

Where did a black dahlia get its color?

Scientists have long known the molecular mechanisms that give a plant called a dahlia its wide spectrum of color. These colors are created by the accumulation of flavonoids, with the over 20,000 varieties of dahlia cultivars ranging from white to purple. But the black dahlia is rare in comparison to the other colors. In 2012, scientists uncovered the reason: the black cultivars had a low concentration of flavones; this, in turn, favors the accumulation of the anthocyanins—and the dark color is formed.

What pigment has recently been found in plants?

A pigment called bilirubin, found in animals, was only recently found in plants. For example, in humans, bilirubin is responsible for the yellow color we see in people who have jaundice (as a result most often of poor liver function). It is formed when the pigment hemoglobin, which makes our blood red, breaks down into heme; that further breaks down from heme into bilirubin. Recently, bilirubin has been found in the orange fuzz on seeds of the white Bird of Paradise—but in this case, it's not hemoglobin breaking down, but chlorophyll molecules. Apparently, the chlorophyll has the same structures as heme—and after breaking down into another molecule, it eventually results in bilirubin.

What is the difference between primary growth and secondary growth?

Primary growth occurs in the tips of stems and roots in plants, thus increasing the length of the stems and roots. Secondary growth allows a plant to increase its diameter. The results of secondary growth form the division of a cylinder of cells around the plant's periphery.

What are meristems?

Meristems (from the Greek term *meristos*, meaning "divided") are unspecialized cells that divide and generate new cells and tissues. Apical meristems, found at the tips of all roots and stems, are responsible for a plant's primary growth, in which the root divides and produces cells inwardly and outwardly. The cells that are produced inwardly grow backward up the root, while the cells that are produced outwardly grow forward in the direction the root is growing (and create what is called a root cap). The vascular cambium and cork cambium are the meristems responsible for a plant's secondary growth.

What are determinate and indeterminate tomatoes?

Unlike many organisms that stop growing when they reach maturity, some plants continue to grow during their entire life span while other plants do stop at maturity. One good example of both types of plants is tomatoes: Unlimited, prolonged plant growth is described as indeterminate tomatoes; for example, the Peacevine cherry tomato is considered to be indeterminate, continuing to grow and produce fruit until a frost. A determinate tomato plant spreads laterally, and its fruits ripen all at once; for example, the Burbank tomato is considered determinate.

SEEDS

What is a seed?

A seed is a mature, fertilized ovule. The seed (of a dicot plant) is made up of a seed coat (for protection), an embryo, and the nutrient-rich tissue called the endosperm (or cotyledon). The seed embryo is actually a miniature root and shoot. Once the seed is protected and enclosed in a seed coat, it ceases further development and becomes dormant. It is interesting to note here that although the seed is not growing, it still "breathes," needing oxygen in order to stay viable enough to eventually grow. When it does begin to grow, the embryonic root (or radicle) is the first organ to emerge from the germinating seed (see below for more about the radicle).

What are the advantages of seed dormancy?

While a seed is dormant—when growth and development do not occur—this is the time for the dispersal of seeds. Thus, the plant can send its seeds into new environments because the dormancy assures survival. Seeds remain dormant until the optimum conditions of temperature, oxygen, and moisture are available for germination and further development.

What temperature conditions are necessary for seed germination?

Overall, the best temperatures for seed germination in most plants are 77 to 86°F (25–30°C). Depending on the variety, some seeds are able to germinate at temperatures ranging from around 40 to 86°F (5–30°C). In fact, some seeds need a certain amount of time in below-freezing temperatures in order to grow, called stratification; for example, parsley seeds must be around 40°F (5°C, usually in a refrigerator) for around two weeks to a month—to simulate a winter environment—so it will germinate better.

At the other extreme, some seeds, such as those of the Rocky Mountain lodgepole pine tree (*Pinus contorta*), require extreme heat for germination. The

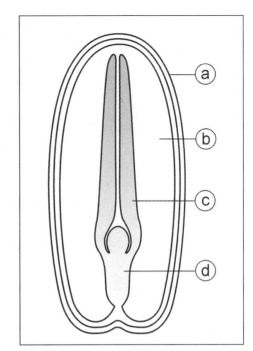

This simple seed diagram shows the a) seedcoat, b) endosperm, c) cotyledons, and d) embryo.

Depending on the plant species, different temperatures may be required in order for the seed to successfully germinate.

153

cones of this pine are covered with a resin that essentially "glues" the scales of the cone. The high temperatures (113–120°F [45–50°C]) associated with moderate to severe forest fires are required to melt the resin and release the seeds.

When a plant is said to be "double dormant," what does that mean?

Plants that are double dormant require a unique sort of layering or stratification in order for their seeds to germinate. The seeds of these plants must have a period of warmth and moisture, followed by a cold spell. Both the seed coat and the seed embryo require this double dormancy if they are to germinate. In nature, this process usually takes two years. Some well-known plants that live the life of double dormancy include lilies, dogwoods, junipers, lilacs, and tree peonies.

ROOTS

What is the first structure to emerge following germination?

The first structure to emerge following germination is the embryonic root (called a radicle). This structure allows the developing seedling to become anchored in the soil and to absorb water. This first root (also called the primary root) develops branch roots called lateral roots, which eventually send out additional lateral roots, creating the multibranched root system.

What is the root cap?

The root cap is a thimblelike mass of parenchyma cells (for more about these cells, see this chapter) that covers and protects the growing root tip as it penetrates the soil. The root cap is pushed forward as the root tip grows longer. The cells on the periphery of the root cap are thrown off as the root cap is pushed forward; new cells are added by the meristem. The root cap also protects the meristem, aids the root as it penetrates the soil, and plays an important role in controlling the response of the root to gravity (called gravitropism, or the response of a plant growth to gravity). From their origin until they are sloughed off, root cap cells live four to nine days, depending on the length of the root cap and the plant species.

What are the functions of the root system?

The major functions of roots are: 1) to anchor the plant in the soil; 2) the storage of energy resources, such as the carrot and sugar beet; 3) absorption of water and minerals from the soil; and 4) to conduct water and minerals to and from the shoot. The roots store the food (energy resources) of the plant, which is either used by the roots themselves or digested; the products of digestion are transported back up through the phloem to the above-ground portions of the plant. Plant hormones are synthesized in certain regions of the roots and transported upward in the xylem to the upper parts of the plant, thus stimulating growth and development.

How deep and wide does the root system of trees penetrate the soil?

How deep a root system penetrates the soil is dependent on moisture, temperature, composition of the soil, and the specific plant. Most roots actively absorb water and minerals. Although the "feeder roots" of most plants are found in the upper 3 feet (1 meter) of the soil, those of many trees are mainly in the upper 6 inches (15 centimeters) of the soil—the part of the soil richest in organic matter. In general, the lateral spread of a tree's root system is four to seven times greater than the spread of the crown of the tree.

What plants have the deepest root systems?

Plenty of root systems go deep underground—some have made it to record depths. The roots of the *Boscia albitrunca* (Shepherd's tree) reach 223.1 feet (68 meters) deep, and the *Acacia erioloba* (camel thorn) roots extend 196.9 feet (60 meters) into the ground, both in the Kalahari Desert; the *Juniperus monosperma* (one-seed juniper) roots go up to 200.1 feet (61 meters) deep in the Colorado Plateau; the *Eucalyptus spp.* in the Australian forests; and the roots of the desert shrub mesquite (*Prosopis juliflora*), which grow nearly 175 feet (53.5 meters) deep near Tucson, Arizona.

What are some causes of root damage?

Roots may be damaged by temperature extremes, drought, nematodes, and other soil microfauna, such as springtails that nibble the succulent roots. When the roots are damaged, cutting back the shoot system helps to re-establish the balance between the root and shoot systems. Unless a gardener is careful, transplanting seedlings and other plants can also damage roots, especially the smaller plants with more fragile root hairs, such as cucumbers.

What are aerial roots?

Aerial roots form on above-the-ground structures, such as a leaf or stem, instead of the roots in soil. They serve different functions in different species; for example, the banyan tree (*Ficus benghalensis*) and red mangrove (*Rhizophora mangle*) have aerial roots called prop roots since they support the plant. The aerial roots of ivy (*Hedera helix*) and Spanish moss (*Tillandsia usneoides*) cling to the surface of an object, providing support for the stem.

Why are banyan trees in Southeast Asia so amazing?

The banyan tree, *Ficus benghalensis*, native to tropical regions of Southeast Asia, is a member of the genus *Ficus*. It is a magnificent evergreen that can reach 100 feet (30.48 meters) in height. As the massive limbs spread horizontally, the tree sends down roots that develop into secondary, pillarlike supporting trunks. Over a period of years, a single tree may spread to occupy as much as 2,000 feet (610 meters) around its periphery.

Do monocot and dicot plant root systems differ?

Yes, the root systems of the monocot and dicot plants do differ. The root system of a monocot plant is a fibrous mass of roots that gives a plant broad exposure to soil water

155

Banyan trees like this one in Southeast Asia drop down aerial roots that become supportive trunks, allowing the tree to spread its canopy over hundreds, even thousands of square feet.

and minerals. The root system of a dicot plant consists of one large taproot with many small secondary lateral roots growing out of it.

What are root hairs?

Root hairs are tiny projections and outgrowths on the outermost layer of the root epidermis. They occur near the tips of roots, where they are abundant; in particular, they increase the surface area of the root system, allowing the roots to absorb water and minerals more efficiently. Root hairs are short-lived, and new ones are produced at approximately the same rate as older ones die. They are extensive, too: In a study on one rye plant, it was estimated that the plant had approximately fourteen billion root hairs, with an absorbing surface area of 480 square yards (401 square meters). If these root hairs were placed end to end, they would extend well over 6,214 miles (10,000 kilometers)!

What are the differences between a bulb, corm, tuber, tuberous root, and rhizome?

Many differences exist, but each of these structures is actually a modified stem that grows below ground. The following lists the major differences between each of these structures:

Bulb—Many times the term "bulb" is applied to any underground storage organ in which a plant stores energy for its dormant period. Dormancy is a device a plant utilizes to get through difficult weather conditions (winter cold or summer drought). A true bulb consists of fleshy scales containing a small basal plate (a modified stem from which the roots emerge) and a shoot. The scales that surround the embryo are modified leaves that contain the nutrients for the bulb during dormancy and early growth. Some bulbs have a tunic (a paper-thin covering) around the scales. The basal plate can also hold the scales together. New bulbs form from the lateral buds on the basal plate. For example, tulips, daffodils, lilies, and hyacinths are examples of bulb flowers.

Corm—A corm is actually a stem that has been modified into a mass of storage tissue. The eye or eyes at the top of the corm are growing points. The corm is covered by dry leaf bases similar to the tunic covering of the bulb, with the roots growing from the basal plate on the underside of the corm. New corms form on top of or beside the old one. Examples of corm-type flowers include gladiolus, freesia, and crocus.

Tuber—A tuber is a solid underground mass of stem like a corm, but it lacks both a basal plate and a tunic. Roots and shoots grow from eyes (growth buds) out of its sides, bottom, and sometimes its top. Some tubers are roundish. Others are flattened and lumpy. Some examples of tubers are gloxinia, caladium, and anemone.

Tuberous root—A tuberous root is a swollen root that has taken in moisture and nutrients and in many ways resembles a tuber. New growth occurs on the base of the old stem, where it joins the root. They can be divided by cutting off a section with an eye-bearing portion from where the old stem was attached. For example, dahlias have tuberous roots.

Rhizomes—A rhizome (or a rootstock) is a thickened, branching storage stem that usually grows laterally along or slightly below the soil surface. Roots develop downward on the bottom surface, while buds and leaves sprout upward from the top of the rhizome. A rhizome is propagated by cutting the parent plant into sections. For example, Japanese, Siberian, and bearded irises, cannas, and trilliums are rhizomes.

SHOOTS, STEMS, AND LEAVES

How does the shoot develop following germination?

Shoots are classified based on whether the cotyledons (seed leaves) are carried above ground or remain below ground. Seed germination during which the cotyledons are carried above ground is called epigeous. The food stored in the cotyledons is digested, and the products are transported to the growing parts of the young seedling. When the seedling becomes established and is no longer dependent upon the stored food in the seed for nutrition, the cotyledons gradually decrease in size, wither, and fall off.

Seed germination during which the cotyledons remain underground is called hypogeous. The seedling uses the stored food from the cotyledons for growth, then the cotyledons decompose. The cotyledons remain in the soil during the entire process.

What are the parts and functions of a stem?

All stems vary in shape and size; in addition, some plants have modified stems. For example, strawberry plants have runners (stolons) that are horizontal stems growing along the surface of the ground. Iris plants also have horizontal stems (rhizomes). Most stems have nodes and internodes: The nodes are the points where the leaves are attached to the stem, while the internodes are the parts of the stem between the nodes. The four main functions of stems are: 1) to support leaves; 2) produce carbohydrates; 3) store materials such as water and starch; and 4) transport water and solutes between roots and leaves and also provide the link between the water and dissolved nutrients of the soil and the leaves.

What are buds on stems?

Buds are called terminal or axillary on a stem. The terminal bud is at the upper part of the stem, where the plant growth is concentrated. It contains developing leaves and a compact series of nodes and internodes. The axillary buds are found in the angles formed by a leaf and the stem and are usually dormant.

What are the differences between thorns, spines, and prickles on a stem or branch?

The differences between thorns, spines, and prickles on plants are as follows: Thorns are modified branches or stems arising from the axils of leaves, their main purpose being to protect the plant from grazing animals. For example, hawthorn trees have true thorns. Spines are modified leaves that harden, giving protection to the plant. For example, spines are found on most cactus plants. Prickles are sharp outgrowths from the outer layer (epidermis) of various plants, especially the leaves and stems. For example, raspberry canes often have prickles.

What are trichomes?

Trichomes are hairlike outgrowths of the plant's outer layer (epidermis). They are often found on stems, leaves ("wooly" or "fuzzy" leaves are covered by trichomes), and reproductive organs. Trichomes have many functions: They increase the rate of water absorption and reduce the rate of water loss due to evaporation in order to keep the leaf surface

How can a gardener make certain plants and fruit trees "bushier"?

Many houseplants, annual plants, and fruit trees exhibit a phenomenon in which the terminal bud produces hormones inhibiting the growth of axillary buds. This allows the plant to grow taller, increasing its exposure to light. Under certain conditions, the axillary buds begin to grow, producing branches. This can occur when the terminal bud is pruned ("pinched back") on certain plants and fruit trees, stimulating the axillary bud growth and producing bushy, full-looking plants.

cool. They also provide a defense against insects, since the "hairiness" of the leaf impedes insect infestation; in addition, certain "hairs" release special chemicals to defend against potential damaging herbivores.

What are leaves?

Leaves are the main photosynthetic organ for plants; they are also organized to maximize sugar production while making sure little water loss occurs. Thus, they are also important in gas exchange and water movement throughout the whole plant. Leaves—outgrowths of the shoot tips—are found in a variety of shapes, sizes, and arrangements. Most leaves have a blade

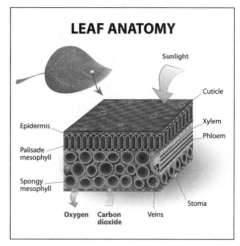

LEAF ANATOMY

Sunlight
Cuticle
Epidermis
Xylem
Phloem
Palisade mesophyll
Spongy mesophyll
Stoma
Oxygen   Carbon   Veins
dioxide

A cutaway of a typical leaf shows its interior structure.

(the flattened portion of the leaf), a petiole (the slender stalk of the leaf), and leaflike stipules (found on some leaves and located at the base of the petiole where it joins the stem). Cross-sections of a leaf also show a variety of features, including the cuticle (the outer covering to minimize water loss), veins (also called vascular bundles, which carry water and nutrients from the soil to the leaf and also carry sugar), stoma (plural stomata; water escapes through the stoma; they open and close), guard cells (modified epidermal cells that contain chloroplasts and control the opening of the stoma), and palisade and spongy mesophyll cells (for photosynthesis).

What is a cuticle in a plant leaf?

The cuticle contains a waxy substance, called cutin, that covers the parts of the plant exposed to the air: the stem and leaves. It is relatively impermeable and provides a barrier to water loss, thus protecting the plant from desiccation.

What is the purpose of the stomata?

Stomata (singular "stoma" from the Greek term *stoma*, meaning "mouth") are specialized pores in the leaves and sometimes in the green portions of the stems, as well as flowers and fruits. Carbon dioxide (CO_2) enters the plant through the stomata, while water vapor escapes through the same pores. The guard cells that border the stomata expand and contract to control the passage of water, carbon dioxide (CO_2), and oxygen (O_2).

What are guard cells in plants?

The guard cells in all plants have the same function: they regulate the exchange of gases and water by opening and closing the stomata pores. However, structurally they are different from dicot and monocot plants: Guard cells in dicots are kidney-shaped, while those in monocots are shaped like dumbbells.

How many leaves are on a mature tree?

Leaves are one of the most conspicuous parts of a tree—and each tree has a certain number of leaves at maturity. For example, a maple tree (genus *Acer*) with a trunk around 3 feet (1 meter) in diameter will have about 100,000 leaves in a good summer (that means no drought conditions or defoliation of trees from creatures like gypsy moths). An oak (genus *Quercus*) will have about 700,000 leaves, while an American elm (*Ulmus americana*) can have more than five million leaves in one season.

What are some examples of modified leaves?

Some plants have leaves that perform functions other than photosynthesis. While the tendrils of some plants are modifications to the stems and provide support for the plant, in other species, such as pea plants (*Pisum sativum*), the tendrils are modified leaves. In carnivorous plants, such as the Venus flytrap (*Dionaea muscipula*) and the pitcher plant (*Sarracenia purpurea*), the leaves attract, capture, and digest the insects with enzymes. Other examples of modified leaves include certain desert plants: Many grow mainly underground, with only a small transparent "window" tip protruding above the soil surface—allowing light to penetrate and reach the site of photosynthesis. The soil covering the leaf protects it from dehydration by the harsh desert winds.

In what ways are leaves economically important?

Many leaves are used for food and beverages, dyes and fibers, medicine, and industry—and the list is long. A few of these plants include, for food, plants, such as cabbage (*Brassica oleracea*), lettuce (*Lactuca sativa*), spinach (*Spinacia oleracea*), and most herbs—including parsley (*Petroselinum crispum*) and thyme (*Thymus vulgaris*)—which are grown for their edible leaves. Bearberry leaves (*Arctostaphylos uva-ursi*) contain a natural yellow dye, while henna leaves (*Lawsonia inermis*) contain a natural red dye. The leaves of palm trees are used to make clothing, brooms, and thatched huts in tropical climates. Aloe leaves (*Aloe vera*) are well known for treating burns and are also used in manufacturing medicated soaps and creams.

How does water move up a tree?

Water is carried up a tree through the xylem tissue in a process called transpiration. At the bottom of the tree, the roots absorb the vast majority of water that a tree needs. Above ground, the constant evaporation from the leaves creates a flow of water from roots to shoots. The properties of cohesion and adhesion allow the water to move up a tree regardless of its height: Cohesion allows the individual water molecules to stick together in one continuous stream, while adhesion permits the water molecules to adhere to the cellulose molecules in the walls of xylem cells. When the water reaches a leaf, it evaporates—thus allowing additional water molecules to be drawn up through the tree.

FLOWERS

What are the main parts of a flower?

Flowers have four main parts—all of which help to identify the flower and to allow the flower to be easily pollinated, often by specific pollinators. If a flower has all of these parts, it is called complete; if it lacks any of them, it is called incomplete. For example, some flowers may not have any petals or sepals. In terms of sexual reproduction in flowers, only stamens and pistils are necessary; some flowers contain only the male structures, while others only the female reproductive structures. Flowers with both structures are called perfect, but if they lack either one or the other they are called imperfect.

The following is a list of the plant parts and some of their components.

Sepals—The sepals are found on the outside of the bud or on the underside of the open flower. They serve to protect the flower bud from drying out. Some sepals ward off predators by displaying spines or secreting chemicals. Collectively, the sepals form the calyx.

Petals—The petals attract pollinators and are usually dropped shortly after pollination occurs. Collectively, the petals form a corolla.

Stamen—The stamen is the male part of a flower. It consists of a filament and anther, where pollen is produced.

Pistil—The pistil is the female part of a flower. It consists of the stigma, style, pollination tube, and ovary, which contains ovules. After fertilization, the ovules mature into seeds.

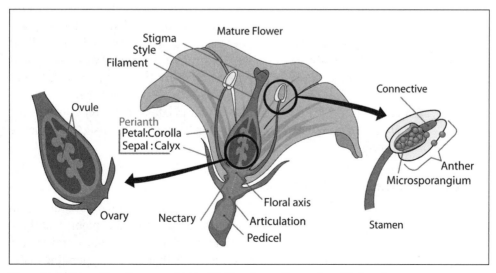

The parts of a flower. Many flowers can be identified based upon the number of their parts.

161

How are many flowers are easily identified based on their parts?

Many flowers can often be identified by their parts—in particular, the number of sepals, petals, and stamens. For example, a stonecrop succulent plant has thirty sepals, thirty petals, and thirty or sixty stamens; buckwheat has three to six sepals, no petals, and six or nine stamens. But it's not really that cut and dried—for example, a jewelweed has three or five unequal sepals, one forming a spur, and five petals, two united and three separate. And with so many species of plants—including trees identified by their seeds—it's no wonder identifying plants is truly a major study!

What are the types of plant pollination?

Effective pollination occurs when viable pollen is transferred to a plant's stigmas, ovule-bearing organs, or ovules (seed precursors). Without pollination, no fertilization would occur. Since plants are immobile organisms, they usually need external agents to transport their pollen from where it is produced in the plant to where fertilization can occur. This situation produces cross-pollination, wherein one plant's pollen is moved by an agent to another plant's stigma. Some plants are able to self-pollinate—transfer their own pollen to their own stigmas. But of the two methods, cross-pollination seems more advantageous, for it allows new genetic material to be introduced.

How does cross-pollination occur?

Cross-pollination agents include insects, wind, birds, mammals, and water. Many times flowers offer one or more "rewards" to attract these agents—sugary nectar, oil, solid food bodies, perfume, a place to sleep, or sometimes the pollen itself. Other times the plant can "trap" the agent into transporting the pollen. Plant structure can accommodate the type of agent used. For example, plants such as grasses and conifers, whose pollen is carried by the wind, tend to have a simple structure lacking petals, with freely exposed and branched stigmas to catch airborne pollen and dangling anthers (pollen-producing parts) on long filaments. This type of anther allows the light round pollen to be easily caught by the wind. These plants are found in areas such as prairies and mountains, where insect agents are rare. In contrast, semi-enclosed, non-symmetrical, long-living flowers such as irises, roses, and snapdragons have a "landing platform" and nectar in the flower base to accommodate insect agents such as the bee. The sticky, abundant pollen can easily become attached to the insect to be borne away to another flower.

How does a plant attract agents of pollination?

Agents of pollination include not only insects, but spiders, birds, and other animals. In general, the plants use color and fragrances as attractants to lure the majority of these agents. For example, a few orchids use a combination of smell and color to mimic the female of certain species of bees and wasps so successfully that the corresponding males will attempt to mate with them. Through this process (pseudocopulation), the orchids achieve pollination. While some plants cater to a variety of agents, other plants are very selective and are pollinated by a single species of insect only. This extreme pollinator specificity tends to maintain the purity of a plant species.

What are nectaries?

Plants secrete a variety of substances from specialized structures called secretory structures—nectaries are the structures that secrete nectar, a sugary compound that attracts insects, birds, or other animals. Most nectaries are associated with flowers and are called floral nectaries. Nectar is 10 to 50 percent sugar, especially sucrose, glucose, and fructose. Plants usually produce small amounts of nectar, which forces foraging animals to visit several flowers before obtaining a full meal. A single insect or bird can, therefore, pollinate tens or hundreds of plants.

PLANTS AND SOILS

How do the different types of soil affect plant growth?

Soil is the weathered outer layer of the Earth's crust and is a mix of tiny rock fragments and organic matter. The three broad categories of soils are clay, sandy, and loam. Clay soils are heavy, with the particles sticking close together; most plants have a hard time absorbing the nutrients in clay soil, and the soil tends to become waterlogged. On the other hand, these soils can be good for a few deep-rooted plants, such as mint, peas, and broad beans. Sandy soils are light and have particles that do not stick together, which is good for many alpine and arid plants, some herbs such as tarragon and thyme, and vegetables such as onions, carrots, and tomatoes. Loam soils are a well-balanced mix of smaller and larger particles that can provide nutrients to plant roots easily. They also drain and retain water well, and thus are considered ideal for plant growth.

What do the numbers on a bag of fertilizer indicate?

The three numbers on a fertilizer bag, such as 15-20-15, refer to the percentages by weight of macronutrients. The first number stands for nitrogen, the second for phosphorus, and the third for potassium. In order to determine the actual amount of each element in the fertilizer, multiply the percentage by the fertilizer's total weight in pounds. For example, in a 50-pound bag of 15-20-15, 7.5 pounds is nitrogen, 10 pounds is phosphorus, and 7.5 pounds is potassium. The remaining pounds are filler.

What is the best soil pH for growing plants?

Literally, pH stands for "potential of hydrogen" and is the term used by soil scientists to represent the hydrogen ion concentration in a soil sample. Soil testing below 7 is said to be acidic; soil testing above pH 7 is alkaline. Because it is based on logarithms with a base of ten, a soil with a pH of 5 is ten times as acidic as soil of pH 6 while a soil test-

What does hydroponics mean?

The term "hydroponics" refers to growing plants in some medium other than soil—usually a liquid solution, with the inorganic plant nutrients (such as potassium, sulfur, magnesium, and nitrogen) continuously supplied to the plants by the solution. It was first pioneered by Julius von Sachs (1832–1897), a researcher in plant nutrition; in 1937, William Frederick Gericke, a scientist at the University of California, defined the word hydroponics (although he asserted that the term was actually suggested by W. A. Setchell of the University of California)—and he proved its worth by growing twenty-five-foot tomato vines in his backyard using hydroponics.

Now hydroponics is mostly used in areas where little or unsuitable soil is available or poor weather conditions, such as growing vegetables in the Middle East or in Antarctica. In addition, because it calls for precise control of nutrient levels and oxygenation of the roots, it is often used to grow plants for research or commercial uses. And although successful for research, hydroponics has many limitations—which is why the average gardener who tries to grow plants hydroponically should have a great deal of patience!

ing pH 4 is one hundred times as acidic as soil testing pH 6. Nutrients such as phosphorous, calcium, potassium, and magnesium are most available to plants when the soil pH is between 6.0 and 7.5. Under highly acidic (low pH) conditions, these nutrients become insoluble and relatively unavailable for uptake by plants. However, some plants such as rhododendrons grow better in acidic soils. High soil pH can also decrease the availability of nutrients. If the soil is more alkaline than pH 8, phosphorous, iron, and many trace elements become insoluble and unavailable for plant uptake. (For more about pH, see the chapter "Basics of Biology.")

PLANT RESPONSES TO STIMULI

What is symbiosis?

Symbiosis is the close association of two or more different organisms. One type of symbiosis is known as mutualism, defined as an association that is advantageous to both parties. The most common (and possibly the most important) mutualistic, symbiotic relationship in the plant kingdom is known as mycorrhiza—a specialized association between the roots of plants and fungi that occurs in the vast majority of both wild and cultivated plants. (For more about symbiosis in animals, see the chapter "Aquatic and Land Animal Diversity"; for more about mycorrhiza, see the chapter "Fungi.")

What is tropism?

Tropism is the movement of a plant in response to a stimulus. The following lists the categories of tropism in plants:

Chemotropism—This is a response to chemicals by plants in which in-curling of leaves may occur.

Gravitropism—Formerly called geotropism, this is a plant's response to gravity, in which the plant moves in relation to gravity. Shoots of a plant are negatively geotropic (growing upward), while roots are positively geotropic (growing downward).

Hydrotropism—This is a plant's response to water or moisture, in which roots grow toward the water source.

Paraheliotropism—This is a response by the plant leaves to avoid exposure to the Sun.

Phototropism—This is a plant's response to light, in which the plant may be positively phototropic (moving toward the light source) or negatively phototropic (moving away from the light source). For example, the main axes of shoots are usually positively phototropic, whereas roots are generally insensitive to light.

Thermotropism—This is a response to temperature by plants and is greatly variable between plant species.

Thigmotropism or *haptotropism*—This is a plant's response to touch by the climbing organs of a plant. For example, the plant's tendrils—such as in a pea plant—may touch and then curl around a wire or string support.

What are turgor movements?

Turgor movements in plants are often reversible and are caused by changes in what is called the "turgor pressure" in specific cells. For example, some plants exhibit different flower positions during the day than at night, such as morning glories and four o'clocks; others, such as clovers and some beans, often display different leaf positions from day to night. One of the main reasons for the leaf reactions is changes in the internal water pressure, usually found in tissue at the base of leaflets. One of the more well-known turgor movements is with a Venus flytrap, which grabs an insect that touches the plant's leaf (trap). This is the result of a turgor movement from rapid changes in cellular water in the plant, allowing the plant to close and digest its prey.

What are the main plant hormones?

Many hormones control plants—all that help coordinate growth, development, and responses to environmental stimuli. The major ones are auxins (helps in growth, such as growing a plant upward rather than laterally), gibberellins (promote stem and leaf elongation), cytokinins (stimulate cytokinesis [dividing of the cytoplasm] and cell division), ethylene (promotes fruit ripening), and abscisic acid (inhibits growth, such as during a drought). (For more about plant hormones and their discoverers, see the chapter "Plant Diversity.")

What is the difference between short-day and long-day plants?

Short-day and long-day plants exhibit a response to the changes in light and dark in a twenty-four-hour cycle (photoperiodism). Short-day plants form flowers when the days become shorter than a critical length; they bloom in late summer or autumn in middle latitudes. For example, chrysanthemums, goldenrods, poinsettias, soybeans, and ragweed are short-day plants. Long-day plants form flowers when the days become longer than a certain, critical length; they bloom in spring and early summer. For example, clover, irises, and hollyhocks are long-day plants. These two responses are also why florists and plant growers at commercial nurseries can adjust the amount of light a plant receives—and force it to bloom out of season.

Can plants be parasitic?

Yes, plants can be parasitic, obtaining nutrients from, and thus harming, other plants. The link between parasites and their hosts is called a haustorium. In many plants, the link is one or more xylem-to-xylem connections between the two plants. The parasitic plant depends largely on the evaporation of water from its leaves as a means of pulling nutrient-containing water from the xylem of its host. In addition, the stomata of many parasitic plants always remain at least partially open, even at night, ensuring a continuous supply of nutrients from the host. For example, cancer root (*Orobanche uniflora*) parasitizes hardwood trees; trees such as sandalwood (genus *Santalum*) obtain their nutrients from nearby grasses.

Many parasitic plants lack chlorophyll and cannot carry out photosynthesis, and thus they depend entirely on their host for nutrients. In some cases, the presence of chlorophyll does not guarantee an independent lifestyle. For example, mistletoe (genus *Phoradendron*) and witchweed (genus *Striga*) are green, yet grow only as parasites. The green portions of these parasites contain only small amounts of certain enzymes required in photosynthesis.

What is allelopathy?

Allelopathy is the release of chemicals by certain plants that inhibit the growth and development of competing plants. The chemicals are usually terpenes or phenols and may

---

What was the historical significance of hemp?

During the early years of colonial America, hemp (*Agave sisalana*)—a fabric that looks and feels like linen—was as common as cotton is now. It was an easy crop to grow, requiring little water, no fertilizers, and no pesticides. It was used for uniforms of soldiers, paper (the first two drafts of the Declaration of Independence were written on hemp paper), and as an all-purpose fabric—including Betsy Ross's flag made of red, white, and blue hemp.

be found in roots, stems, leaves, fruits, or seeds. An example of this relationship among plants is the black walnut tree (*Juglans nigra*): A chemical compound in the leaves and green stems of the black walnut tree is leached by rainfall into the soil; it is then hydrolyzed and oxidized into another compound called juglone—a compound shown to be very toxic to many plants as well as an inhibitor of seed germination. For example, tomatoes and alfalfa will wilt when grown near black walnuts, and their seedlings will die if their roots contact the tree's roots. Similarly, white pine (*Pinus strobus*) and black locust (*Robinia pseudoacacia*) are often killed by black walnuts growing in their vicinity.

PLANT USES

How is maple syrup harvested?

Maple syrup is harvested from the trunks of sugar maple trees (*Acer saccharum*). Production of maple syrup requires daily temperatures to fluctuate between freezing and thawing. During cold nights (below freezing), starch made during the previous summer and stored in wood is converted to sugar. During the day when temperatures rise above freezing, a positive pressure is created in the xylem's sapwood. When a maple sap harvester drives tubes (called spiles) into the sapwood, the positive pressure pushes the sugary sap out of the tree trunk at a rate of 100 to 400 drops per minute—which is why around February in the northern regions with sugar maples, metal pails are often seen hanging from the trees. The flow stops when temperatures drop below freezing.

How did the navel orange originate?

Navel oranges are oranges without seeds. In the early nineteenth century an orange tree in a Brazilian orchard produced seedless fruit, while the rest of the orchard produced seeded oranges. This naturally occurring mutation gave rise to what we now refer to as the navel orange. A bud from the mutant tree was grafted onto another orange tree; the branches that resulted were then grafted onto other trees, soon creating orchards of navel orange trees. Every time you eat a navel orange, remember that it was derived from pieces of the tree that first produced the mutated fruit!

How long can an orange tree produce oranges?

An average orange tree will produce fruit for fifty years, but eighty years of productivity is not uncommon, and a few trees are known to be still producing fruit after more than a century. An orange tree may attain a height of 20 feet (6.1 meters), but some trees are as much as 30 feet (9.1 meters) high. Orange trees grow well in a variety of soils but prefer subtropical settings.

How are seedless grapes grown?

Since seedless grapes cannot reproduce in the same manner that grapes usually do (for example, dropping seeds), growers have to take cuttings from the plants, root them, then plant the cuttings. Seedless grapes come from a naturally occurring mutation in which the hard seed casing fails to develop. Although the exact origin of seedless grapes is unknown, they might have been first cultivated thousands of years ago in present-day Iran or Afghanistan. Currently, 90 percent of all raisins are made from Thompson seedless grapes.

Do seedless watermelons occur naturally?

Seedless watermelons are relatively new in the fruit world, first introduced in 1988 after fifty years of research. A seedless watermelon plant requires pollen from a seeded watermelon plant. Farmers frequently plant seeded and seedless plants close together and depend on bees to pollinate the seedless plants. The white "seeds," also known as pods, found in seedless watermelons serve to hold a fertilized egg and embryo. Because a seedless melon is sterile and fertilization cannot take place, pods do not harden and become a black seed, as occurs in seeded watermelons.

What are some specific examples of how plants are economically important?

Materials of plant origin are found in a wide variety of industries, including paper, food, textile, and construction—a list too long to mention here. But familiar examples are available in abundance: Chocolate is made from cocoa seeds, specifically seeds of the species *Theobroma cacao*. Foxglove (*Digitalis purpurea*) contains cardiac glycosides used to treat congestive heart failure. The berries obtained from the plant *Piper nigrum* produce black pepper; the berries are dried, resulting in black peppercorns, which can then be cracked or ground. Tea can be made from the leaves of *Camellia sinensis*. Fibers taken from the stem of flax plants (*Linum usitatissimum*) have been used to make linen, while the seeds are commonly consumed and are a source of linseed oil. Paper money is even made from flax fibers!

The cardiac glycosides digitoxin contained in purple foxglove can be lethal if eaten, but in carefully administered doses it is used to treat congestive heart failure.

How much wood is needed to make one ton of paper?

In the United States, wood pulp is usually used in paper manufacturing. Pulp is measured by cord or weight. Although the fiber used in making paper is derived overwhelmingly from wood, many other ingredients are needed as well. One ton of paper typically requires two cords of wood, 55,000 gallons (208,000 liters) of water, 102 pounds (46 kilograms) of sulfur, 350 pounds (159 kilograms) of lime, 289 pounds (131 kilograms) of clay, 1.2 tons of coal, 112 kilowatt hours of power, 20 pounds (9 grams) of dye and pigment, and 108 pounds (49 kilograms) of starch. Other ingredients may also be necessary.

What wood is used to make baseball bats?

Wooden baseball bats are made from white ash (*Fraxinus Americana*). This wood is ideal for producing bats because it is tough and light and can thus help drive a ball a great distance. A tree roughly seventy-five years old and 15.7 inches (40 centimeters) in diameter can produce approximately sixty bats.

What are frankincense and myrrh?

Frankincense is an aromatic gum resin obtained by tapping the trunks of trees belonging to the genus *Boswellia*. The milky resin hardens when exposed to the air, forming irregularly shaped granules—the form in which frankincense is usually marketed and sold. Also called olibanum, frankincense is used as an ingredient in many different products, including pharmaceuticals, perfumes, fixatives, fumigants, and incense. Myrrh comes from a tree of the genus *Commiphora*, native to the northeastern region of Africa and the Middle East. Myrrh is also a resin obtained from trees; it is used in pharmaceuticals, perfumes, and toothpastes.

How is commercial cork cultivated?

Commercial cork is the outer bark of the cork oak (*Quercus suber*) grown in the western Mediterranean. The first layer is commercially useless; thus, it is removed from the tree and discarded when the tree is approximately ten years old. When the tree is twenty to twenty-five years old and has a diameter of approximately 15.75 inches (40 centimeters), a usable cork layer 1.2 to 3.9 inches (3–10 centimeters) thick can be harvested. A similar layer can be harvested approximately once every ten years until the tree is approximately 150 years old.

The cork of a cork oak breaks away at the cork cambium and can be peeled off without harming the tree. Cork consists of densely packed cells (about one million cells per

What is replacing natural cork for wine stoppers?

Plastic "corks" are replacing natural cork for wine stoppers—and for a good reason. During the 1980s and early 1990s, bad cork was traced to the fungal contaminant called 2,4,6-trichloroanisole, TCA. Not only is it not good to drink such a fungus, but TCA also flattens the taste of the wine—removing the flavors the winemaker worked so hard to produce.

cubic centimeter) that contain the plant wax suberin, making cork impermeable to liquids and gases. Half of its volume is trapped air; therefore, it is four times lighter than water. It is virtually indestructible, fire-resistant, and durable; resists friction; and absorbs vibration and sound.

What was one of the most famous criminal cases involving forensic botany?

Forensic botany is the identification of plants or plant products used to produce evidence for legal trials. One of the first criminal cases to use forensic botany was the famous 1935 trial of Bruno Hauptmann (1899–1936), who was accused, and later convicted, of kidnapping and murdering the son of Charles and Anne Morrow Lindbergh. The botanical evidence presented in the case centered on a homemade wooden ladder used during the kidnapping and left at the scene of the crime. After extensive investigation, the plant anatomist Arthur Koehler (1885–1967) showed that parts of the ladder were made from wooden planks taken from Hauptmann's attic floor.

What is herbal medicine?

Herbal medicine treats disease and promotes health using plant materials. For centuries, herbal medicine was the primary method of administering medically active compounds. The following lists some herbs used for medicinal purposes:

Herb	Botanical Name	Common Use
Aloe	*Aloe vera*	Skin, gastritis
Black cohosh	*Cimicifuga racemosa*	Menstrual problems, menopause
Dong quai	*Angelica sinensis*	Menstrual problems, menopause
Echinacea	*Echinacea angustifolia*	Colds, immune health
Evening primrose oil	*Oenothera biennis*	Eczema, psoriasis, menopause
Feverfew	*Tanacetum parthenium*	Migraine headaches
Garlic	*Allium sativum*	Cholesterol, hypertension
Ginkgo biloba	*Ginkgo biloba*	Cerebrovascular insufficiency, memory problems
Ginseng	*Panax ginseng, Panax quinquifolius, Panax pseudoginseng, Eleutherococcus senticosus*	Energy, immunity, libido
Goldenseal	*Hydrastis candensis*	Immune health, colds
Hawthorne	*Crateaegus laeviagata*	Cardiac function
Kava kava	*Piper methysticum*	Anxiety
Milk thistle	*Silybum marianum*	Liver disease
Peppermint	*Mentha piperita*	Dyspepsia, irritable bowel syndrome
Saw palmetto	*Serona repens*	Prostate problems
St. John's Wort	*Hypericum perforatum*	Depression, anxiety, insomnia
Tea tree oil	*Malaleuca alternifolia*	Skin infections
Valerian	*Valeriana officinalis*	Anxiety, insomnia

What are some common culinary herbs?

Herbs are often used to enhance flavors in food. They usually come from the leaves of nonwoody plants and sometimes the fruits or bulbs. The following lists some of the more familiar culinary herbs:

Common Name	Scientific Name	Part Used
Basil	*Ocimum basilicum*	Leaves
Bay leaves	*Laurus nobilis*	Leaves
Cumin	*Cuminum cyminum*	Fruit
Dill	*Anethum graveolens*	Fruit, leaves
Garlic	*Allium satiavum*	Bulbs
Mustard	*Brassica alba; Brassica nigra*	Seed
Onion	*Allium cepa*	Bulb, some leaves
Oregano	*Origanum vulgare*	Leaves
Parsley	*Petroselinum crispum*	Leaves
Peppermint	*Mentha piperita*	Leaves
Sage	*Salvia officinalis*	Leaves
Tarragon	*Artemesia dracunculus*	Leaves
Thyme	*Thymus vulgaris*	Leaves

What is an example of an herb that has been used throughout history?

Herbs have had a long and varied history, used mostly for culinary and medicinal purposes but also for celebrations and ceremonies. Many herbs had multiple uses. For example, the herb dill (*Anethum graveolens*) has long been used for medicinal purposes: The Egyptians used it as a soothing medicine; Greeks habitually used the herb to cure the hiccups. During the Middle Ages, dill was prized for the protection it purportedly provided against witchcraft. Magicians and alchemists used dill to concoct spells, while a commonly known "wives' tale" stated that dill added to wine could enhance passion. Colonial settlers brought dill to North America, where it became known as "meetin' seed," because children were given dill seed to chew during long sermons in church. Today most of us use dill in our cuisine, including in dill pickles, baked fish, and bread.

What is wormwood?

Artemisia absinthium, known as wormwood, is a hardy, fragrant perennial that grows to heights of 2 to 4 feet (.6 to 1.2 meters). Wormwood is native to Europe but has been widely naturalized in North America. Absinthe, a liquor, is distilled and flavored using this plant. The plant was banned in the United States in the early 1900s because it was considered habit forming and hazardous to one's health, but it is now easily found in most herb nurseries.

Which plants have been used to create dyes?

Natural materials, including many plants, were the source of all dyes until the late nineteenth century. Blue dye was historically rare and was obtained from the indigo plant (*Indigofera tinctoria*). Another color difficult to obtain for dye was red; the madder plant

(*Rubia tinctorum*) was an excellent source of red dye and was used for the famous "red coats" of the British Army. Other, more common, natural dyes derived from plant sources are summarized in the following chart:

Common Name	Scientific Name	Part of Plant Used	Color
Black walnut	*Juglans nigra*	Hulls	Dark brown, black
Coreopsis	*Coreopsis*	Flower heads	Orange
Lilac	*Syringa*	Purple flowers	Green
Red cabbage	*Brassica oleracea-capitata*	Outer leaves	Blue, lavender
Turmeric	*Curcuma longa*	Rhizome	Yellow
Yellow onion	*Allium cepa*	Brown, outer leaves	Burnt orange

What plant native to Central and South America can be used as both a poison and a healing remedy?

Chondrodendron tomentosum, a plant that produces the poison called curare, has properties that are both healing and poisonous. In both Central and South America, the plant has been used by many different Indian tribes to develop a poisonous mixture. The poisonous stems and roots of the plant are crushed and cooked until taking on a syrupy consistency. Indian tribes often dipped the tips of arrows and other weapons into the poisonous paste before battle. However, the root of the vine also has healing properties. In Brazil, especially, it is used as a diuretic and fever reducer and is commonly used to treat tissue inflammation, kidney stones, bruises, contusions, and edema.

What part of the wheat plant (*Triticum aestivum*) is used to make flour?

Wheat is a grass whose fruit, the grain or kernel, contains one seed. The endosperm and embryo of the wheat plant are surrounded by the pericarp, or fruit wall, and the remains of the seed coat. More than 80 percent of the volume of the wheat kernel is made up of the starchy endosperm, and white flour is made by milling this starchy endosperm.

Wheat bran constitutes approximately 14 percent of the wheat kernel and is found in the covering layers and the outermost layer of the kernel. (It is the embryo of the wheat plant and represents approximately 3 percent of the wheat kernel.) Although nearly two dozen species of wheat exist, the most important ones for commercial use are common wheat (*Triticum aestivum*, sometimes referred to as *Triticum vulgare*) and durum wheat (*Triticum durum*). The varieties of common wheat account for 90 percent of the wheat grown worldwide. Durum wheat accounts for 5 to 7 percent of the wheat grown, and all other species account for the remainder of the wheat grown.

What plants are commonly used in the perfume industry?

Perfumes are made of a mixture of a large variety of scents. Although many perfumes are created synthetically, the expensive designer scents still use natural essential oils extracted from plants. The perfume industry uses all parts of the plant to create a unique blend of scents. Some commonly used plant materials for essential-oil extraction are:

Plant Part	Source
Bark	Indonesia and Ceylon cinnamons and cassia
Flowers	Rose, carnation, orange blossoms, ylang-ylang, violet, and lavender
Gums	Balsam and myrrh
Leaves and stems	Rosemary, geranium, citronella, lemongrass, and a variety of mints
Rhizomes	Ginger
Roots	Sassafras
Seeds and fruits	Orange, lemon, and nutmeg
Wood	Cedar, sandalwood, and pine

What parts of plants are sources for spices?

Spices are aromatic seasonings derived from many different parts of plants, including the bark, buds, fruit, roots, seeds, and stems. Some common spices and their sources are as follows:

Why were tomatoes often called "love apples" and considered aphrodisiacs?

Tomatoes belong to the nightshade family; they were cultivated in Peru and introduced to Europe by Spanish explorers. Tomatoes were introduced to Italy from Morocco, so the Italian name for the fruit was *pomi de Mori*, (meaning "apples of the Moors"). The French called the tomato *pommes d'amore*, (meaning "apples of love"). This latter name may have referred to the fact that tomatoes were thought to have aphrodisiac powers, or it may have been a corruption of the Italian name. When tomato plants were first introduced to Europe, many people viewed them with suspicion, since poisonous members of the nightshade family were commonly known. Although the tomato is neither poisonous nor an aphrodisiac, it took centuries for it to fully overcome its undeserved reputation.

Spice	Scientific Name of Plant	Part Used
Allspice	*Pimenta dioica*	Fruit
Black pepper	*Piper nigrum*	Fruit
Capsicum peppers	*Capsicum annum; Capsicum baccatum; Capsicum chinense; Capsicum frutescens*	Fruit
Cassia	*Cinnamomum cassia*	Bark
Cinnamon	*Cinnamomum zeylanicum*	Inner bark
Cloves	*Eugenia caryophyllata*	Flower
Ginger	*Zingiber officinale*	Rhizome
Mace	*Myristica fragrans*	Seed
Nutmeg	*Myristica fragrans*	Seed
Saffron	*Crocus sativus*	Stigma
Turmeric	*Curcuma longa*	Rhizome
Vanilla	*Vanilla planifolia*	Fruit

What is the most expensive spice in the world?

The world's most expensive spice is saffron, from the Arabic word *za'faran*, meaning "yellow." It comes from the stigmas of plants called autumn crocus (*Crocus sativus*) and is native to eastern Mediterranean countries and Asia Minor. The spice is not only expensive today, but was highly sought after by the ancient civilizations of Egypt, Assyria, Phoenicia, Persia, Crete, Greece, and Rome.

The blooming period for the crocus is approximately two weeks, and the flowers must be picked while they are in full bloom, before any signs of wilting are visible. Once picked, the three-part stigmas are removed from the petals before the petals wilt; this is a time-consuming process that can only be done by hand as the stigmas are very fragile. Then the stigmas are roasted and sold either as whole threads (whole stigmas) or powder. In order to harvest 1 pound (0.45 kilogram) of the spice, between 70,000 flowers must be picked to yield 210,000 stigmas. The price of saffron varies—depending on the type, such as Persian or Italian saffron—but can, to date, run about $15 for 0.035 ounces (1 gram); in another place, $2,700 per pound (454 grams); and still another place, it could run up to $10,000 per pound.

AQUATIC AND LAND ANIMAL DIVERSITY

HISTORICAL INTEREST IN ANIMAL DIVERSITY

How many different animals have ever existed?

Biologists have described and named more than one million species of animals, and some believe that several million to ten million more species remain to be discovered, classified, and named. Over time, deep into the Earth's past history, even more different animals have existed, with some estimates into the hundreds of millions. The greatest numbers of animals live in the oceans, with fewer in the freshwater areas and even fewer on land.

Who developed the first method for classifying mammals?

The French naturalist and anatomist George Léopold Chrétien Frédérie Dagobert Cuvier (1769–1832) was the first to use a specific method to classify mammals. This method also established a system of zoological classification, dividing the animals into four categories based on their structures. His four categories were Vertebrata, Mollusca, Articulata, and Radiata. Cuvier is also considered by many to be the founder of comparative anatomy.

Who are considered the "fathers" of zoology, modern zoology, and experimental zoology?

Greek philosopher and scientist Aristotle (384–322 B.C.E.) is considered the "father of zoology." His contributions to the field include vast quantities of information about the variety, structure, and behavior of animals; the analysis of the parts of living organisms; and the beginnings of the science of taxonomy. During Aristotle's time, only 500 species were known—a list he divided into eight classes. Swiss naturalist Konrad von Gesner is often credited as the "father of modern zoology." In 1551, he wrote the first volume to his three-volume *Historia Animalium* (*The History of Animals*) that served as a standard reference work throughout Europe in the sixteenth and seventeenth centuries. Swiss naturalist and philosopher Abraham Trembley (1710–1784) is considered the founder of experimental zoology. Much of his research involved studying the regeneration of hydras.

ANIMALS IN GENERAL

What are the main characteristics of animals?

Animals are an *extremely* diverse group of organisms, with all of them sharing a number of characteristics. In general biological terms, animals are multicellular eukaryotes that are heterotrophic, ingesting and digesting food inside the body. Their cells lack the cell walls that provide support in the bodies of plants and fungi. The majority of animals have muscle systems and nervous systems, responsible for movement and rapid response to stimuli in their environment.

Most animals reproduce sexually; one reason for this versus asexual reproduction is variation, with each offspring being the product of both parents, giving them a better chance of survival. In most animal species, a large, nonmotile (movable) egg is fertilized by a small, flagellated sperm; this forms what is called a diploid zygote (or the fertilized egg). The transformation of the zygote into an animal's specific form depends on special regulatory genes in the cells of the developing embryo. (For more about cells and reproduction, see the chapters "Cellular Basics" and "Animal Behavior and Reproduction.")

Can animals be grouped according to body symmetry?

Yes, animals are often divided into two groups according to their symmetry—the arrangement of body structures in relation to the axis of the body. For example, the bodies of most primitive animals such as jellyfish, sea anemones, and starfish have radial symmetry—a body in the form of a wheel or cylinder, with similar structures arranged as spokes from a central axis. Animals with bilateral symmetry have right and left halves that are mirror images of each other; they also have top (dorsal) and bottom (ventral) portions and a front (anterior) end and back (posterior) end. More sophisticated animals fall into this category, such as flatworms. Some organisms even exhibit both—such as the echinoderms that have bilateral symmetry as larvae and revert to radial symmetry as adults.

How do the structures of most animals develop?

The structures of most animals develop from three embryonic tissue layers. The outer layer (ectoderm) gives rise to the outer covering of the body and the nervous system. The inner layer (endoderm) forms the lining of the digestive tube and other digestive organs. The middle layer (mesoderm) gives rise to most other body structures, including muscles, skeletal structures, and the circulatory system.

What are the most common animal phyla?

Scientists often categorize animal phyla into about thirty-five divisions—whether they are found in the oceans, on land, or in between. The nine major phyla include Porifera (sponges), Cnidaria (hydra and jellyfish), Mollusca (mollusks), Echinodermata (starfish and sea urchins), Nematoda (roundworms), Annelida (segmented worms), Platyhelminthes (flatworms), Arthropoda (insects, crustaceans, and arachnids), and Chordata (fish, amphibians, reptiles, birds, and mammals).

What are the major characteristics of chordates?

All chordates—which include fish, amphibians, reptiles, birds, and mammals—share several features, the major ones being a notochord, dorsal nerve cord, and pharyngeal gill pouches. The notochord, a supporting rod made of cartilage, runs along the dorsal part of the body. It is always found in embryos, but in most vertebrates it is replaced during late embryonic development by a backbone of bony or cartilagelike vertebrae. The tubular dorsal nerve cord, near the notochord, is also formed during development of the embryo. In most vertebrates, the nerve cord eventually becomes a hollow cord and is protected by the backbone. The pharyngeal gill pouches appear during embryonic development on both sides of the throat region (the pharynx), but in some species, it does not develop. In human embryos, these gill pouches show a series of folds (thus the word pouches) that look similar to those of early fish embryos. But while in fish they would develop into gills, in humans (and other mammals) they develop into the ear's Eustachian tube and middle ear, tonsils, parathyroid, and thymus.

What is the difference between an invertebrate and a vertebrate?

Invertebrates are animals that lack a backbone. In fact, almost all animals (99 percent) are invertebrates—biologists believe millions more may be yet undiscovered. Vertebrates are animals with backbones (vertebral column); with more than one million identified animals on the planet, vertebrates represent only 42,500—around 25 percent—of those animals.

What are the largest and heaviest invertebrates?

The largest invertebrate to date is the colossal squid, or the Antarctic or Giant Cranch squid—it even has its own genus: Mesonychoteuthis (*Mesonychoteuthis hamiltoni*). It is estimated that the squid measures about 50 feet (15 meters) in length and lives at depths of at least 7,218 feet (2,200 meters) in the Southern Ocean. It also has what is thought to be the largest eyes in the animal world, measuring 11 inches (27 centimeters) across.

The heaviest invertebrate is highly debated. Some scientists believe a bivalve called the giant clam, *Tridacna gigas*, found in the coral reefs of the South Pacific and Indian

Oceans, holds the record. This invertebrate has an average weight of around 440 pounds (200 kilograms), although reports have been heard of a giant clamshell found in the twentieth century that weighed close to 750 pounds. It has a shell around 47 inches (120 centimeters) in length and is thought to have a life span of more than 100 years.

What are the major features shown by all vertebrates?

Animals in the subphylum Vertebrata are distinguished by several features. Most prominent is the endoskeleton of bone or cartilage, centering around the vertebral column (spine or backbone). Composed of separate vertebrae, a vertebral column combines flexibility with enough strength to support even a large body. Other vertebrate features include: 1) complex dorsal kidneys; 2) a tail (lost via evolution in some groups) extending between the anus; 3) a

Giant clams are one of the heaviest of the invertebrates. Weighing hundreds of pounds, they are thought to live as long as a century.

closed circulatory system with a single, well-developed heart; 4) a brain at the top end of the spinal cord, with ten or more pairs of cranial nerves; 5) a cranium (skull) protecting the brain; 6) paired sex organs in both males and females; and 7) two pairs of movable appendages (this would be fins in the fish that evolved into legs in land vertebrates).

What are the largest and smallest vertebrates?

Of the more than 60,000 vertebrates currently known to humans, finding the largest doesn't seem to be too much of a problem. To date, scientists believe the largest vertebrate known is the marine mammal called the blue whale (*Balaenoptera musculus*); it is also considered the largest known animal on Earth (and one of the loudest animals on Earth). Mature blue whales can reach around 75 to 100 feet (23–30.5 meters) in length; they can weigh up to 150 tons (136 metric tons), although because of whaling hunts, the largest of the whales are thought to have dwindled in number, with the average being about 75 to 80 feet (23–25 meters) long. Even baby blue whales are bigger than most animals, averaging, at birth, about 25 feet (7.6 meters) long.

For obvious reasons, determining the smallest vertebrates can be very difficult, and it is often debated how to measure different animals, such as a frog versus a fish. One of the latest contenders was found in 2012—a new species of frog from New Guinea called *Paedophryne amauensis* that measured around one-third of an inch (7.7 millimeters) in size. It was also found with a "bigger cousin"—the frog *Paedophryne swiftorum*—that averaged only about 0.33 inch (8.5 millimeters) in size. Still other scientists point to a creature called a *Paedocypris progenetica* found in 2006—a fish that measures about 0.31 inch (7.9 millimeters) long and lives in the acid swamps of Sumatra.

The following chart lists the largest and smallest vertebrates known to date:

Name	Average Length and Weight
Largest vertebrates	
Sea mammal Blue whale	75–100 ft (23–30.5 m) long; (*Balaenoptera musculus*) weighs around 150 tons (135 metric tons)
Land mammal African bush elephant	Bull is 10.5 ft (3.2 m) tall (*Loxodonta africana*) at shoulder; weighs 5.25–6.2 tons (4.8–5.6 metric tons)
Living bird North African ostrich	8–9 ft (2.4–2.7 m) tall; (*Struthio c. camelus*) weighs 345 lb (156.5 kg)

Fish
    Whale shark
      (*Rhincodon typus*)
	41 ft (12.5 m) long; weighs 16.5 tons (15 metric tons)

Reptile
    Saltwater crocodile
      (*Crocodylus porosus*)
	14–16 ft (4.3–4.9 m) long; weighs 900–1,500 lb (408–680 kg)

Rodent
    Capybara
      (*Hydrochoerus hydrochaeris*)
	3.25–4.5 ft (1–1.4 m) long; weighs 250 lbs (113.4 kg)

Smallest vertebrates	

Sea mammal
    Commerson's dolphin
      (*Cephalorhynchus commersonii*)
	Weighs 50–70 lbs (236.7–31.8 kg)

Land mammal
    Bumblebee bat or Kitti's hog-nosed bat
      (*Craseonycteris thong longyai*)
    or the pygmy shrew (*Suncus erruscus*)
	Bat is 1 in (2.54 cm) long; weighs 0.062–0.07 oz (1.6–2 g); 1.5–2 in (3.8–5 cm) long, weighs 0.052–0.09 oz (1.5–2.6 g)

Bird
    Bee hummingbird
      (*Mellisuga helenea*)
	2.25 in (5.7 cm) long; weighs 0.056 oz (1.6 gm)

Fish
    Dwarf pygmy goby
      (*Trimmatam nanus*)
	0.35 in (8.9 mm) long

Reptile
    Gecko
      (*Spaerodactylus parthenopion*)
	0.67 in (1.7 cm) long

Name	Average Length and Weight
Rodent	
Pygmy mouse	4.3 in (10.9 cm) long; weighs 0.24–0.28 oz
(*Baiomys taylori*)	(6.8–7.9 g)

What was the first group of vertebrates?

The first vertebrates were fish that appeared 500 million years ago around the beginning of the Cambrian period (on the geologic time scale). They are called agnathans (from the Greek *a*, meaning "without," and *gnath*, meaning "jaw")—small, jawless fish up to about 8 inches (20 centimeters) long. They have also been called ostracoderms ("shell skin") because their bodies were covered with bony plates, most notably a head shield protecting the brain.

What is the largest group of vertebrates?

The largest group of vertebrates is fish, a diverse group that includes almost 21,000 species, more than all other kinds of vertebrates combined. Most members of this group are osteichythes, or "bony fish," which includes fish well known to fishermen and fish-lovers alike, such as bass, trout, and salmon.

Do animals ever help each other?

Populations of organisms within an environment may engage in a variety of relationships with each other. For example, in a relationship known as mutualism, each species provides a benefit to the other; for instance, it can occur between two animal species like large coral reef fish and smaller species like the wrasses that swim into their mouths and eat the parasites that may have taken up residence there.

What is the difference between ectotherms and endotherms?

Ectotherms, also known as cold-blooded animals, warm their bodies by absorbing heat from their surroundings. These animals have large variations in normal body temperature due to their changing environment, with the most common ectotherms being invertebrates, fish, reptiles, and amphibians. The body temperature of endotherms, also known as warm-blooded animals, depends on the heat produced by the animal's metabolism. Mammals, birds, some fish, and some insects are endotherms. Their normal body temperature is fairly constant, even when vast temperature differences in their environment occur.

What is the meaning of the phrase "ontogeny recapitulates phylogeny"?

Ontogeny is how an organism develops from fertilized egg to adult; phylogeny is the evolutionary history of a group of organisms. The phrase "ontogeny recapitulates phylogeny" originated with German biologist and naturalist Ernst Haeckel (1834–1919), and it means that as an embryo of an advanced organism grows, it will pass through stages that look very much like the adult phase of less advanced organisms. Although further research demonstrated that early stage embryos are not representative of our evolutionary ancestors, Haeckel's general concept does reveal some clues about evolutionary history. In particular, animals with recent common ancestors tend to share more similarities during development than those that do not. For example, a dog embryo and a pig embryo will look more alike through most stages of development than a dog embryo and a salamander embryo.

AQUATIC ANIMALS

How many animals exist in the oceans?

Because the oceans are so vast—taking up around 70 percent of the Earth's "surface"—the number of marine animals is not truly known. What is known is that the numbers are immense; for example, of living species, to date, about 800 species of cephalopods (including squids and octopi), around 28,000 species of fish, and estimates of 50,000 to 120,000 species of mollusks exist.

What is a sponge?

A sponge is member of the phylum Porifera—one of the most primitive animals in the world. Overall, approximately 5,000 species of marine (saltwater) sponges and 150 species of freshwater sponges exist. These living invertebrates may be brightly colored—green, blue, yellow, orange, red, or purple—or they may be white or drab. The bright colors are due to the various bacteria or algae that live on or within the sponge.

A sponge's body contains holes that lead to an inner water chamber. The organisms pump water through those pores and expel it through a large opening at the top of the chamber. As water passes through the body, the sponge gathers nutrients, oxygen is absorbed, and waste is eliminated. Sponges are distinctive in possessing choanocytes (special flagellated cells whose beating drives water through the body cavity) that characterizes them as suspension feeders (also known as filter feeders). A marine sponge that is 4 inches (10 centimeters) tall and 0.4 inch (1 centimeter) in diameter pumps about 23 quarts (22.5 liters) of water through its body in one day. To obtain enough food to grow by 3 ounces (100 grams), a sponge must filter about 275 gallons (1,000 kilograms) of seawater!

What animals are members of the phylum Cnidaria?

Cnidarians include the corals, jellyfish, sea anemones, and hydras. The name Cnidaria refers to the stinging structures that are characteristic of some of these animals. These organisms have a vase-shaped body plan and a digestive cavity with only one opening to the outside; this opening is surrounded by a ring of tentacles used to capture food and defend against predators. The tentacles—and sometimes the outer body surface—contain longer, harpoonlike structures called stingers or nematocysts; the stinging cells are called cnidocytes.

What are the two distinct body forms of cnidarians?

The two forms of cnidarians are called the polyp stage and the medusa (plural, medusae), or jellyfish, stage. Polyps generally live attached to a hard surface and bud to produce more polyps and, in some cnidarians, to produce the medusa stage of the life cycle. These medusae, or jellyfish, drift with the ocean currents or swim by pulsating their umbrella-shaped bodies. They also release sperm and eggs into the water; after external fertilization, the embryo develops into a larva that eventually settles to the ocean bottom, becoming another polyp and completing the life cycle. Not all cnidarians go through both polyp and medusa stages; some, such as corals and sea anemones, exist only as polyps.

What is the longest jellyfish?

The longest (and thus, probably the largest) jellyfish is the *Cyanea capillata*, a "jelly" that has no skeleton and lives in the cold, northern ocean regions, such as the North Atlantic, Arctic, and North Pacific Oceans. Called the lion's-mane jellyfish, it is one of the largest invertebrates and normally ranges in bell size (the jellyfish's bulbous body) from about 5 to 6 feet (1.5–1.8 meters) across, with some reports of bell diameter as wide as 9 feet (2.7 meters). The tentacles add to their size, and some have been known to be up to 100 feet (30 meters) long.

What are some interesting features of jellyfish?

Jellyfish live close to the ocean shorelines, spending most of their time floating near the water's surface. They have bell-shaped, see-through (called gelatinous) bodies that

Jellyfish are marine animals that have an umbrella-like top that is used to propel them through water, and trailing tenticles that can sting and capture prey.

are between 95 percent and 96 percent water; they have a muscular ring around the margin of a bell-shaped body that contracts rhythmically to propel them through the water. Jellyfish are carnivores, subduing their prey with stinging tentacles and drawing the paralyzed animal into the digestive cavity. For humans, such a sting from most types of jellyfish is extremely painful.

How are coral reefs formed, and how fast are they built?

Coral reefs grow only in warm, shallow water. The calcium carbonate skeletons of dead corals serve as a framework upon which layers of successively younger animals attach themselves. Such accumulations, combined with rising water levels, slowly lead to the formation of reefs that can be hundreds of meters deep and long. For example, the major reef builder in Florida and Caribbean waters, *Montastrea annularis* (star coral), requires about one hundred years to form a reef just 3 feet (1 meter) high.

The coral animal, or polyp, has a columnar form; its lower end is attached to the hard floor of the reef, while the upper end is free to extend into the water, with the whole colony consisting of thousands of such individual polyps. Two kinds of corals exist, hard and soft, depending on the type of skeleton secreted. The polyps of hard corals deposit around themselves a solid skeleton of calcium carbonate (chalk), so most swimmers see only the skeleton of the coral; the animal is in a cuplike formation into which it withdraws during the daytime. (For more about coral reefs, see the chapter "Environment and Ecology.")

What is a hydra?

A hydra, a well-known member of phylum Cnidaria, is a tiny (0.4 inch or 1 centimeter in length) organism found in freshwater ponds. It exists as a single polyp that sits on a basal disk that it uses to glide around—it can also move by somersaulting. It usually has six to ten tentacles, which it uses to capture food, and they reproduce both sexually and asexually (budding). Hydras are named after the multiheaded monster of Greek mythology that was able to grow two new heads for each head cut off. When a hydra is cut into several pieces, each piece is able to regrow all the missing parts and become a whole animal.

What are the major groups of mollusks?

The four major groups of mollusks (phyla Mollusca) are: 1) chitons; 2) gastropods, including snails, slugs (mostly marine, but some freshwater), and nudibranches. This is the largest and most diverse group of mollusks (around 40,000 different species); 3) bivalves, including clams, oysters, and mussels; and 4) cephalopods, including squids and octopuses. Although mollusks vary widely in external appearance, some share the following body plan: a muscular foot, usually used for movement; a mass containing most of their internal organs; and a mantle, or a fold, of tissue that drapes over the mass and secretes a shell (that is, in organisms that have a shell).

How are pearls created?

Pearls are formed in saltwater oysters and freshwater clams. A curtainlike tissue called the mantle is within the body of these mollusks. Certain cells on the side of the mantle toward the shell secrete nacre, also known as mother-of-pearl, during a specific stage of the shell-building process.

A pearl is the result of an oyster's reaction to a foreign body, such as a piece of sand or a parasite, within the oyster's shell. The oyster neutralizes the invader by secreting thin layers of nacre around the foreign body, eventually building it into a pearl. The thin layers are alternately composed of calcium carbonate, argonite, and conchiolin. Humans can also have a hand in the formation of pearls—in particular, they intentionally place irritants within an oyster—resulting in what we often see in a jewelry display as "cultured pearls."

What are the major groups of echinoderms?

According to one common classification, six principle groups of echinoderms (phyla Echinodermata, from the Greek terms *echina*, meaning "spiny," and *derma*, meaning "skin") exist: 1) class Crinoidea (sea lilies and feather stars); 2) class Asteroidea (sea stars, also called starfish); 3) class Ophiuroidea (basket stars and brittle stars); 4) class Eichinoidea (sea urchins and sand dollars); 5) class Holothuroidea (sea cucumbers); and 6) class Concentricycloidea (sea daisies that live on waterlogged wood in the deep ocean and were first discovered in 1986).

Do all starfish (sea stars) have five arms?

No, this type of echinoderm has a variety of arms. They are members of the class Asteroidea, and their bodies consist of a central disk; from that disk, from five to more than twenty arms radiate.

FISH

What were some early studies about fish?

Fish have been a mainstay of food for thousands of years. Because of this, they were often studied along the way. For example, in 1656, Italian mathematician and physiologist Giovanni Alfonso Borelli (1608–1679) showed that a fish moved primarily by moving its tail, not its fins. In 1738, Swedish naturalist Petrus (Peter) Artedi (1705–1735) wrote the book *Petri Arted, seuci, medici, ichthyologia sive opera omnia de piscibus*—about fish and fish taxonomy—giving him the title of "the father of ichthyology" (which is the study of fish). And between 1788 and 1804, French scientist Bernard Lacepede (1756–1825) wrote forty-four volumes for his *Histoire Naturelle*—initially started by the French scientist Georges-Louis Leclerc, Comte de Buffon (1707–1788)—with eight of the volumes devoted to serpents and fish.

What fish is called a "living fossil"?

In a South African fish market in 1908, and in a small trawler fishing off the coast of west Africa in 1938, a certain type of fish was caught—a sea cave-dwelling, 5-foot (16.5-meter) ferocious predatory fish with limblike fins and a three-lobe tail called an African coelacanth. This was what is now called a "living fossil" (a term coined by naturalist Charles Robert Darwin), based on the fact that the fish closely resemble their more than three-hundred-million-year-old fossil ancestors—animals thought to have gone extinct around seventy million years ago. It took until the early 1950s for another live specimen to be caught, and since then, several more have been found.

185

In 2013, an international team of researchers delved even deeper into the fish's past—decoding the genome of a coelacanth—with its 2.8 billion units of DNA, about the same size as a human genome. They found that the genes in the fish are evolving more slowly than other organisms—possibly because of a lack of predators and/or that the fish do not *have* to change. This may be due to a characteristic of the fish's habitat: They live off the Eastern African coast (and another species off the coast of Indonesia) at dimly lit ocean depths of about 500 feet (496.7 meters)—regions that have not changed for thousands, if not millions, of years.

Does a connection between fish and humans exist?

Besides the obvious answer—yes, many people eat fish—researchers are also currently working on one puzzling possible connection: Did humans evolve from fish? In 2013, scientists decoded the genome of the coelacanth, a "living fossil" that resembles fossil fish that existed about 300 million years ago. Between the coelacanth and the lungfish (an air-breathing freshwater fish)—both with lobed fins that look like limbs—scientists are trying to see which animal may be closer to the first ancestral fish that used their fins to walk on land. Whatever the creatures were, they gave rise to the tetrapods—all the vertebrate animals from reptiles and birds—and to mammals, such as humans. Scientists label lobe-finned fish like the coelacanth and lungfish as sarcopterygians ("fleshy fins"); tetrapods, including humans, are descended from the sarcopterygians—and thus, the coelacanth is probably more closely related to people than to other fish!

How many fish species inhabit the Earth—from freshwater to oceans?

According to recent estimates, more than 28,000 fish species inhabit freshwater and ocean waters. They are divided into thirty-six orders and 400 families in one of the more common classification systems. The majority of fish species are bony fish (class Osteichthyes, often called true fish), with skeletons made mostly or only of bones; others have skeletons of cartilage (class Chondrichthyes).

What general characteristics do all fish have in common?

All fish have the following characteristics: 1) gills that extract oxygen from water; 2) an internal skeleton with a skin that surrounds the dorsal nerve cord; 3) single-loop blood circulation in which the blood is pumped from the heart to the gills and then to the rest of the body before returning to the heart; and 4) nutritional deficiencies, particularly some amino acids that must be consumed and cannot be synthesized.

How do fish swimming in a school change their direction simultaneously?

About 80 percent of the approximately 28,000 fish species travel in schools. Fish travel in schools for both protection and for efficiency. Safety in numbers (in a school) is a form of predator avoidance, because trying to catch one fish in a large, moving school can be difficult for a predator. Secondly, fish that travel in schools have less drag (fric-

How much electricity does an electric eel generate?

An electric eel (*Electrophorus electricus*) has current-producing organs made up of electric plates (modified muscle cells) on both sides of its vertebral column running almost its entire body length. The charge—350 volts on average, but as great as 550 volts—is released by the central nervous system. The shock consists of four to eight separate charges, each of which lasts only two- to three-thousandths of a second. These shocks, used as a defense mechanism, can be repeated up to 150 times per hour without any visible fatigue to the eel. The most powerful electric eel, found in the rivers of Brazil, Colombia, Venezuela, and Peru, produces a shock of 400 to 650 volts.

tion) and therefore use less energy for swimming. Also, when fish spawn, a school ensures that some eggs will evade predators and live to form another school.

The movements of a school of fish, which confuse predators, happen because the fish detect pressure changes in the water. The detection system, called the lateral line, is found along each side of the fish's body. Along the line are clusters of tiny hairs inside cups filled with a jellylike substance. If a fish becomes alarmed and turns sharply, it causes a pressure wave in the water around it. This wave pressure deforms the "jelly" in the lateral line of nearby fish. The deformation moves the hairs that trigger nerves, and a signal is sent to the brain telling the fish to turn.

What are sharks?

Chondrichthyes are fish that have a cartilaginous skeleton rather than a bony skeleton and include sharks, skates, and rays. Of those animals, one of the most well known, and often feared, is the shark. The 375 species of sharks currently known about range in length from 6 inches (15 centimeters) to 49 feet (15 meters)—only around a dozen are considered to be dangerous to humans. The relatively rare great white shark (*Carcharodon carcharias*) is the largest predatory fish, with some specimens reaching 20 feet, 4 inches (6.2 meters) long and weighing 5,000 pounds (2,270 kilograms). The largest shark is the whale shark (*Rhincodon typus*) that measures around 60 feet (20 meters) long; the smallest is the Caribbean Ocean's deepwater dogfish shark (*Etmopterus perryi*) that measures 8 inches (20 centimeters) long. The fastest shark is thought to be the shortfin mako (*Isurus oxyrinchus*), reported to swim 20 miles (32 kilometers) per hour.

Why do people fear sharks?

One of the major reasons why humans are afraid of sharks is obvious: We can't see underwater, and a shark can come seemingly out of nowhere. While around thirty species are known to have attacked humans, in the United States, shark attacks average about sixteen per year, with around one fatality every two years (around the world, shark attacks number around fifty to seventy per year). In one study, the researchers found that most shark attacks occur near shore, which is not surprising since most people who enter the water stay close to the shore. Some researchers believe that shark attacks on humans will only increase over time, as more people live—and play—along ocean coastlines.

What is unusual about shark's teeth?

Sharks were among the first vertebrates to develop teeth. The teeth are not set into the jaw, but rather sit on top of it—thus, they are not firmly anchored and are easily lost. The teeth are arranged in six to twenty rows, with the ones in front doing the biting and cutting. Behind these teeth, others grow. When a tooth breaks or is worn down, a replacement moves forward. In fact, one shark may eventually develop and use more than 20,000 teeth in its lifetime.

AQUATIC MAMMALS

What freshwater mammal is venomous?

The male duck-billed platypus (*Ornithorhynchus anatinus*) has venomous spurs located on its hind legs. When threatened, the animal will drive the spurs into the skin of a potential enemy, inflicting a painful sting. The venom released is relatively mild and generally not harmful to humans, but for most animals that try to bother the platypus, it makes them hesitate to attack the platypus again.

What is the difference between porpoises and dolphins?

Marine dolphins (family Delphinidae) and porpoises (family Phocoenidae) together comprise about forty species. The chief differences between dolphins and porpoises occur in the snout and teeth: True dolphins have a beaklike snout and cone-shaped teeth, while true porpoises have a rounded snout and flat or spade-shaped teeth.

What are the fastest and slowest whales?

The orca or killer whale (*Orcinus orca*) is the fastest-swimming whale. In fact, it is the fastest-swimming marine mammal, with speeds that reach 31 miles (50 kilometers) per hour. Right whales (*Eubalaena japonica*) are one of the slowest whale species, typically traveling at 1.2 to 2.5 miles (2–4 kilometers) per hour.

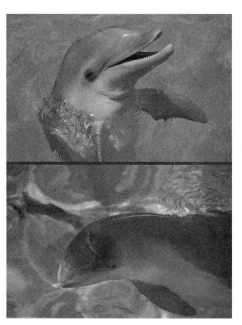

A dolphin (top) is most easily distinguished from a porpoise (bottom) by its beaklike snout versus the rounded snout of the porpoise.

188

What are manatees?

In the winter, the marine mammal called the West Indian manatee (*Trichechus manatus*) moves to the more temperate parts of Florida, such as the warm headwaters of the Crystal and Homosassa rivers in central Florida or the tropical waters of southern Florida. When the air temperature rises to 50°F (10°C), it will wander back along the Gulf Coast and up the Atlantic Coast as far as Virginia. Long-range offshore migrations to the coast of Guyana and South America have been documented. In 1983, when the population of manatees in Florida was reduced to several thousand, the state gave it legal protection from being hunted or commercially exploited. However, many animals continue to be killed or injured by environmental problems (such as water pollution or "red tide," also known as algal blooms) and the encroachment of humans. Entrapment in locks and dams, collisions with barges and power boat propellers, and so on cause at least 30 percent of manatee deaths, which total 125 to 130 annually. According to researchers, it is difficult to tell how many manatees are in all of Florida's waters, but in 2013, estimates put the animal's population at about 5,000. (For more about manatees and algal blooms, see the chapter "Environment and Ecology.")

IN BETWEEN SEA AND LAND

AMPHIBIANS

What are amphibians?

The word "amphibian," from the Greek *amphi* ("both") and *bios* ("life"), refers to the animal's double life on land and in water. The usual life cycle of amphibians begins with eggs laid in water, which develop into aquatic larvae with external gills; in a development that shows its evolution, the fishlike larva develops lungs and limbs and becomes an adult—thus, amphibians have made a partial transition to terrestrial life. The living amphibians include newts, salamanders, frogs, and toads. Although lungfish made a partial transition to living out of the water, amphibians were the first to struggle onto land and become adapted to a life of breathing air while not constantly surrounded by water. They were the first vertebrates to have true legs, tongues, ears, and a voicebox—with branches of certain amphibians eventually giving rise to the reptiles.

What fossil amphibian was once mistaken for a human skeleton?

In 1731, Swiss geologist Johann Scheuchzer (1672–1733) found a fossil of what he thought was a human skeleton—what he called *Homon diluvii testis*, or "man, a witness of the flood," in reference to the Bible. What he had actually discovered was the fossil of a twenty million-year-old extinct giant salamander.

Why must amphibians stay near water or moist environments?

Although amphibians breathe air and walk on crude legs—for example, alligators—they cannot stray too far from water. This is because they need to lay their eggs in water, and their larval stage needs water. For example, frogs and toads start their early life as eggs deposited in water (or near water), then develop into tadpoles—they must live in water to survive. At the adult stage, amphibians can either be amphibious (live on land and in the water) or totally aquatic, with some exceptions: For example, some species of frogs live only on land (but in a moist area), and some are more common in arid regions. For example, some desert amphibians store water in their bladders; they wait for cyclic rains (such as the Arizona "monsoon" season), then use the resulting water puddles to reproduce.

Amphibians like this frog must stay near the water to live because their eggs can only survive in a wet environment.

Do amphibians hibernate?

Yes, some amphibians hibernate—those that live in the colder climates of the north hibernate during the winter. Those that live in the more arid, desert climates do not hibernate, but estivate—become inactive—during the hottest times of the year.

What are the major groups of amphibians?

More than 4,000 species of amphibians are known to live on Earth and are found on every continent except Antarctica, which are divided into three major groups. The following lists those groups, their order (in terms of classification), and the known number (to date) of living species:

Group	Order	No. of Species
Frogs and toads	Anura (Salientia)	3,800
Salamanders and newts	Caudata (Urodela)	360
Caecilians	Apoda (Gymnophiona)	160

What's the difference between a frog and a toad?

Frogs and toads are different animals, but distinguishable from each other. Frogs and toads are of the same order, either Anura or Salientia (depending on the classification system). They breed and have their young in similar ways, and they both hunt for food at twilight or at night (although some toads are active during the day).

If you can get close enough before either creature dives into the pond as you approach, though, you can see the differences: Frogs have much smoother skin than toads; toads almost always have warts; frogs live near moist areas, but toads can live in drier places; frogs lay their eggs in clumps, whereas toads tend to lay them in long strands in a pond; toads are more fat-bodied, while frogs tend to be slimmer; toads have an oval,

---

What frog is carrying a deadly fungus?

In 2013, scientists discovered that the African clawed frog—a species found all over the world—is carrying a deadly amphibian disease and is thus threatening hundreds of species of frogs and salamanders. Initially, the frogs were shipped around the world for use in human pregnancy tests from the 1930s to 1950s; if a female frog began ovulating within about ten hours after being injected with a woman's urine, the woman was probably pregnant. Because other methods were developed to determine if a woman is pregnant, the frogs were released into the wild and are now spreading a fungus called *Batrachochytrium dendrobatidis* (Bd). Most of the surviving African clawed frogs have found a way to survive the fungus, but other amphibians are not as lucky. The disease infects the skin, causing it to thicken to around forty times more than normal—toppling the creature's electrolyte balance and literally causing a heart attack.

---

raised glandular area behind the eye, but frogs do not; and toads have an L-shaped ridge between and in back of their eyes, while frogs do not.

What are the caecilians?

The caecilians are mostly legless, wormlike amphibians found everywhere around the world, but mostly in tropical regions. Many are eyeless, have smooth skins, and burrow into the ground like earthworms. Not much is known about this group, and they are the least studied of the living amphibians. But it is thought their ancestors may have played an important part in the development of early amphibians.

WORMS

What are the three phyla of worms?

The many types of worms in the world are divided into three phyla: Flatworms belong to the phylum Platyhelminthes and are flat, elongated animals that have bilateral symmetry and primitive organs (they include the planarians, flukes, and tapeworms). The phylum Nematoda includes the roundworms, or unsegmented worms, also with bilateral symmetry, but have little in terms of sensory organs. Most of them are parasitic, including the parasite *Trichinella* that causes trichinosis from uncooked pork. The third phylum is Annellia, or segmented worms, including sandworms, tube worms, earthworms, and leeches. They also have bilateral symmetry, a digestive tract that is a tube within a tube—complete with a crop, gizzard, and intestine—and a closed circulatory system.

What are the most common tapeworm infections in humans?

The long, flat bodies of tapeworms have long been associated with humans—especially in association with eating certain meals that contain the tapeworm. For example, the **191** beef tapeworm (*Taenia saginata*) comes from eating rare beef and is the most common of tapeworms in humans; the pork tapeworm (*Taenia solium*) comes from eating rare pork, but is less common than the beef tapeworm; and the fish tapeworm (*Diphyllobothrium latum*) is found most often in the Great Lakes region of the United States when eating rare or poorly cooked fish. It is also the largest tapeworm and can grow to a length of 66 feet (20 meters); to compare, the beef tapeworm may only reach around 33 feet (10 meters).

How numerous are roundworms?

Roundworms, or nematodes, are members of the phylum Nematoda (from the Greek term *nematos*, meaning "thread") and are numerous in two respects: 1) number of known and potential species; and 2) the total number of these organisms in a habitat. Approximately 12,000 species of nematodes have been named, but it has been estimated that if all species were known, the number would be closer to 500,000. Nematodes live in a variety of habitats ranging from the sea to soil. Six cubic inches (100 cubic centimeters) of soil may contain several thousand nematodes, a square yard (.85 square meters) of woodland or agricultural soil may contain several million of them, and good topsoil may contain billions per acre.

In what ways are earthworms beneficial?

Earthworms are extremely beneficial to gardeners and growers by making the soil more fertile. An earthworm literally eats its way through soil and decaying vegetation. As it moves about, the soil is turned, aerated, and enriched by nitrogenous wastes. (In fact, English naturalist Charles Robert Darwin once calculated that a single earthworm could eat its own weight in soil every day.) Much of what is eaten is then excreted on the Earth's surface in the form of "casts." The worms then rebury these casts with their burrowing process.

Earthworms serve an important function in the ecosystem by aerating and fertilizing soil, which is beneficial to all plants.

What are giant tube worms?

Giant tube worms were discovered near the hydrothermal (hot water) ocean vents in 1977 as the submersible Alvin was exploring the ocean floor of the Galapagos Ridge (located 1.5 miles [2.4 kilometers] below the Pacific Ocean surface). Growing to lengths of 5 feet (1.5 meters), *Riftia pachyptila Jones*, named after worm expert Meredith Jones of the Smithsonian Museum of Natural History, lack both mouth and gut and are topped with feathery plumes composed of over 200,000 tiny

Why are leeches important in the field of medicine?

Leeches have been used in the practice of medicine since ancient times. During the 1800s, leeches were widely used for bloodletting because of the mistaken idea that body disorders and fevers were caused by an excess of blood—thus, leech collecting and cultures were practiced on a commercial scale during this time.

The medical leech, *Hirudo medicinalis*, is used even today to remove blood that has accumulated within tissues as a result of injury or disease. Leeches have also been applied to fingers or toes that have been surgically reattached to the body. The sucking by the leech unclogs small blood vessels, permitting blood to flow normally again through the body part. The leech releases hirudin, secreted by the salivary glands, which is an anticoagulant that prevents blood from clotting and dissolves preexisting clots. Other salivary ingredients dilate blood vessels and act as an anesthetic. A medicinal leech can absorb as much as five to ten times its body weight in blood. Complete digestion of this blood takes a long time, and these leeches feed only once or twice a year in this manner.

tentacles. The phenomenal growth of these worms is due to their internal food source—symbiotic bacteria, over one hundred billion per ounce of tissue, that live within the worms' troposome tissues. To these tissues, the tube worms transport absorbed oxygen from the water, together with carbon dioxide and hydrogen sulfide. Using this supply, the bacteria in turn produce carbohydrates and proteins that the worms need to thrive.

What is the largest leech known?

Most leeches are between 0.75 inch and 2 inches (2–6 centimeters) in length, but some "medicinal" leeches reach 8 inches (20 centimeters). The giant of all leeches is the Amazonian *Haementeria ghilanii* (from the Greek term *haimateros*, meaning "bloody") that can reach up to 12 inches (30 centimeters) in length.

AQUATIC AND LAND ARTHROPODS

What are arthropods?

Members of the phylum Arthropoda are characterized by jointed appendages and an exoskeleton of chitin. More than one million species of arthropods are currently known to science, and many biologists believe that millions more will be identified. Arthropods are the most biologically successful group of animals on Earth today. This is because they are the most diverse organisms and live in a greater range of habitats than do the members of any other animal phylum, from land to the oceans. One major reason for their abundance is one of their classes—the Insecta class, the organisms we all fondly (or not too fondly) call insects; they are the most successful organisms on Earth in terms of diversity, geographic distribution, number of species, and number of individuals. In fact, to date, more than 900,000 different kinds of insects are known (and amazingly, it is estimated that ten quintillion (10,000,000,000,000,000,000) individual insects are alive at any given time!

What are the major groups of arthropods?

More than one million arthropod species have been described, with insects making up the vast majority of them. Zoologists estimate that the arthropod population of the world numbers about a billion million (10^{18}) individuals; in fact, two out of every three organisms known on Earth are arthropods, and the phylum is represented in nearly all habitats of the biosphere. About 90 percent of all arthropods are insects, and about half of the named species of insects are beetles. Several groups of arthropods contain a mix of animals that

Brilliantly colored jewel bugs are true bugs, not beetles. A major difference between bugs and beetles is that beetles have mandibles to chew their food, whereas bugs use a proboscis to pierce their food and suck out the contents.

inhabit water environments and animals that live on land. They include the subphyla of Chelicerata (horseshoe crabs, spiders, scorpions, ticks, and mites); Crustacea (lobsters, crabs, shrimps, isopods, copepods, and barnacles); and Unirania (grasshoppers, roaches, ants, bees, butterflies, flies, beetles, centipedes, and millipedes).

Is there a difference between beetles and bugs?

Yes, there is a definite difference between beetles and bugs—a subject that is confusing to most of us, especially since we often use the term "bug" for almost any type of tiny, flying creature! The biggest difference is that they belong to two distinct insect groups—in most biological classifications, bugs are actually called "true bugs," and are listed under the Order Hemiptera, while beetles are under the Order Coleoptera. There are some similarities, as both true bugs and beetles have the overall characteristics of the insect world: six jointed legs, antennae, a relatively hard exoskeleton, and three main body parts: the head, thorax, and abdomen. But there are differences, too, which is the reason for the two orders:

True bugs—For example, assassin bugs, stink bugs, and bed bugs—have piercing-sucking mouthparts (many of them suck nectar and sap); membrane-thin wings, if they have any at all; and the juveniles look like miniature adult bugs, except they don't have wings quite yet (if at all). (For more about bugs, see below.)

Beetles—For example, Asian ladybugs, weevils, and dung beetles—eat a wide range of materials, from plants, leaves, and bark to animals; they have small, thick sheaths (called elytra) that fold out when they are ready to fly; and their life cycle runs from larva to adult. This order makes up around forty percent of all insects species around the world—and it is estimated that they represent between twenty-five to thirty percent of all animal species on Earth.

What marine and freshwater invertebrates are some of the most important of all animals?

Copepods, tiny crustaceans, are the link between the photosynthetic life in the ocean or pond and the rest of the aquatic food web. They are primary consumers grazing on algae in the waters of the oceans and ponds. These organisms, among the most abundant multicellular animals on Earth, are then consumed by a variety of small predators, which are eaten by larger predators, and so on. Virtually all animal life in the ocean depends on the copepods, either directly or indirectly. Although humans do not eat copepods directly, our sources of food from the ocean would disappear without the copepods.

What arthropods can affect humans (in the United States)?

Many arthropods can affect humans, especially in terms of unhealthy bites, itches, scratches, diseases, and allergies. The following lists only a few of these creatures that live in various spots in the United States:

Arthropod	Effect on Human Health
Black widow spider (*Latrodectus mactans*)	Venomous bite; most in South
Brown recluse or violin spider (*Loxosceles reclusa*)	Venomous bite
Scorpion (*Centruroides exilicauda*)	Venomous bite; most in South and Western states
Chiggers (*Trombiculid mites*)	Dermatitis; most in South
Itch mite (*Sarcoptes scabiei*)	Scabies
Deer tick (*Ixodes dammini*)	Bite can transmit lyme disease
Dog tick, wood tick (*Dermacentor species*)	Bite can transmit Rocky Mountain spotted fever in some places in the U.S.
Mosquitoes	Bite can transmit diseases (for example, West Nile virus, encephalitis, filarial worms)
Horseflies, deerflies	Female has painful bite
Houseflies	Many transmit bacteria and viruses
Fleas	Dermatitis
Bees, wasps, ants	Venomous stings (single stings not dangerous unless person is allergic)

What are the only crustaceans that don't move around?

Barnacles are the only sessile (permanently attached to one location) crustaceans. They were described by the nineteenth-century naturalist Louis Agassiz (1807–1873) as "nothing more than a little shrimplike animal standing on its head in a limestone house and licking food into its mouth." Accumulations of barnacles may become so great that the speed of a ship may be reduced by up to 40 percent, necessitating dry-docking the ship to remove the barnacles.

What are "sea monkeys"?

"Sea monkeys" are often sold as sea creatures by a variety of companies. But in reality, they are crustaceans—a type of brine shrimp called *Artemia salina*. Other types of brine shrimp *not* sold as "sea monkeys" are a very important part of the marine food chain—and as food for fish and people. The sea monkeys you buy also have some strange characteristics. For example, they breathe through their feet; can survive for years without water by hibernating; and are born with one eye, but eventually develop two more.

What are the largest and smallest aerial spider webs?

The largest aerial webs are spun by the tropical orb weavers of the genus *Nephila*, which produce webs that measure up to 18.9 feet (6 meters) in circumference. The smallest webs are produced by the species *Glyphesis cottonae*; their webs cover an area of about 0.75 square inch (4.84 square centimeters).

Do male mosquitoes bite humans?

No, male mosquitoes live on plant juices, sugary saps, and liquids from decomposition. They do not have a biting mouth that can penetrate human skin, but female mosquitoes do. In some species, the females, who lay as many as 200 eggs, need blood to lay their eggs—which leads to those painful bites that make humans and other animals itch.

Why are insects considered the most successful group of animals?

With more than one million described species (and perhaps millions more not yet identified), class Insecta is the most successful group of animals on Earth. In the United States alone, about 91,000 different

Black Widow spiders are easily distinguished by the bright-red hourglass marking on their abdomens. Their bite is venomous and very painful, although not usually lethal for healthy adults. Nevertheless, if you are bitten, obtain treatment as soon as possible.

Do bat-eating spiders exist?

Yes, bat-eating spiders live everywhere in the world except Antarctica. About 90 percent of these invertebrates live in the warmer regions of the globe and include some web-building spider species such as the *Argiope savignyl* and a tarantula species *Poecilotheria rufilata*; both are known to capture and kill small bats. Some tropical orb-weaving spiders—with a leg span of 4 to 6 inches (10–15 centimeters)—catch bats in webs that can reach 5 feet (16.5 meters) in diameter. Still other spiders have been seen capturing and killing a small bat, such as the huntsman spider (*Heteropoda ventoria*) in India. But they are not the only creatures the spiders eat—some larger species have also been known to capture and eat fish, frogs, and even snakes and mice.

species have been described, with an estimated 73,000 species not yet described. In fact, the largest numbers of species in the U.S. fall into the four insect Orders: Coleoptera (beetles) at about 23,700; Diptera (flies) at about 19,600; Hymenoptera (ants, bees, wasps) at about 17,500; and Lepidoptera (moths and butterflies) at about 11,500.

More species of insects have been identified than of all other groups of animals combined. What insects lack in size, they make up for in sheer numbers. If we could weigh all the insects in the world, their weight would exceed that of all the remaining terrestrial animals. About 200 million insects are alive at any one time for each human. And why are they successful? Flight is one key to the great success of insects. An animal that can fly can escape many predators, find food and mates, and disperse to new habitats much faster than an animal that must crawl about on the ground.

Does a connection exist between early flowering plants and insects?

Yes, a connection is possible. Early flowering plants—although a debate is still up as to when they first arose, from 140 to 190 million years ago to 215 million years ago—are thought to be connected to the rise of insects. Modern insects such as bees and wasps rely on flowers for pollen and nectar, and if the plants evolved 215 million years ago, it may be why both became very successful on land. (For more about early plants, see the chapter "Plant Diversity.")

What is a "bug," biologically speaking?

The biological meaning of the word "bug" is significantly more restrictive than in common usage. People often refer to all insects as "bugs," even using the word to include such organisms as bacteria and viruses as well as glitches in computer programs. In the strictest biological sense, a "bug" is a member of the order Hemiptera, also called true bugs. Members of Hemiptera include bedbugs, squash bugs, clinch bugs, stink bugs, and water striders.

What is the largest group of insects that has been identified and classified?

The largest group of insects that has been identified and classified is the order Coleoptera (beetles, weevils, and fireflies), with some 350,000 to 400,000 species. Beetles are the dominant form of life on Earth, as one of every five living species is a beetle.

What are beneficial insects?

Beneficial insects include bees, wasps, flies, butterflies, moths, and others that pollinate plants. Many fruits and vegetables depend on insect pollinators for the production of seeds. Insects are an important source of food for birds, fish, and many animals. In some countries, such insects as termites, caterpillars, ants, and bees are eaten as food by people. Products derived from insects include honey and beeswax, shellac, and silk. Some predators such as mantises, ladybugs or lady beetles, and lacewings feed on other harmful insects. Other helpful insects are parasites that live on or in the body of harmful insects. For example, some wasps lay their eggs in caterpillars that damage tomato plants.

What animals fluoresce?

Fluorescence is luminescence caused by a natural pigment molecule; pigments are molecules that absorb some colors of light while reflecting others. For example, green pigments (like those in leaves) absorb red and blue wavelengths of light but reflect green wavelengths. After being energized by photons (units of light energy), the electrons of some pigment molecules actually give off light as they fall back to their normal state. Chlorophyll molecules that play a role in photosynthesis have this ability, as do molecules found in several organisms, such as certain jellyfish. The body of the insect *Photinus pyralis*, better known as a lightning bug, has the enzyme luciferase, which generates the chemical reaction that leads to a drop in the energy state of electrons—a reaction similar to that which occurs in a chlorophyll molecule—allowing the insect to "glow" on and off during its mating season.

What are some of the most destructive insects in the world?

The most destructive insect is the desert locust (*Schistocera gregaria*), the locust of the Bible; their habitat ranges from the dry and semi-arid regions of Africa and the Middle East through Pakistan and northern India. This short-horn grasshopper can eat its own weight in food a day, and during long migratory flights a large swarm can consume 20,000 tons (18,144,000 kilograms) of grain and vegetation a day.

What is the gypsy moth?

The gypsy moth (*Porthetria dispar*) lays its eggs on the leaves of oaks, birches, maples, and other hardwood trees. When the yellow hairy caterpillars hatch from the eggs, they devour the leaves in such quantities that the tree becomes temporarily defoliated. Some trees can withstand two years of such defoliation, but other trees often die in the infestation year. In 1869, Professor Leopold Trouvelot (1827–1895) originally brought gypsy moth egg masses from France to Medford, Massachusetts. His intention was to breed the gypsy moth with the silkworm to overcome a wilt disease of the silkworm. He placed the egg masses on a window ledge, and evidently the wind blew them away. Only a decade later, the caterpillars covered trees in that vicinity, and in twenty years trees in eastern Massachusetts were being defoliated. Since that time, it has spread (in cycles) to at least twenty-five states in the United States.

What are "killer bees"?

Africanized honeybees—the term entomologists prefer rather than killer bees—are a hybrid bee originating in Brazil, where African honeybees were imported in 1956. The breeders, hoping to produce a bee better suited to producing more honey in the tropics, instead found that African bees soon hybridized with and mostly displaced the familiar European honeybees. Although they produce more honey, Africanized honeybees (*Apis mellifera scutellata*) are more dangerous than European bees because they attack intruders in greater numbers. Since their introduction, they have been responsible for approximately 1,000 human deaths.

In addition to such safety issues, concern is growing regarding the effect of possible hybridization on the U.S. beekeeping industry. In October 1990, the bees crossed the Mexican border into the United States; in 1996, these bees were found in parts of Texas, Arizona, New Mexico, Nevada, and California; by 2009, they had crept up even further into those states and added southern Oklahoma, Louisiana, and Florida—even Puerto Rico and the U.S. Virgin Islands—to the list. Their migration northward has slowed—probably because they are a tropical insect and cannot live in colder climates.

How many flowers produce 1 pound of honey?

Bees must gather 4 pounds (1.8 kilograms) of nectar, which requires the bees to tap about two million flowers, in order to produce 1 pound (454 grams) of honey. The honey is gathered by worker bees, whose life span is three to six weeks, long enough to collect about a teaspoon of nectar.

What is a "daddy longlegs"?

The name applies to a harmless, nonbiting, long-legged arachnid. Also called a harvestman, it is often mistaken for a spider, but it lacks a spider's segmented body shape. Although it has the same number of legs (eight) as a spider, the harvestman's legs are far longer and thinner. These very long legs enable it to raise its body high enough to avoid ants or other small enemies. Harvestmen are largely carnivorous, feeding on a variety of small invertebrates such as insects, spiders, and mites. They never spin webs as spiders do. They also eat some juicy plants and in captivity can be fed almost anything edible, from bread and milk to meat, and also need to drink frequently. (The term "daddy longlegs" is also another name for a cranefly—a thin-bodied insect with long legs—and a snoutlike structure with which it sucks water and nectar.)

LAND ANIMALS

Early Land Animals

What were the earliest animals to walk on land?

The only way scientists know about the earliest land animals on Earth is through fossils found in ancient rock. And from these fossils, we know that almost all species of animals have changed dramatically since the first animals walked on land. The changes became necessary as the Earth has gone though such environmentally changing events as volcanic eruptions, the movement of the continents, changes in climate, and massive extinctions. (For more about early animals, see the chapter "Basics of Biology.")

But what animals were responsible for first walking on land remains a highly debated topic. One possibility was suggested in 2013, in which researchers suggested that a toothy creature called the *Ichthyostega* represented one of the first transitions between fish and terrestrial animals. It lived about 374 to 359 million years ago and is thought to have lived in the shallow water of swampy areas—either along coastlines or waterways. The best guess is that the animal was looking for something to eat and was lured by the possible food on land. The four-legged creature probably dragged itself out of the water and on land by using its front legs and dragging its back legs, much like how modern seals move. The creature also had a string of bones in its chest that may have been the precursors to the sternum that holds the ribcage—necessary for supporting the weight of its chest as it moved on land.

How does a mastodon differ from a mammoth?

Although the words are sometimes used interchangeably, the mammoth and the mastodon were two different species. The mastodon seems to have appeared first, and a side branch may have led to the mammoth. The mastodon lived in Africa, Europe, Asia, and North and South America, appearing in the Oligocene era (38 to 25 million years ago) and surviving until less than one million years ago. It stood a maximum of 10 feet (3 meters) tall and was covered with dense, woolly hair; its tusks were straight forward and nearly parallel to each other. The mammoth evolved less than two million years ago and died out about 10,000 years ago; they lived in the cooler regions of North America, Europe, and Asia. Like the mastodon, the mammoth was covered with a long, coarse layer of outer hair to protect it from the cold. It was somewhat larger than the mastodon, standing 9 to 15 feet (2.7–4.5 meters); the mammoth's tusks tended to spiral outward, then up.

The reasons for both creatures' demise are still a matter of speculation. When it comes to the mammoths, the evidence is more "recent": After finding several mammoths frozen in such places as Siberia, scientists believe that the gradual warming of the Earth's climate—and thus a change in the animals' environments—was the primary factor in the mammoth's extinction. But they suggest that early humans may have killed many mammoths as well—perhaps hastening the extinction process.

Did dinosaurs and humans ever coexist?

One of the most famous of all the ancient animals were the dinosaurs—creatures that only lived on land (so far, no dinosaurs are known to have lived in the air or water)—first appearing in the early Triassic period (about 220 million years ago) and all but disappearing at the end of the Cretaceous period (about sixty-five million years ago). It is thought that modern humans (*Homo sapiens sapiens*) appeared only about 25,000 years ago. Although we know movies that show humans and dinosaurs existing together are only Hollywood fantasies, one caveat remains: Many scientists believe that modern birds are the ancestors of the dinosaurs—and thus, in some ways, we can say that the dinosaurs never truly became extinct! (For more about dinosaur extinctions, see the chapter "Heredity, Natural Selection, and Evolution.")

Why are cockroaches so amazing?

The word "cockroach" can often send chills up a human's spine. They are considered pests by most, but in reality, they've been on Earth for much longer than humans. The earliest cockroach fossils are about 350 million years old; by about 220 million years ago, cockroaches were so prolific—and are thought to have been the first organism to master flight—that many paleontologists nickname that time the "Age of Cockroaches."

Modern cockroaches (order Blattaria or Blattodea) are nocturnal scavenging insects that eat not only human food, but book bindings, ink, paper, and seemingly everything in between. Of the about 4,500 species of cockroaches, only thirty species are associated with human habitats, and of those, only about four are considered definite pests. The most well-known one in the United States is the American cockroach (*Periplaneta americana*) that can measure about 1.2 inches (30 millimeters) long—although when you see them invade your kitchen at night, they seem to look much larger! The world's heaviest cockroach is the Australian giant burrowing cockroach—a creature that can measure 3.5 inches (9 centimeters) long. They are related to the termite; in fact, recent genetic evidence suggests that termites evolved directly from true cockroaches. And they are amazing creatures that have adapted well to life on Earth; they can even go for a month without food, for two weeks without water, and the female mates once and can stay pregnant for all of her life.

REPTILES AND BIRDS

What is the difference between a reptile and an amphibian?

Reptiles are clad in scales, shields, or plates, and their toes have claws; amphibians have moist, glandular skins, and their toes lack claws. Reptile eggs have a thick, hard, or parch-

Crocodillians like this Florida alligator have been around since the dinosaurs. They are osteoderms, meaning they have bony deposits in their skin, which forms an effective armor on their backs and tails.

mentlike shell that protects the developing embryo from moisture loss, even on dry land. The eggs of amphibians lack this protective outer covering and are always laid in water or in damp places. Young reptiles are miniature replicas of their parents in general appearance, if not always in coloration and pattern. Juvenile amphibians pass through a larval (usually aquatic) stage before they metamorphose (change in form and structure) into the adult form. Reptiles include alligators, crocodiles, terrapins, tortoises, turtles, lizards, and snakes; amphibians include salamanders, toads, and frogs.

Were reptiles the first group of vertebrates to become truly terrestrial?

Yes, reptiles were the first group of vertebrates to become true land animals. They had several adaptations: Legs were arranged to support the body's weight more effectively than in amphibians, allowing reptile bodies to be larger and to be able to run. Reptilian lungs were more developed, with a greatly increased surface area for gas exchange than the saclike lungs of amphibians. The three-chambered heart of reptiles was more efficient than the amphibian heart. In addition, the skin was covered with hard, dry scales to minimize water loss. However, the most important evolutionary adaptation was the amniotic egg, in which an embryo could survive and develop on land. The eggs were surrounded by a protective shell that prevented the developing embryo from drying out.

What is the most successful and diverse group of terrestrial vertebrates?

Birds, members of the class Aves, are the most successful of all terrestrial vertebrates. Twenty-eight orders of living birds with almost 10,000 species are distributed over almost the entire Earth. The success of birds is mostly due to the development of the feather, providing insulation from the cold, allowing them to fly, and as protective coloration, not only as camouflage, but for territorial displays for mating.

Are birds related to dinosaurs?

According to many scientists, birds are essentially modified dinosaurs with feathers. American paleontologists Robert T. Bakker (1945–) and John H. Ostrom (1928–2005) did extensive research on the relationship between birds and dinosaurs in the 1970s. They suggested that the bony structure of small dinosaurs was very similar to *Archaeopteryx*, the first animal classified as a bird (it had true feathers)—thus dinosaurs and birds probably evolved from the same ancestors. Ostrom also suggested that dinosaurs may have been warm-blooded and thus more active and similar to birds. The evidence since that time has become stronger, as paleontologists continue to discover more dinosaur fossils—some that show even more evidence of birdlike feathers. In fact, in 2013, several researchers found yet another dinosaur fossil with evidence of feathers and suggested that dinosaurs were not "overgrown lizards," but that feathered dinosaurs were actually the norm.

What birds have some of the largest wingspans?

An African wading bird called the Marabou stork (*Leptoptilos crumeniferus*) has the largest wingspan of any bird (to date), measuring about 13.2 feet (4 meters). The albatross family has the next largest wingspans, including the wandering albatross (*Diomedea exculans*), the royal albatross (*Diomedea epomophora*), and the Amsterdam Island albatross (*Diomeda amsterdiamensis*), with a spread from 8 to 11 feet (2.5–3.3 meters). The next in line is the Trumpeter Swan (*Cygnus buccinator*), with a wingspan of about 11 feet (3.3 meters), which is the largest (and heaviest) bird in North America.

Will wild birds reject baby birds that have been touched by humans?

Contrary to popular belief, birds generally will not reject hatchlings touched by human hands. The best thing to do for newborn birds (probably with the exception of large birds such as raptors—call a wildlife rehabilitator for help in that instance) that have fallen or have been pushed out of the nest is to locate the nest as quickly as possible, put on some gloves, gently pick up the baby, and carefully put them back into the nest (especially if other nestlings are in the nest). If you cannot find the nest, leave the bird alone and watch from a distance—usually the mother will return and get the baby back to the nest in a relatively short time. If no mother returns within a few hours (or less if it is snowing or raining), contact your local wildlife rehabilitator for more information.

MAMMALS

What characterizes a mammal?

The class Mammalia includes more than 5,000 species in about twenty-six orders (depending on which classification system you use), which are found on all continents and in all oceans and seem to have a disproportionately large ecological role compared to their abundance—mainly because their metabolism is so high, they have to eat more. In general, most mammals have several characteristics—especially three not found in other animals: The females produce milk using modified sweat glands (called mammary glands, which can vary from two to a dozen or more); have hair (mostly made of a protein called keratin, similar to what makes up human fingernails and mostly developed to insulate the mammal against extreme environments); and have three middle ear bones (the malleus, incus, and stapes, allowing mammals to better hear sounds).

Do any mammals fly?

Bats (around 1,200 species) are the only truly flying mammals, although several gliding mammals are referred to as "flying" (such as the flying squirrel and flying lemur).

This Gambian epauletted fruit bat is one of 1,200 species of bat, the only true flying mammal on our planet.

The "wings" of bats consist of double membranes of skin stretching from the sides of the body to the hind legs and tail that are actually skin extensions of the back and belly. The wing membranes are supported by the elongated fingers of the forelimbs (or arms).

Do any cats live in the desert?

The sand cat (*Felis margarita*) is the only member of the cat family tied directly to desert regions. Found in the deserts of North Africa, the Arabian Peninsula, Turkmenistan, Uzbekistan, and western Pakistan, the sand cat has adapted to extremely arid desert areas. The padding on the soles of its feet is well suited to the loose sandy soil, and it can live without drinking freestanding water. Having sandy or grayish-ochre dense fur, its body length is 17.5 to 22 inches (45–57 centimeters). Mainly nocturnal (active at night), the cat feeds on rodents, hares, birds, and reptiles. The Chinese desert cat (*Felis bieti*) does not live in the desert as its name implies, but inhabits the steppe country and mountains. Likewise, the Asiatic desert cat (*Felis silvestris ornata*) inhabits the open plains of India, Pakistan, Iran, and Asiatic Russia.

What are the smallest and largest bear species?

The smallest bear in the world—and also considered the least studied—is the sun (or honey) bear that lives in the human lowland tropics of Southeast Asia. They have short, sleek, and dense black fur (for protection from dirt and insects), are good tree climbers, are omnivores (eating anything they can find in the rain forest, especially honey and bee larvae), and can weigh about 100 pounds (45 kilograms). The largest bear species is the Polar Bear (or "sea bears") that live around the Arctic Circle, including the North Pole, and northern parts of Europe, Asia, and North America (with about 60 percent of the population in Canada). Their fur looks white, but it can also be yellow or even greenish depending on the light. They are one of the largest land predators in the world, can reach about 10 feet (305 centimeters) in height and weight between 800 and 1,600 pounds (360–720 kilograms), although females weigh less. The largest recorded polar bear was a male that weighed over 2,200 pounds (990 kilograms) and was over 12 feet (365 centimeters) tall.

Do camels store water in their humps?

The hump or humps do not store water, since they are fat reservoirs. The ability to go long periods without drinking water, up to ten months if plenty of green vegetation and dew is available to feed on, results from a number of physiological adaptations of the animal. For example, camels can lose up to 40 percent of their body weight with no ill effects; they can also withstand a variation of their body temperature by as much as 14°F (8°C).

What is one of the most successful groups of mammals known?

Contrary to what you may think—that humans are the most successful—other animals beat us, in particular the bat. These creatures have more than 1,200 species that comprise about one-fifth of all mammal species, and other than owls, hawks, spiders, and snakes, they have few natural enemies.

How long do animals, in particular mammals, live?

Of the mammals, humans and fin whales live the longest. The following lists the maximum life span for various longer-lived animal species:

Animal (Latin name)	Maximum Life Span (in years)
Marion's tortoise (*Testudo sumeirii*)	152+
Quahog (*Venus mercenaria*)	c. 150
Common box tortoise (*Terrapene Carolina*)	138
European pond tortoise (*Emys orbicularis*)	120+
Spur-thighed tortoise (*Testudo graeca*)	116+

Why are nine-banded armadillos so amazing?

Nine-banded armadillos—found in the U.S. and Central and South America—are not well known to most people, but they have fascinating features for a mammal. For example, they are always born as same-sex identical quadruplets, and hungry armadillos can eat up to 40,000 ants at a single "meal." They are also excellent in the water, holding their breath for up to ten minutes by inhaling air into their lungs, stomach, and intestines, all of which make them buoyant, too. They also can feel by walking along the bottom of a river or shallow lake—all by letting their breath out.

Animal (Latin name)	Maximum Life Span (in years)
Fin whale (*Balaenoptera physalus*)	116
Human (*Homo sapiens sapiens*)	116 (although this is probably closer to 125)
Deep-sea clam (*Tindaria callistiformis*)	c. 100
Killer whale (*Orcinus orca*)	c. 90
European eel (*Anguilla anguilla*)	88
Lake sturgeon (*Acipenser fulvescens*)	82
Freshwater mussel (*Margaritana margaritifera*)	80 to 70
Asiatic elephant (*Elephas maximus*)	78
Andean condor (*Vultur gryphus*)	72+
Whale shark (*Rhiniodon typus*)	c. 70
African elephant (*Loxodonta africana*)	c. 70
Great eagle-owl (*Bubo bubo*)	68+
American alligator (*Alligator mississipiensis*)	66
Ostrich (*Struthio camelus*)	62.5
Horse (*Equus caballus*)	62
Orangutan (*Pongo pygmaeus*)	c. 59
Hippopotamus (*Hippopotamus amphibious*)	54.5
Chimpanzee (*Pan troglodytes*)	51
Gorilla (*Gorilla gorilla*)	50+
Domestic goose (*Anser a. domesticus*)	49.75
European brown bear (*Ursus arctos arctos*)	47
Blue whale (*Balaenoptera musculus*)	c. 45
Goldfish (*Carassius auratus*)	41
Common toad (*Bufo bufo*)	40
Roundworm (*Tylenchus polyhyprus*)	39
Giraffe (*Giraffa camelopardalis*)	36.25
Domestic cat (*Felis catus*)	34
Canary (*Serinus caneria*)	34
American bison (*Bison bison*)	33
Bobcat (*Felis rufus*)	32.3
Sperm whale (*Physeter macrocephalus*)	32+
American manatee (*Trichechus manatus*)	30
Domestic dog (*Canis familiaris*)	29.5
Lion (*Panthera leo*)	c. 29
Theraphosid spider (*Mygalomorphae*)	c. 28
Tiger (*Panthera tigris*)	26.25
Giant panda (*Ailuropoda melanoleuca*)	26
American badger (*Taxidea taxus*)	26
Common wombat (*Vombatus ursinus*)	26
Bottle-nosed dolphin (*Tursiops truncates*)	25
Domestic chicken (*Gallus g. domesticus*)	25
Gray squirrel (*Sciurus carolinensis*)	23.5
Aardvark (*Orycteropus afer*)	23
Coyote (*Canis latrans*)	21+
Domestic goat (*Capra hircus domesticus*)	20.75
Queen ant (*Myrmecina graminicola*)	18+

Animal (Latin name)	Maximum Life Span (in years)
Common rabbit (*Oryctolagus cuniculus*)	18+
Walrus (*Odobenus rosmarus*)	16.75
Domestic turkey (*Melagris gallapave domesticus*)	16
American beaver (*Castor Canadensis*)	15+
Land snail (*Helix spiriplana*)	15
Guinea pig (*Cavia porcellus*)	14.8
Hedgehog (*Erinaceus europaeus*)	14
Golden hamster (*Mesocricetus auratus*)	10
Millipede (*Cylindroiulus landinensis*)	7
House mouse (*Mus musculus*)	6
Common octopus (*Octopus vulgaris*)	2 to 3
Monarch butterfly (*Danaus plexippus*)	1.13
Bedbug (*Cimex lectularius*)	0.5 or 182 days
Black widow spider (*Latrodectus mactans*)	0.27 or 100 days
Common housefly (*Musca domesticus*)	0.04 or 17 days

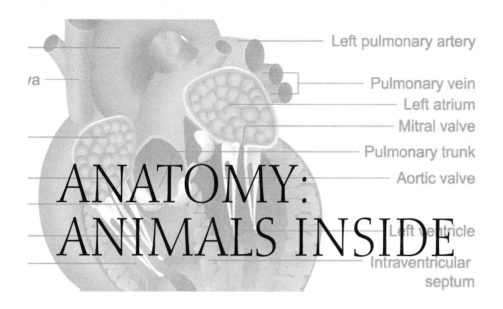

Left pulmonary artery
Pulmonary vein
Left atrium
Mitral valve
Pulmonary trunk
Aortic valve
Left ventricle
Intraventricular septum

ANATOMY: ANIMALS INSIDE

ANIMAL ANATOMY BASICS

What is anatomy?

Anatomy is the study of an organism's (plant or animal) structure or any of its parts, including the major organs, tissues, and cells. In other words, it refers to the internal structure (and how it is organized within) of an organism. Major branches of anatomy include comparative anatomy (the similarities and differences between structures and organizations of organisms), histology (study of organisms' tissues), cytology (study of cells), and human anatomy (anthropotomy).

To compare, and since function is included in this field, anatomy is directly related to physiology, or the mechanical, physical, and biological functions of organisms and the parts involved in those functions. For purposes of this text, the current chapter focuses on the animals' physical attributes as anatomy; the following chapter ("Physiology: Animal Function and Reproduction") on physiology includes more detail about how animals function.

What are the four levels of structural organization in animals?

Every animal has four levels of hierarchical organization: cell, tissue, organ, and organ system. Each level in the hierarchy is of increasing complexity, and all organ systems work together to form an organism. (For more about cells and tissues, see the chapter "Cellular Basics.")

What were some of the earliest studies in human anatomy?

The Greek physician Diocles of Carystus (c. 350 B.C.E.), a student of Aristotle, wrote the first anatomy books (along with the first book of herbal remedies). Around the same time, Greek physicians Herophilus (c. 320 B.C.E.) and Erasistratus (c. 304 B.C.E.)—both **209** of whom lived in Alexandria—performed dissections in public and thus described such human organs as the liver, ovaries, prostate gland, and spleen. They also determined that the place of reasoning was in the brain, not the heart.

What were some of the earliest studies in animal anatomy other than humans?

No one can truly pinpoint when people began to study animal anatomy. Many scientists believe it came from all the religious sacrifices of animals and how these structures compared to humans. Early Egyptians, and other cultures, also had to have some crude internal (for example, organs and veins) knowledge of certain animals, especially those creatures that were embalmed and buried with the human dead. One of the first people who probably knew about anatomy of animals was Greek philosopher Aristotle (c. 330 B.C.E.), who dissected animals and described many of their physical attributes. One positive outcome of such dissections, though, was that they eventually led to more knowledge about human anatomy.

Who were Hippocrates and Galen?

Aegean physician Hippocrates (c. 455–360 B.C.E.) was the first to define the profession of physicians, making up the Hippocratic oath—urging physicians to separate medicine from religion—and is thus considered by some as the "father of medicine." Some of the first studies of human anatomy were carried out by Roman physician and anatomist Galen (c. 130—200 C.E.). Not only was he the first to use the pulse as a diagnostic aid, but he also wrote all medical knowledge that existed at the time into one systematic treatment—one that would be used by physicians until the end of the Middle Ages.

What was one of the first books on human anatomy?

In 1543, Flemish anatomist Andreas Vesalius (1514–1564) wrote the first accurate work on human anatomy, titled *De humani corporis fabrica* (*On the structure of the human body*). Many scientists call it one of the books that changed the world—complete with exceptional illustrations (usually attributed to the painter Stephen van Calcar) and showing the true anatomy of the human body, all reproduced with extreme accuracy (previous hand-copied books had many inaccuracies). The book also caused a scandal: after it was published, accusations of body snatching and heresy were directed against Vesalius, mainly by physicians who were not as progressive or had older ideas of the human body, much of it based on Roman physician and anatomist Galen.

TISSUE AND CELLS

What are epithelial tissues?

A tissue (from the Latin *texere*, meaning "to weave") is a group of similar cells that perform a specific function. For example, the epithelial tissue, also called epithelium (from the Greek *epi*, meaning "on," and *thele*, meaning "nipple"), covers every surface, both

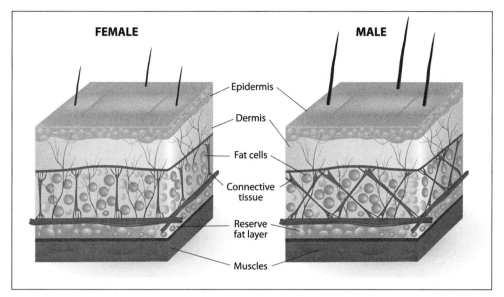

The main difference between female and male skin is that a man' skin is thicker and tougher (more connective tissue). Also, men produce more sebum (oil), which is why they tend to have more pimple problems at puberty than women do.

external and internal, of the body. The outer layer of the skin, the epidermis, is one example of epithelial tissue; other examples are the lining of the lungs, kidney tubules, and the inner surfaces of the digestive system—including the esophagus, stomach, and intestines—and the lining of parts of the respiratory system. In general, the epithelium forms a barrier, allowing the passage of certain substances while impeding the passage of other substances.

How often is the epithelium replaced?

Epithelial cells are constantly being replaced and regenerated during an animal's lifetime, which is why some people say in jest, "As humans, we become a 'different person' every so often." For example, the epidermis (outer layer of the skin) is renewed every two weeks; the epithelial lining of the stomach is replaced every two to three days; and the liver, a gland consisting of epithelial tissues, easily regenerates after portions are removed surgically.

What are connective tissues?

Six major types of connective tissue are in our bodies. The following lists and briefly describes those tissues:

- *Loose connective tissue* is a mass of widely scattered cells in a loose weave of fibers, many with strong protein fibers called collagen. These tissues are found beneath the skin and between organs and are actually like a "binding and packing material" that provides support to hold other tissues and organs in place.

211

- *Adipose tissue* consists of adipose cells in loose connective tissue; each cell stores a large droplet of fat, which swells when fat is stored and shrinks when fat is used to provide energy. In general, these tissues pad and insulate an animal's body.

- *Blood* is loose connective tissue; the surrounding material is a liquid called plasma. Blood consists of red blood cells, erythrocytes, white blood cells, leukocytes, and thrombocytes or platelets, which are pieces of bone marrow cell. Plasma also contains water, salts, sugars, lipids, and amino acids. Blood is approximately 55 percent plasma and 45 percent elements, and its main duty is to transport substances from one part of the body to another, which also plays an important role in the immune system of the animal. (For more about blood, see below.)

- *Collagen* (sometimes called fibrous or dense connective tissues) has a matrix of densely packed collagen fibers. The first type of collagen is regular dense connective tissues, which are lined up in parallel; tendons, which bind muscle to bone, and ligaments, which join bones together, are examples of dense regular connective tissue. The other type of collagen is irregular dense connective tissues, which are strong tissues covering various organs, such as kidneys and muscles.

- *Cartilage* (from the Latin *cartilago*, meaning "gristle") is connective tissue with an abundant number of collagen fibers in a rubbery matrix. It is both strong and flexible and provides support and cushioning, such as the cartilage found between the discs of the vertebrae in the spine.

- *Bone* is a rigid connective tissue that has a matrix of collagen fibers embedded in calcium salts. It is the hardest tissue in the body, although it is not brittle. Most of the skeletal system is comprised of bone, providing support for muscle attachment and protecting the internal organs.

What are the various types of muscle tissues in vertebrates?

Muscle tissue, consisting of bundles of long cells called muscle fibers, provides the capability of movement for the organism or for the movement of substances within the body of the organism. Vertebrates have three types of muscle tissue; the following lists those muscle types and their definition:

Smooth muscle tissue—Smooth muscle tissues are organized into sheets of long cells shaped like spindles, with each cell containing a single nucleus. For example, smooth muscle tissues line the walls of the digestive tract (stomach and intestines), blood vessels, urinary bladder, and irises of the eyes. Smooth muscle contraction is considered to be involuntary, since it occurs without intervention (or conscious thought) of the animal.

Skeletal muscle tissue—These tissues consist of numerous, very long muscle fibers that lie parallel to each other. Since the muscle fibers are formed by the fusion of several muscle cells, each long fiber has many nuclei. Muscle fibers have alternating light and dark bands, giving the appearance of a striped or striated fiber. Skeletal muscles allow animals to move, lift, and utter sounds. For example, tendons attach skeletal muscles to the bone, and when skeletal muscles contract, they cause the bone to move at the joint. Skeletal muscles are voluntary since the animal consciously contracts them. (It is interesting to note: Exercise does not increase the number of muscle cells because adult animals have a fixed number of skeletal muscle cells; however, exercise does enlarge existing skeletal muscle cells.)

Cardiac muscle tissue—As the name suggests, cardiac muscle tissues are found in the hearts of vertebrates and consist of small, interconnected cells, each with a single nucleus. They are striated muscle fibers similar to skeletal muscle tissues, but are involuntary like smooth muscle tissues, since the animal does not consciously move them. The ends of the tissue cells form a tight latticework, allowing signals to spread from cell to cell—and thus cause the heart to contract.

What is the function of nerve tissue?

Nerve tissue serves as the communication system of an animal. Nerve tissue allows an animal to receive stimuli from its environment and to relay an appropriate response. (For more about nerves and nerve cells, see the chapter "Physiology: Animal Function and Reproduction.")

ORGANS AND ORGAN SYSTEMS

What is an organ?

An organ is a group of several different tissues working together as a unit to perform a specific function or functions. Each organ performs functions that none of the component tissues can perform alone, with this cooperative interaction of different tissues being a basic feature of animals. The heart is an example of an organ: It consists of cardiac muscle wrapped in connective tissue; heart chambers, which are lined with epithelium tissue; and nerve tissue, which controls the rhythmic contractions of the cardiac muscles.

How many organs are in the human body?

About seventy-eight organs make up the human body—each one with a different function (or sometimes the same, as with two kidneys), size, function, and sometimes shape. The largest organ, with respect to its size and weight, in the human body is the skin. Not everyone has organs in the same place, either; for example, sometimes organs such as the kidney may be located closer to the pelvis, or only one kidney may be present. These differences can be due to genetics, difference in the growth of cells of that organ, or even disease.

What are some of the major organs in animals and humans?

Not all animals have the same organs, but some organs are common to most organisms (mostly mammals) inside their bodies. The following lists only a few organs (from various organ systems):

Brain—The brain is the central control area of the body and responsible for all the actions of the body parts. For humans, it is the third largest organ and averages about one hundred billion cells. (For more about the brain, see the chapter "Physiology: Animal Function and Reproduction.")

Heart—The heart pumps blood to every part of the animal's body, which then delivers energy to every cell. For humans, it is the fifth largest organ and weighs more in the male, 0.69 pounds (about 315 grams), than in females, 0.58 pounds (about 265 grams). (For more about the heart, see this chapter on the circulatory system.)

Kidneys—The kidneys separate waste material by filtering it from the bloodstream. Humans have two kidneys, with the average weight of both being around 0.64 pounds (290 grams); both the organs filter our blood fifty times a day. If one does not function, the other will enlarge and take on the duty of both organs.

Liver—The liver receives blood filled with digested food from our gut (stomach). It will store some of the food (which turns into a waste called urine) and send the other part to the rest of the body's cells through the bloodstream. In humans, the liver is about 3.44 pounds (1,560 grams), making it the second largest organ in the body (skin is the largest organ).

Lungs—Lungs allow organisms to exhale carbon dioxide and inhale oxygen—which also means carbon dioxide is taken out of the blood cells and oxygen is put into the red blood cells. For humans, they are the fourth largest organ in the body, and if you spread out all the parts of the lungs, it would cover over 295 feet (90 meters).

Pancreas—This is actually a gland in the body that produces several vital hormones, including insulin and somatostatin. It also has two functions (called a dual-function gland), as it has the features of both an endocrine and exocrine gland. In humans, it only weighs about 0.22 pounds (98 grams).

Skin—The largest organ in the human body, the skin on most organisms acts the same: to protect and hold in the internal organs, help with heat regulation, and protect from diseases and environmental interactions. For humans, the skin weighs about twenty-four pounds (10.886 kilograms), but varies according to height, weight, and shape of the body.

Spleen—This is the organ that forms red and white blood cell pulp, thus helping it make blood and at the same time, increase the immunity of the organism. In humans, it is the heaviest of all the organs (using comparable size and weight), weighing about 0.37 pounds (170 grams).

Thyroid—The thyroid in an organism controls the production of thyroxine and several other hormones and thus is tied to many functions of other organs in the body. In humans, it is the largest gland and is located on either side of the neck (often referred to as a butterfly shape across your windpipe).

Gallbladder—Most vertebrates have a gallbladder that acts as a storage area for bile and acts the same way for most animals. For example, the gallbladder in humans is often thought of as the "forgotten" organ, but it is responsible for an important function: it helps with fat digestion and helps to concentrate the bile our liver produces. This sac, about 4 inches (10.2 centimeters) long, is located under the liver; when you eat fats, a hormone called cholecystokinin that is produced in the small intestines signals the gallbladder to release bile (a fluid made by your liver to digest fat) that helps with fat breakdown and absorption in the small intestines.

What is an organ system?

An organ system is a group of organs working together to perform a vital body function. The vast majority of vertebrate animals have twelve major organ systems; the following lists those systems, their components, and their functions:

Organ System	Components	Functions
Cardiovascular	Heart, blood, and blood vessels	Transports blood throughout the body, and circulatory system, supplying nutrients and carrying oxygen to the lungs and wastes to kidneys
Digestive	Mouth, esophagus, stomach, intestines, liver, gallbladder, and pancreas	Ingests food and breaks it down into smaller chemical units
Endocrine	Pituitary, adrenal, thyroid, and other ductless glands	Coordinates and regulates the activities of the body
Immune	Lymphocytes, antibodies; macrophages, and also thymus, tonsils and spleen	Removes foreign substances; fights diseases
Integumentary	Skin, hair, nails, and sweat glands	Protects the body
Lymphatic	Lymph nodes, lymphatic capillaries, lymphatic vessels, spleen, and thymus	Captures fluid and returns it to the cardiovascular system
Muscular	Skeletal muscle, cardiac muscle, and smooth muscle	Allows body movements
Nervous	Nerves, sense organs, brain, and spinal cord	Receives external stimuli, processes information, and directs activities
Reproductive	Testes, ovaries, and related organs	Carries out reproduction
Respiratory	Lungs, trachea, and other air passageways	Exchanges gases—captures oxygen (O_2) and disposes of carbon dioxide (CO_2)
Skeletal	Bones, cartilage, ligaments, tendons	Protects the body and provides support for locomotion and movement
Urinary (or Excretory)	Kidneys, ureters, bladder, and urethra	Fluid balance, excretion of urine; removes wastes from the bloodstream
Vestibular	Ear system	Contributes to balance and special orientation

What is a vestigial organ?

Years ago, scientists believed that many organs in our body had no real function at all and needed to be removed. The most common examples are the gallbladder and the appendix; now scientists know that both have a major function in the body.

DIGESTION

What are the steps of food processing for animals?

The first step in digestion is for animals to ingest food. The food is then broken down via the digestive process into molecules; once the food is digested, it is absorbed through the digestive tract to provide energy for the organism. The final step of food processing is elimination—the undigested material is passed out of the digestive tract.

How are animals classified based on what they eat?

Often times, animals are classified based on whether they eat plants, other animals, or a combination of both. Animals that eat only plant matter are called herbivores (from the Latin *herba*, meaning "green crop," and *vorus*, meaning "devouring"), such as cattle, deer, and many aquatic species that eat algae. Animals that eat other animals are called carnivores (from the Latin *carne*, meaning "flesh," and *vorus*, meaning "devouring"), such as lions, sharks, snakes, and hawks. And finally, animals that eat both plants and other animals are called omnivores (from the Latin *omnis*, meaning "all," and *vorus*, meaning "devouring")—including humans, bears, crows, and raccoons.

How do the teeth (called dentition) of animals reflect their diet?

Depending on the animal's main diet, it has specialized teeth to eat with. For example, herbivores have sharp incisors to bite off blades of grass and other plant matter, along

---

How do zombie worms eat?

Yes, the ocean contains worms—the so-called zombie worms (genus *Osedax*)—that seem to like the taste of whale bones. Scientists have discovered that these creatures may like to munch the bones, but without any mouth parts or a digestive tract, for that matter—how the animals eat has long been a mystery. In 2013, scientists finally made the discovery—the bone-drilling worms actually produce acid in large quantities. The acid is produced by what are called proton pumps, or protein-containing structures that exist at the front end of the worm's body (to compare, human kidneys use similar proton pumps to process wastes). How they process the dissolved bone as food is still a question, but one suggestion is that symbiotic bacteria may be involved.

with flat premolars and molars for grinding and crushing plants. Carnivores have pointed incisors and enlarged canine teeth to tear off pieces of meat; their premolars and molars are jagged to aid in chewing flesh. Omnivores have nonspecialized teeth to accommodate a diet of both plant material and animals.

How do continuous feeders differ from discontinuous feeders?

Continuous feeders, also known as filter feeders, are aquatic animals that constantly feed by having water filled with food particles (for example, small plankton or fish) entering through the mouth. In addition, they do not need a storage area, such as a stomach, for food. Discontinuous feeders must hunt for food on a regular basis; they need a storage area, such as a stomach, to house food until it is digested.

What are the two general types of digestive systems?

Depending on the organism, the digestive system can include the mouth, alimentary canal or gastrovascular cavity, esophagus, stomach, small intestine, large intestine, and anus. The mouth is the opening through which food is ingested. In animals with a second opening for elimination, the digestive system contains an alimentary canal, essentially a tube allowing for the passage of food from the mouth to the anus. In contrast, animals with only one opening have a gastrovascular cavity that serves as the site of digestive activities.

How do humans digest their food?

Humans have a special digestive system that allows the body to digest (break down large food particles into smaller molecules) and absorb nutrients. For example, fats are broken down into glycerol and fatty acids and proteins into amino acids; vitamins and minerals are small enough to be absorbed without being digested. Overall, the average length of the human digestive tract—from the mouth to the anus, also called the alimentary canal—is about 30 feet (9.9 meters) and is lined mostly with smooth muscles (involuntary) that push the food through the body in a process called peristalsis. The following lists some of the major parts of the human alimentary canal:

Mouth—The human mouth, tongue, and teeth are responsible for breaking the food down with mechanical action. As omnivores, human teeth include incisors for cutting, canines for tearing, and molars for grinding. As the food is in the mouth, salivary glands release saliva (the scientific name of salivary amylase), which then chemically breaks down the starch in the food.

Esophagus—The esophagus is where the food goes after a person swallows. It is directed into the esophagus by the epiglottis, or a flap of cartilage in the back of the pharynx (throat), which essentially stops food from going into the windpipe (and why we sometimes choke on food).

Stomach—Food is churned mechanically in the stomach, an organ that secretes gastric juice, or a mix of special enzymes and hydrochloric acid. The cardiac sphincter

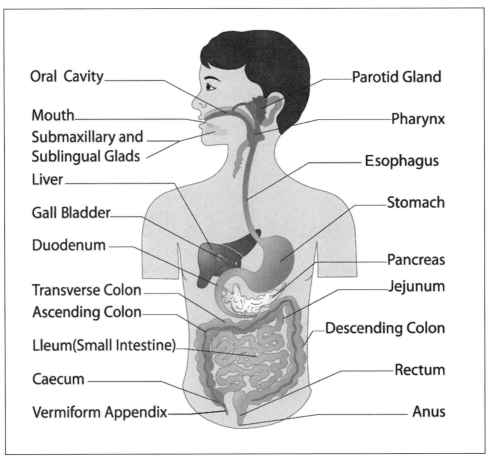

The human digestive system

acts as a stop to prevent food in the stomach from backing up into the esophagus and thus burning it; the pyloric sphincter at the bottom of the stomach helps keep the food in the stomach long enough to be digested.

Small intestine—The upper part of the small intestine—about the first 12 inches (30 centimeters)—is called the duodenum. Bile from the liver and stored in the gall-bladder is released into the small intestine, allowing fats to break down and assisting other digestive enzymes to work. The intestine is lined with millions of villi, which absorb all the nutrients released from the digested foods.

Large intestine—The large intestine is also called the colon and has three functions: the removal of undigested waste (egestion), bacteria that produces vitamins, and removal of excess water. All together, both intestines reabsorb 90 percent of the water that enters the mouth—but if too much water is removed, constipation results, or if not enough water is removed, diarrhea results. The waste is finally released at the end of the digestive tract, or the anus.

What are gallstones?

Gallstones are produced in the gallbladder—part of our digestive system—and are small deposits that most often form when substances in the bile harden. It is estimated that about 10 to 15 percent of adults have gallstones. The usual symptoms of a pain around the abdomen, back, or just under the right arm occur most often if a blockage is in the gallbladder's duct that releases the bile—most often caused by a gallstone. The one positive part about having gallstones is that for most people, the removal of the gallbladder has little effect, as bile has other ways of getting to your small intestines.

How do earthworms digest their food?

Earthworms have a long straight tube as their digestive tract. As they burrow into the ground, they create tunnels that aerate the soil (much to the happiness of gardeners); their mouths are also ingesting the decaying organic matter along with the soil. The food moves to their esophagus, then to the crop, where it is stored. The gizzard—a sand- and soil-filled structure with thick, muscular walls—then grinds up the food. From there, the food passes through the intestines (that also holds a large fold in its upper surface, called the typhlosole, that increases the intestine's surface area), and chemical digestion and absorption occurs.

How do grasshoppers digest their food?

Grasshoppers have a digestive tract similar to the earthworm, with a crop and gizzard, but have some major differences. The grasshopper has certain, specialized mouth parts that allow it to taste, bite, and crush food. In addition, its gizzard has plates made of chitin (for more about chitin, see this chapter) that helps grind up its food, and its digestive tract removes the waste products (uric acid) from its body.

Why do cows have four stomachs?

Cows—mammals called ruminants—have four stomachs in order to process their low-quality diet of grass; they eat rapidly and do not completely chew much of their food before swallowing. The liquid part of their food enters what is called the reticulum first, while the solid part of their food enters what is called the rumen, where it softens. Bacteria in the rumen break the food material down as a first step in digestion. Ruminants later regurgitate the partially liquefied plant parts into their mouth, where they continue to munch it in a process known as "chewing their cud." Cows chew their cud about six to eight times per day, spending a total of five to seven hours in rumination. The chewed cud goes directly into the other chambers of the stomach, where various microorganisms assist in further digestion. And to give their intestines time to absorb nutrients, these herbivores have a longer small intestine than most mammals.

How do rodents, rabbits, and hares digest plant material?

Unlike cows that have a rumen to digest cellulose (plant material), herbivores such as rabbits and hares have a cecum, a large pouch that aids in digestion with the assistance of microorganisms. The cecum is located at the junction between the small and large intestines. It is impossible for these animals to regurgitate the contents of their stomachs (like ruminants) because the cecum is located beyond the stomach. Instead, these animals pass their food through their digestive tract a second time by ingesting their feces. When feces pass through the digestive tract, it is possible for these animals to absorb the nutrients produced by the microorganisms in the cecum.

Rabbits and hares, such as this mountain hare, have a cecum in their intestines that helps them to digest the plant matter that makes up their diet.

RESPIRATION

What is respiration?

Respiration is the exchange of gases (oxygen and carbon dioxide) between an animal and its environment. The three phases to the process of respiration (gas exchange) are 1) breathing, when an animal inhales oxygen and exhales carbon dioxide; 2) transport of gases via the blood (circulatory system) to the body's tissues; and 3) at the cellular level, when the cells take in oxygen from the blood and, in return, add carbon dioxide to the blood. Different types of animals have different respiratory organs for gas exchange, with the four types of respiratory organs being the skin, gills, tracheae, and lungs.

How do certain animals breathe through their skin?

Many invertebrates and some vertebrate animals, including amphibians, breathe through their skin—a process known as cutaneous respiration. Most of these animals are small, long, and flattened, like earthworms and flatworms. All animals that rely on their skin for respiration live in moist, damp places in order to keep their body surfaces moist. Capillaries (the small blood vessels) bring blood rich in carbon dioxide and deficient in oxygen to the skin's surface, where gaseous exchange takes place by diffusion.

How do certain animals use gills to breathe?

Gills may be external extensions of the body surface such as those found in aquatic insect larvae and some aquatic amphibians. Diffusion of oxygen occurs across the gill surface into capillaries, while carbon dioxide diffuses out of the capillaries into the environment. Fish and some other marine animals also have internal gills: Water enters the animals through the mouth, then flows over the gills in a steady stream and out through gill slits. Although some animals with gills spend part of the time on land, they all must spend some time in moist, wet environments in order for the gills to function.

How do certain animals use lungs to breathe?

Lungs are internal structures found in most terrestrial animals in which gas exchange occurs. The lungs are lined with moist epithelium (outer cells) to avoid becoming dried out. Some animals, including certain lungfish, amphibians, reptiles, birds, and mammals, have special muscles to help move air in and out of the lungs; other animals have lungs connected to the outside surface with special openings and do not require special muscles to move air in and out of their lungs.

How do many insects breathe?

Insects have a system of internal tubes (called tracheae) that lead from the outside to internal regions of the body by what are called spiracles, where gases are exchanged. Some insects rely on muscles to pump the air in and out of the tracheae, while in other insects, the process is a passive exchange of gases. In addition, some insects, such as spiders, have "lungs" in addition to tracheae—hollow, leaflike structures through which the blood flows. These lungs hang in an open space that is connected to a tube, with the other side of the tube in open contact with the air.

How do humans breathe?

Humans take in oxygen and give off carbon dioxide—in fact, humans consume more than 23,670 ounces (2,959 cups or 700 liters) of oxygen a day, exhaling it as carbon dioxide (although around 15 to 18 percent of what we breathe out is still oxygen). The way oxygen and carbon dioxide gases are exchanged is relatively simple: The gases move between the small sacs in the lungs (alveoli) and the blood by diffusion—the oxygen diffusing from the alveoli into the blood, and the carbon dioxide from the blood to the alveoli. This also involves pressures (called concentration), in which the oxygen in the alveoli is kept at a higher pressure than in the blood, and carbon dioxide pressure in the alveoli is kept at a lower level than in the blood—all done by our breathing in and out.

What are some human lung statistics?

The human body has two lungs—both divided into lobes: the left has two lobes, and the right has three lobes. As a person breathes, the air travels down the windpipe, eventually branching out into the mucus-lined bronchi; from there, the bronchi split into tens of thousands of even smaller tubes called bronchioles that connect to tiny sacs called alveoli. The number of airways in the average human is immense, with both lungs containing about 1,500 miles (2,400 kilometers) of airways, with a total surface area of about the size of a tennis court, plus about 300 to 500 alveoli. The lung capacity changes depending on several factors, too—for example, females usually have 20 to 25 percent lower capacity than human males, and logically, taller people have more capacity than shorter people. And if you have lived at sea level all your life, your lung capacity is probably less than a person who has lived in a high-mountain region all their life (because the mountain air is thinner and has less oxygen, the body compensates by expanding the capacity of the lungs and thus, takes in more oxygen).

What is the respiration rate of various animals?

The respiratory rate of an animal depends on many factors, including the creature's metabolism and even how the animal is built. The following lists the breaths per minute for various animals:

Animal	Breaths per minute
Diamondback snake	4
Horse	10
Human	12
Dog	18
Pigeon	25–30
Cow	30
Giraffe	32
Shark	40
Trout	77
Mouse	163

What are the breath-holding capabilities of mammals?

It's not easy for most humans to hold their breath underwater, but some animals seem to be the champions. The following lists the average time in minutes that certain animals can hold their breath—in other words, not take oxygen into their system. It is interesting to note that some humans can hold their breath for longer; the underwater record by 2012 was 22 minutes and 22 seconds—humans can hold their breath longer underwater than on land because of what is called the "diving reflex," when the body slows down the heart and metabolism to conserve oxygen and energy when in cold water.

Mammal	Average time in minutes
Human	1
Polar bear	1.5
Pearl diver (human)	2.5
Sea otter	5
Platypus	10

Mammal	Average time in minutes
Muskrat	12
Hippopotamus	15
Sea cow (manatee)	16
Beaver	20
Porpoise	15
Seal	15–70 (depends on the seal type)
Fin whale	20
Greenland whale	60
Sperm whale	75–90
Bottlenose whale	120

How do air-breathing mammals dive underwater for extended periods of time?

Seals and whales are able to dive underwater for extended periods of time because they are able to better store oxygen in their systems. While humans store 36 percent of their oxygen in their lungs and 51 percent in their blood, seals store only approximately 5 percent of their oxygen in their lungs and 70 percent in their blood. They also store more oxygen in the muscle tissue—25 percent compared with only 13 percent in human muscle tissue. While underwater, these mammals' heart rates and oxygen consumption rates decrease, allowing some species to remain underwater for up to twenty minutes at a time.

In 2013, researchers finally uncovered how diving mammals evolved their underwater endurance. They found that marine mammals, such as the sperm whale, all have high concentrations of oxygen-binding protein called myoglobin—the protein that gives red meat its color—in concentrations so high that the muscles of these mammals are almost black in color. The scientists knew that the protein tends to stick together in such high concentration and cuts down on their function to carry oxygen. They found that deep-diving mammals—and even aquatic mammals such as beavers—had more of an electric charge on the surface of their myoglobin. The researchers further speculated that such an electric charge causes "electro-repulsion"—similar to trying to put two magnets with the same pole together (it's not possible)—thus preventing the stickiness of the protein in the mammals and allowing for more oxygen in their muscular systems for those deep, long dives.

CIRCULATORY SYSTEM

What are the functions of the circulatory system?

The primary function of the circulatory system is to transport oxygen and nutrients to all the cells of an organism. The circulatory system also transports wastes from the cells to waste-disposal organs such as the lungs for carbon dioxide and the kidneys for other metabolic wastes. In addition, the circulatory system plays a vital role in maintaining homeostasis—in other words, to keep your entire system balanced.

What are the differences between an open and a closed circulatory system?

In an open circulatory system, found in many invertebrates (for example, spiders, crayfish, and grasshoppers), the blood is not always contained within the blood vessels. Periodically, the blood leaves the blood vessels to bathe the tissues with blood and then returns to the heart; thus, no interstitial body fluid is separate from the blood. A closed circulatory system, also called a cardiovascular system, is found in all vertebrate animals and many invertebrates; in a closed system, the blood never leaves the blood vessels.

What are the main components of the circulatory system?

The components of the circulatory system are vessels, heart, and blood (for more about the heart and blood, see this chapter). The following are the three types of transport vessels in a closed circulatory system:

Arteries and arterioles—The arteries and arterioles always transport blood *away* from the heart (via rhythmic contractions known as the pulse) to the various organs in the body—an area under enormous pressure. Thus, their walls are made of thick, elastic, smooth muscles.

Veins and venules—Veins return blood to the heart after it circulates through the body. They are relatively thin-walled vessels that lack muscular tissues but are located within skeletal muscle—which means the blood is propelled upward and back to the heart as the body moves. They also contain one-way "valves" that prevent the backflow of blood within the vein.

Capillaries—The smallest vessels are the capillaries—mostly microscopic vessels that form an elaborate network that conveys blood between arteries and veins (their walls are only one cell in thickness). They branch from the ends of small arteries and carry oxygen-rich blood to all tissues of the body—allowing for the diffusion of nutrients and wastes between cells and the blood.

What are the main components of human blood?

Blood is considered a complex tissue—a group of similar cells—suspended in a fluid medium for easy transport through blood vessels in many organisms, including humans. The following lists the four main components of human (and many other organisms) blood (for more about cells, see the chapter "Cellular Basics"):

Plasma—Plasma is the liquid portion of the blood, and its principal component is water. Within the water are necessary elements that allow humans to live—dissolved salts, nutrients, gases, and molecular wastes. It also contains clotting factors, hormones, enzymes, and antibodies.

Red blood cells—The red blood cells—also called by the scientific name erythrocytes—are the most abundant cell type in the blood fluid. These dish-shaped cells measure about 8 micrometers in diameter; they lack nuclei and cannot reproduce. But they have an extremely important job: carrying hemoglobin (a red oxygen-carrying pigment; when it is bound to oxygen, it is called oxyhemoglobin) and oxygen throughout the body. They only live about 120 days—forming in the bone marrow and recycled in the liver.

White blood cells—White blood cells are also formed in the bone marrow. They die off fighting infections and are one of the components of pus. These cells come in two types: the phagocytes, which engulf and destroy bacteria that enter the bloodstream through a break in the skin and gather in large numbers at the sites of bacterial infection anywhere in the body, and the lymphocytes (in particular the B lymphocyte) that produce antibodies specifically designed to fight off certain types of antigens (foreign proteins) that enter the bloodstream via almost any route.

Platelets—Platelets—also called by the scientific name thrombocytes—are small, noncellular components (actually cell fragments) that form in the bone marrow from

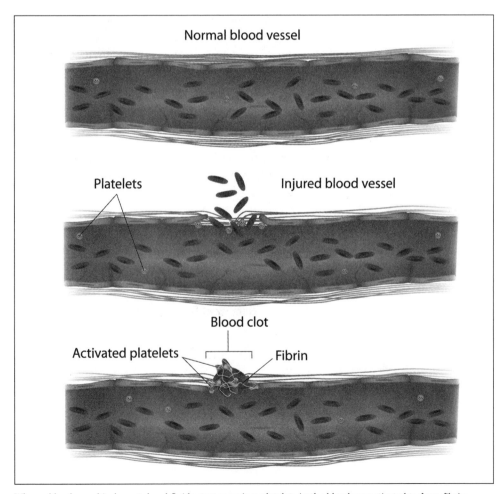

When a blood vessel is damaged and fluids start escaping, platelets in the blood are activated to form fibrin, which traps red blood cells and lymph cells, forming a clot.

megakaryocytes (large bone marrow cells) and are the main reason why blood clots so well.

How does human blood clot?

For most humans, the ability to clot blood is vital—if not, even a small cut could mean bleeding to death. Simply put, blood clotting (or thrombus) is actually a mass of protein fibers that trap lymph and red blood cells, eventually hardening into a cap—what we call a scab—that protects the damaged area. The clotting reaction is thought to be platelets in the area of a cut releasing chemicals into the bloodstream, starting the formation of a clot through a series of enzyme-controlled reactions.

What is hemophilia?

Hemophilia is a rare, hereditary bleeding disorder in which the blood does not clot normally at the site of a wound and the person bruises easily. It occurs more in males than females and is thought to be caused by the lack of a blood clotting factor: Type A, the most common type (around 90 percent of the cases), is caused by a deficiency of factor VIII (one of the proteins that helps blood to form clots) and Type B, which is caused by a deficiency of factor IX. Simply put, the condition is caused by a defect in one of the clotting factor genes that lies on the X chromosome and most often occurs in males since that gene can be passed from mother to son. A son—with XY chromosomes—lacks a second X chromosome to make up for the defective gene, whereas a daughter—with XX chromosomes—is likely to be a carrier but unlikely to actually have the disorder. In order to have hemophilia, females must have the abnormal gene on both X chromosomes, which is a very rare occurrence.

What are blood groups?

Genetically determined blood group systems number in the dozens among humans, but the AB0 and Rh systems are the most important ones used to type blood for human blood transfusions. The following lists the number of blood groups for certain species of animals:

Species	Number of Blood Groups
Pig	16
Cow	12
Chicken	11
Horse	9
Sheep	7
Dog	7
Rhesus monkey	6
Mink	5
Rabbit	5
Mouse	4
Rat	4
Cat	2

What is the ABO blood type mean?

Humans have over fifty different blood antigens—the antigens of the A-B-O (or ABO) group are the most common. This classification system is used by most hospitals and blood donation groups to distinguish blood types for compatibility for blood transfusions. The antigens are known as antigens A and B—both responsible for producing the blood types A, B, AB, and O (also called the universal donor blood type).

Organ and tissue transplants—including blood transfusions—can only be safe if certain markers (antigens) of the donor and recipient are the same or very similar; if not, the recipient's body (the immune system) will reject the blood or tissue as though they were disease-causing organisms by producing antibodies. For example, people with type A have antibodies in their blood against type B; those with type B have antibodies against type A. People with type AB have no anti-A or anti-B antibodies, but those who have type O have both anti-A and anti-B antibodies. People with AB blood are called universal recipients, as they can receive any of the blood types. And usually the most sought-after blood types for donation come from people with type O blood: called universal donors, their blood can be given to people with any of the other blood types.

Do all animals have blood?

Some invertebrates, such as flatworms and cnidarians, lack a circulatory system that contains blood. These animals possess a clear, watery tissue that contains some phagocytic cells, a little protein, and a mixture of salts similar to seawater. Invertebrates with an open circulatory system have a fluid that is more complex and is usually referred to as hemolymph (from the Greek term *haimo*, meaning "blood," and the Latin term *lympha*, meaning "water"); their hemoglobin is not concentrated in cells within the hemolymph, but rather is found floating in the hemolymph. Invertebrates with a closed circulatory system have blood that is contained within blood vessels. Other examples of circulatory systems are found in squids, octopi, and crustaceans—animals that also have oxygen-carrying molecules in their plasma, but their bodies use a copper-based molecule hemocyanin to carry oxygen instead.

What does Rh blood type mean?

Rh stands for Rhesus—after the rhesus macaque (*Macaca mulatta*), or the rhesus monkey (associated with the development of the Rh factor and serum)—and is one of the major blood group systems in humans. The Rh system classifies blood as Rh-positive or Rh-negative, based on the presence or absence of Rh antibodies found in the blood. Similar to the ABO blood type, when it comes to blood transfusions, the Rh often matters: People with Rh-positive blood can have a transfusion of Rh-negative blood without a problem, but people with Rh-negative blood will have a transfusion reaction if they receive Rh-positive blood.

Why do small animals not have a circulatory system?

Smaller animals such as hydras do not have a separate circulatory system since their cells are able to efficiently exchange materials (nutrients, gases, and wastes) through diffusion. The cells of these animals are close to the surface and thus can exchange nutrients effectively and efficiently without the need for a circulatory system.

What animals have different colored blood than humans?

The color of blood is related to the compounds that transport oxygen. For example, hemoglobin containing iron is red and is found in all vertebrates and a few invertebrates. Annelids (segmented worms) have either a green pigment, chlorocruorin, or a red pigment, hemerythrin. Some crustaceans (arthropods having divided bodies and generally having gills) have a blue pigment, hemocyanin, in their blood.

How do skin cells keep your blood from leaking out?

The skin is comprised of multiple layers of cells. The outermost layers are made of dead cells full of keratin. Sebaceous glands coat these dead cells with an oily secretion that makes them water resistant. However, they are not waterproof; about one pint of fluid from deeper tissues leaks through the skin's surface every day and evaporates. This excretion is in addition to perspiration produced by excessive heat or strenuous activity. The very strong junctions holding the cells together, known as desmosomes, prevent large of amounts of fluid leakage across the skin barrier. The linkages are so efficient that the epidermal skin cells tend to slough off in sheets rather than individually.

Who first demonstrated that blood circulates?

English physician William Harvey (1578–1657) was the first person to demonstrate that blood circulates in the bodies of humans and other animals. Harvey's hypothesis was that the heart is a pump for the circulatory system, with blood flowing in a closed circuit. Harvey conducted his research on live organisms as well as dissection of dead organisms to demonstrate that when the heart pumps, blood flows into the aorta. He observed that when an artery is slit, all the blood in the system empties. Finally, Harvey demonstrated that the valves in the veins serve to make sure blood returns to the heart.

How does blood circulate throughout the human body?

For most humans, blood takes a certain path to circulate throughout the body. The deoxygenated blood first enters through the vena cava—a major artery—into the right atrium of the heart. From there, it passes through a one-way valve called the tricuspid valve (right atrioventricular, or AV) into the right ventricle. Then, a strong muscular contraction forces it out through the pulmonary artery and into the lungs. The blood passes through capillaries in the lungs, where gas exchange oxygenates the blood; it then returns via the pulmonary veins to the heart, entering through the left atrium and on to the left ventricle, which then contracts, sending the oxygenated blood out of the heart to the body's organs via the aorta.

What are the various types of circulation?

Several major types of circulation occur in the body. The circulation of blood through the heart is called the coronary circulation. Blood moving through the body organs is called the systemic circulation; this also often includes renal circulation (kidneys) and hepatic circulation (liver). The circulation of blood through the pulmonary artery, lungs, and pulmonary vein is called the pulmonary circulation.

How does the human heart work?

The human heart—actually a muscle that is located beneath the sternum, or the upper left side of a person's torso—is about the size of a clenched fist. On the average, the human heart beats about seventy to seventy-five times per minute (resting) and pumps about five quarts (just over five liters) of blood per minute. Within the heart, two atria receive blood from the body cells, while two ventricles pump blood out of the heart.

The heart also has a natural "pacemaker" called the sinoatrial (SA) node, the part of the heart that times the contraction by generating and sending electrical signals to what is called the atrioventricular node. The impulses are sent to certain fibers, causing the ventricles to contract; these electrical impulses are also what are detected by the electrocardiogram (EKG) many people receive to check for irregularities in the

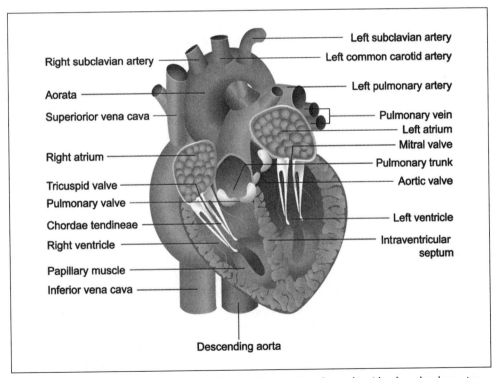

The remarkable muscle in your chest known as the heart (cross section shown above) has four chambers—two ventricles and two atria—that work non-stop to pump blood through your body.

heartbeat. Several factors can affect the heart's pacemaker, including two sets of nerves that speed up or slow down the heart, the release of hormones such as adrenaline, and the body's overall temperature.

What is blood pressure, and why is it important?

Blood pressure is the lowest in the veins and the highest in the arteries when the ventricles of the heart contract. Currently, the blood pressure for normal, resting adults is 120/80 (although this is often debated, depending on the doctor). This means that the systolic number (the measurement of the pressure when the ventricles contract) is 120 and the diastolic number (the measurement of the pressure when the heart relaxes) is 80. One of the main problems with too high a blood pressure—from the narrowing of the arteries because of either the buildup of fatty deposits on the walls or plaque—is that it can cause serious damage to the heart and blood vessels. Such damage can not only cause the heart to lose its ability to pump well, but it can also cause blood vessels to lose their elasticity and ability to carry blood efficiently.

How does a human's heartbeat compare with those of other mammals?

The resting heartbeat of a human often differs with those of other mammals, mainly because certain animals have different needs in terms of their circulatory systems. The following lists some mammals and their average resting heart rates:

Mammal	Resting Heart Rate (beats per minute)
Human	75
Horse	48
Cow	45–60
Dog	90–100
Cat	110–140
Rat	360
Mouse	498

EXCRETORY SYSTEM

How do certain excretory systems work in various animals?

The excretory system is responsible for removing waste products from an organism. It also plays a vital role in regulating the water and salt balance in the organism. Many animals, such as sponges, jellyfish, tapeworms, and other small organisms, do not have distinct excretory organs. Rather, they rid their bodies of waste through diffusion. Larger, more complex animals require specialized, often tubular, organs to rid their bodies of waste. For example, flatworms such as planarians have tubules that collect wastes and expel them to the outside via pores. Segmented worms such as earthworms have nephridia (tubules with a ciliated opening) in each segment. Fluid from the body cavity is propelled through the nephridia, and wastes are expelled through a pore to the outside while certain substances are reabsorbed. In addition, insects have a unique excretory system that consists of Malpighian tubules: Waste products enter the Malpighian tubes from the body cavity, then water and other useful substances are reabsorbed while uric acid passes out of the body.

How do vertebrates dispose of their wastes?

Unlike many other organisms discussed above, vertebrate animals have kidneys to dispose of their metabolic wastes. For example, in humans the kidneys are two bean-shaped organs, both about the size of a fist, which are located in the middle of the back on either side of the spine. Around 20 percent of the blood pumped by the heart goes through the organs, and every day, the average human processes about 200 quarts of blood through the kidneys with about 2 quarts of urine (as waste and extra water) traveling down the ureters to the bladder, where it is stored until you go to the bathroom. Overall, the kidneys have many tasks, from keeping the concentrations of substances in your body in balance and removing wastes from the body (like urea and toxic substances) to helping to maintain your calcium levels and regulate blood pressure.

What are kidney stones?

Most people who have had a kidney stone (called renal lithiasis) attack will tell you it is a very painful experience. The stones are usually small, hard deposits that form inside the kidneys and are made of minerals and acid salts. They often affect many places along the urinary tract, from the kidneys to the bladder, and can form in both or a single kidney. They seem to form when the urine becomes concentrated and/or when the urine contains more crystal-forming substances, such as calcium or uric acid, than can be diluted by the fluid in your urine. The stones are most commonly classified as calcium, struvite, uric acid, and cystine stones—all of which form for various reasons.

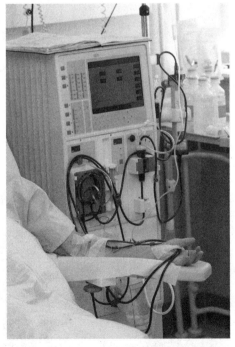

How does kidney dialysis remove waste products from the body?

Damaged or diseased kidneys are not capable of removing toxic waste substances from the body. Kidney dialysis removes the

Dialysis treatments work by filtering out toxins in the blood when a person's kidneys are not up to the task.

231

nitrogenous waste and regulates the pH of the blood when the kidneys do not work. Blood is pumped from an artery through a series of tubes made of a permeable membrane and a special solution; urea and excess salts diffuse out of the blood as it circulates through the dialyzing machine and are then discarded. Necessary ions diffuse from the solution back into the blood, and the cleansed blood is then returned to the body.

What type of nitrogenous waste do animals excrete?

Ammonia, urea, and uric acid are nitrogenous waste products—the result of the breakdown of various molecules, including nucleic acids and amino acids. Since it is highly toxic, excretion of pure ammonia is possible only for aquatic animals because ammonia is very soluble in water. Urea and uric acid are excreted by terrestrial animals. Urea is approximately 100,000 times less toxic than ammonia, so it may be stored in the body and eliminated with relatively little water loss. In addition, uric acid requires very little water for disposal and is often excreted as a paste or dry powder—for example, guano from seabirds and bats is excreted as solid white droppings.

Do marine fish drink water?

Marine bony fish such as tuna, flounder, and halibut drink seawater almost constantly to replace water lost by osmosis and through their gills. It is estimated that they drink an amount equal to 1 percent of their body weight each hour, an amount comparable to a human drinking 1.5 pints, or nearly 3 cups (700 milliliters), of water every hour around the clock. The gills eliminate most of the excess salts obtained by drinking large quantities of seawater. The fish excrete small quantities of urine that is isotonic to their body fluids. By contrast, cartilaginous fish (e.g., sharks and rays) do not need to drink water to maintain the balance of water (osmotic balance) in their bodies. They reabsorb the waste product urea, creating and maintaining a blood urea concentration that is one hundred times higher than that of mammals. Their kidneys and gills thus do not have to remove large quantities of salts from their bodies.

Do freshwater fish drink water?

Freshwater fish never drink water separate from ingesting food. These fish are prone to gain water since their body fluids are hypotonic (containing a lesser concentration of salts) to the surrounding water. They drink water through their gills to maintain the correct balance of salts in their bodies and excrete large quantities of diluted urine daily. It is estimated that freshwater fish eliminate a quantity of urine equal to one-third of their body weight each day.

What are the two types of euryhaline fish?

The two types of euryhaline (from the Greek terms *eurys*, meaning "broad," and *hals*, meaning "salt") fish can tolerate a wide range of salinity. They include flounder, sculpin, and killifish that live in estuaries, intertidal areas where the salinity of the water fluctuates throughout the day. The second type includes salmon, shad, and eels, which spend part of their life cycles in freshwater and the balance in seawater.

Are any other animals able to drink seawater?

Besides fish, certain birds and reptiles that live near the sea are also able to drink seawater. These animals have nasal salt glands near their eyes through which they excrete the excess quantities of salt solution. For example, penguins have such a special adaptation because they take in a great deal of saltwater as they swim and catch food. Although they cannot ingest saltwater, they are able to drink such water because of the supraorbital gland (just above their eye socket) that filters excess salt from their blood; from there, a concentrated fluid similar to brine is secreted from the nasal passages. This is also one reason why you see penguins shake their beaks—they are trying to shake off the excess salt around their nasal passages.

What is the composition of human urine?

Human urine is composed mostly of water containing organic wastes, as well as some salts. The composition of urine can vary according to diet, time of day, and diseases—but overall, the makeup of urine is 95 percent water and 5 percent solids (such as urea, creatinine, uric acid, and trace amounts of such substances as enzymes, carbohydrates, and hormones).

SKELETAL SYSTEM

What is the skeletal system?

The skeletal system is a multifunctional system which provides support, allows an animal to move, and protects the internal organs and soft parts of an animal's body. The following lists the three main types of skeletal systems:

Hydrostatic skeleton—This consists of fluid under pressure and is most common in soft, flexible animals such as hydras, planarians, and earthworms and other segmented worms. For example, hydras and planarians have a fluid-filled gastrovascular cavity; the body cavity, or coelom, of an earthworm is also fluid-filled.

Exoskeleton—Many aquatic and certain terrestrial animals have a rigid, hard exoskeleton. For example, mollusks have an exoskeleton made of calcium carbonate that grows with the animal during its entire lifetime. Another type of exoskeleton common among insects and arthropods is made from chitin, a strong flexible type of polysaccharide. While it provides excellent protection and allows for a large variety of movements, it does not grow with the animal. Thus, when an animal outgrows its skeleton, it sheds and replaces it with a larger one in a process known as molting.

Endoskeleton—An endoskeleton consists of bone and cartilage that grows with the animal throughout its life. It stores calcium salts and blood cells and consists of hard or leathery supporting elements situated among the soft tissues of an animal. Although most common among vertebrates, certain invertebrates such as sponges, sea stars, sea urchins, and other echinoderms have an endoskeleton of hard plates beneath their skin. This skeletal system allows for a wider range of movement than the other two skeletal types.

What is chitin?

Chitin—a white, amorphous, semitransparent mass that is insoluble in common solvents like water and alcohol—is found in the exoskeletons of insects and other arthropods. It is referred to as a glucosamine polysaccharide with the formula of $C_{30}H_{50}O_{19}N_4$. The basic units of this substance are linked together by certain reactions that make up long chains; hydrogen bonds link the chains together and help make chitin rigid and strong.

What is arthritis?

The word arthritis is derived from the Greek words *arth* ("joint") and *itis* ("inflammation")—when your joints become inflamed. To humans, it can mean suffering from joint pain and stiffness. More than one hundred types of arthritis exist, some that affect the skin, muscles, bones, and internal organs as well as the joints. Another type of arthritis is rheumatoid arthritis, in which the joint pain is caused by the immune system's attack on the membrane lining the joints. Together these types of arthritis are called rheumatic diseases, and they are considered to be one of the most common chronic health problems for humans.

What is osteoporosis?

Osteoporosis, from the Greek meaning "porous bones," is a disease of the bone that causes an increase in the risk of fractures and breaks. In particular, the bone mineral density goes down, and the reduction causes the bones to deteriorate. In advanced osteoporosis, the bone is literally porous, with small holes altering the bone structure, allowing the bones to break more easily. In women after menopause, osteoporosis is called primary type 1 or postmenopausal osteoporosis; after the age of seventy-five, it is often

How much weight can an ant carry?

Ants are incredibly strong in relation to their size—and they don't even have a skeleton. They can carry objects ten to twenty times their own weight, with some ants able to carry objects up to fifty times their own weight. In fact, ants are able to carry objects over great distances and even climb trees while carrying such objects. In some cases, this is comparable to a 100-pound (90 kilogram) person picking up a small car, carrying it 7 to 8 miles (11 to 13 kilometers) on his back, then climbing the tallest mountain while still carrying the car!

The top part of a turtle shell is called the carapace, and the bottom is the plastron.

called primary type 2 or senile osteoporosis. Overall, more women then men suffer from the disease.

What are the upper and lower shells of a turtle called?

The turtle (order Testudines) uses its shell as a protective device. The upper shell is called the dorsal carapace, and the lower shell is called the ventral plastron; the shell's sections are referred to as the scutes. The carapace and the plastron are joined at the sides of the turtle.

How many vertebrae are in the necks of various mammals?

Not all mammal neck vertebrae (cervical vertebrae) are created equal. Although most mammals have only seven neck bones, the manatee and the two-toed sloth have only six neck bones; a giraffe's long neck has seven neck bones, the same as many other mammals, but the vertebrae are greatly elongated; the ant bear has eight neck bones, and the three-toed sloth has nine neck bones.

PHYSIOLOGY: ANIMAL FUNCTION AND REPRODUCTION

PHYSIOLOGY BASICS

What is physiology?

Physiology is the term used to describe the study of the function of living organisms, or the science that examines the processes and mechanisms by which living organisms (plants and animals; for our discussion in this chapter, we will concentrate on animals) function under various conditions. The term "physiology" was first used by the Greeks as early as 600 B.C.E. to describe a philosophical inquiry into the nature of things. It was not until the sixteenth century that the term was used in reference to vital activities of humans. During the nineteenth century, its usage was expanded to include the study of all living organisms using chemical, physical, and anatomical experimental methods (for more about the anatomy, or physical attributes, of animals, see the chapter "Anatomy: Animals Inside").

Who is considered the "founder of physiology"?

As an experimenter, Claude Bernard (1813–1878) enriched physiology by introducing numerous new concepts into the field. The most famous of these concepts is that of the French *milieu intérieur* or, loosely translated, the body's "internal environment." This means that the complex functions of the various organs are closely interrelated and are all directed to maintaining a constancy of the animal's internal conditions, even with external changes. All internal cells exist in a combination of blood and lymph (thus, aqueous) environment that bathes the cells and provides a medium for the simple exchange of nutrients and waste material.

What is homeostasis, and when was the term first used?

American physiologist Walter Bradford Cannon (1871–1945), who elaborated on Claude Bernard's concept of the *milieu intérieur*, used the term "homeostasis" to describe the body's ability to maintain a relative constancy in its internal environment. It depends on the coordination of thousands of chemical reactions within the cells all at the same time. For example, the maintenance of a human's constant body temperature is a form of homeostasis.

What was the first professional organization of physiologists?

The first organization of physiologists was the Physiological Society, founded in 1876 in England. In 1878, the *Journal of Physiology* began publication as the first journal dedicated to reporting results of research in physiology. The American counterpart, the American Physiological Society, was founded in 1887; their sponsored publication, the *American Journal of Physiology*, was first published in 1898.

ENDOCRINE SYSTEM

Who discovered the first known animal hormone?

The British physiologists William Bayliss (1860–1924) and Ernest Starling (1866–1927) discovered secretin in 1902. They used the term "hormone" (from the Greek word *horman*, meaning "to set in motion") to describe this chemical substance—one that stimulated an animal's organ at a distance from the chemical's site of origin. Their famous experiment using anesthetized dogs demonstrated that diluted hydrochloric acid, mixed with partially digested food, activated a chemical substance in the duodenum. The activated substance secretin was released into the bloodstream and came into contact with cells of the pancreas; in turn, in the pancreas, it stimulated secretion of digestive juice into the intestine through the pancreatic duct.

What is the function of the endocrine system?

The endocrine system is one of two major regulatory systems that release chemicals in the body (the other one is the nervous system; see below). Hormones, or chemicals made and secreted by endocrine glands or neurosecretory cells, are the main messengers of the endocrine system. Hormones are transported in the blood to all parts of the body and interact with target cells (cells that contain hormone receptors), and they regulate metabolic rate, growth, maturation, and reproduction.

What are the "fight-or-flight" hormones?

Epinephrine and norephinephrine are released by the adrenal glands in times of stress. The familiar feelings of a pounding, racing heart, increased respiration, elevated blood pressure, and goosebumps on the skin are responses to stressful circumstances.

What are some vertebrate endocrine glands and their hormones?

Vertebrates have ten major endocrine glands. The following lists those glands, their target tissues, and their principal function in the body:

Endocrine gland; hormone	Target tissue	Principal function
Posterior pituitary		
Antidiuretic hormone (ADH)	Kidneys	Stimulates water reabsorption by kidneys
Oxytocin	Uterus, mammary glands	Stimulates uterine contractions and milk ejection
Anterior pituitary		
Growth hormone (GH)	General	Stimulates growth, especially cell division and bone growth
Adrenocorticotropichormone (ACTH)	Adrenal cortex	Stimulates adrenal cortex
Thyroid-stimulating hormone (TSH)	Thyroid gland	Stimulates thyroid
Luteinizing hormone (LH)	Gonads	Stimulates ovaries and testes
Follicle-stimulating hormone (FSH)	Gonads	Controls egg and sperm production
Prolactin (PRL)	Mammary glands	Stimulates milk production
Melanocyte-stimulating hormone (MSH)	Skin	Regulates skin color in reptiles and amphibians; unknown function in humans
Thyroid		
Calcitonin	Bone	Lowers blood calcium level
Parathyroid		
Parathyroid hormone (PTH)	Bone, kidneys, digestive tract	Raises blood calcium level
Adrenal medulla		
Epinephrine (adrenaline) and norepinephrine (noradrenaline)	Skeletal muscle, cardiac muscle, blood vessels	Initiates stress responses; raises heart rate, blood pressure, metabolic rates; constricts certain blood vessels
Adrenal cortex		
Aldosterone	Kidney tubules	Stimulates kidneys to reabsorb sodium and excrete potassium
Cortisol	General	Increases blood glucose
Pancreas		
Insulin	Liver	Lowers blood glucose level; stimulates formation and storage of glycogen
Glucagon	Liver, adipose tissue	Raises blood glucose level
Ovary		
Estrogens	General; female reproductive structures	Stimulates development of secondary sex characteristics in females and uterine lining

Endocrine gland; hormone	Target tissue	Principal function
Progesterone	Uterus, breasts	Promotes growth of uterine lining; stimulates breast development
Testes		
Androgens (testosterone)	General; male reproductive structures	Stimulates development of male sex organs and spermatogenesis
Pineal gland		
Melatonin	Gonads, pigment cells	Involved in daily and seasonal rhythmic activities (circadian cycles); influences pigmentation in some species

How do steroid hormones differ from nonsteroid hormones?

Steroid hormones such as estrogen and testosterone enter target cells and directly interact with the DNA in the nucleus. Nonsteroid hormones such as adrenaline generally do not enter the target cell but instead bind to a receptor protein found on external cell membranes. This then causes a sequence of metabolic effects.

How are anabolic steroids harmful to those who use them?

Anabolic (protein-building) steroids are drugs that mimic the effects of testosterone and other male sex hormones. They can build muscle tissue, strengthen bone, and speed muscle recovery following exercise or injury. They are sometimes prescribed to treat some types of anemia as well as osteoporosis in postmenopausal women. Anabolic steroids have become a lightning rod of controversy in competitive sports. The drugs are banned from most organized competitions because of the dangers they pose to health and to prevent athletes from gaining an unfair advantage. Adverse effects of anabolic steroids include hypertension, acne, edema, and damage to the liver, heart, and adrenal glands. Psychiatric symptoms can include hallucinations, paranoid delusions, and manic episodes. In men anabolic steroids can cause infertility, impotence, and premature balding. Women can develop masculine characteristics such as excessive hair growth, male-pattern balding, disruption of menstruation, and deepening of the voice. Children and adolescents can develop problems in growing bones, leading to short stature.

What is a goiter?

A goiter is often noticeable on a person's neck—it is an enlargement of the thyroid gland caused by hypothyroidism (too little thyroxin hormone in the person's system). An insufficient dietary intake of iodine is a common cause of goiter, although it may be caused by a cold nodule or tumor on the thyroid, leading to an insufficient amount of thyroxin to enter the system. Goiters are more often treated by supplementation of the thyroid hormone.

What is the difference between Type I and Type II diabetes?

Diabetes mellitus is a hormonal disease that occurs when the body's cells are unable to absorb glucose from the blood. Type I is insulin-dependent diabetes mellitus (IDDM), and Type II is noninsulin-dependent diabetes mellitus (NIDDM). Insulin is completely deficient in Type I diabetes. In Type II diabetes insulin secretion may be normal, but the target cells for insulin are less responsive than normal.

What are neurosecretory cells?

Neurosecretory cells—found in vertebrates and invertebrates—are specialized nerve cells that produce and secrete hormones. Well-known examples of neurosecretory cells are oxytocin- and vasopressin-secreting neurons in the hypothalamus.

What is the hypothalamus, and why is it important to the endocrine and nervous systems?

The hypothalamus is a collection of structures that, as a group, are actually the bridge between the endocrine and nervous systems. For example, it is part of the nervous system when, in times of stress, it sends electrical signals to the adrenal gland to release adrenaline. Then it acts like a nerve when it secretes what are called gonadotropic-releasing hormones from neurosecretory cells (GnRH), thus stimulating the anterior pituitary gland, causing the secretion of follicle-stimulating hormone (FSH; stimulates gonads to produce sperm and ova) and luteinizing hormone (LH; stimulates ovaries and testes). It also acts like an endocrine gland when it produces oxytocin and antidiuretic hormones that it stores in the posterior pituitary gland. Finally, the hypothalamus is actually the place that contains the body's temperature control and the centers that regulate hunger and thirst.

NERVOUS SYSTEM

What is the nervous system?

The nervous system is an intricately organized, interconnected system of nerve cells that relays messages to and from the brain and spinal cord of an organism in vertebrates. It receives sensory input, processes the input, then sends messages to the tissues and organs for an appropriate response. In vertebrates, the nervous system is made up of two parts: the central nervous system, consisting of the brain and spinal cord; and the peripheral system, consisting of nerves that carry signals to and from the central nervous system.

How does the nervous system of invertebrates differ from that of vertebrates?

The least complex nervous system is the nerve net of cnidarians—a network of neurons located throughout the radial symmetrical body of organisms such as hydras (animals that lack a head and a brain). The neurons are in contact with one another and with muscle fibers within epidermal cells. Invertebrates that display bilateral symmetry—such as planarians, annelids, and arthropods—all have a brain (a concentration of neurons at the anterior or head end), one or more nerve cords, and the presence of a central nervous system. In contrast, vertebrates all have a central nervous system and a peripheral nervous system.

What are neurons, nerve dendrites, and axons?

Neurons are specialized cells that produce and conduct "impulses," or nerve signals. Neurons consist of a cell body containing a nucleus and two types of cytoplasmic extensions, dendrites and axons. Dendrites are thin, highly branched extensions that receive signals. Axons are tubular extensions that transmit nerve impulses away

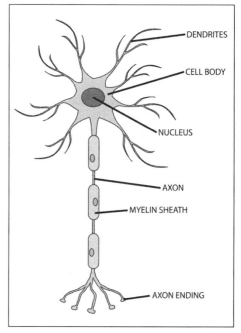

The nerve cell is a specialized cell that is capable of relaying electrical signals from various parts of the body to the brain.

from the cell body, often to another neuron. Other cells in nerve tissue nourish the neurons, insulate the dendrites and axons, and promote quicker transmission of signals. (For more about cells, see the chapter "Cellular Basics.")

How many different types of neurons are found in nerve tissue?

The three main types of neurons are as follows: Sensory neurons conduct impulses from sensory organs (eyes, ears, surface of the skin) into the central nervous system. Motor neurons conduct impulses from the central nervous system to muscles or glands. Interneurons (or association neurons) are neither sensory neurons nor motor neurons. They permit elaborate processing of information to generate complex behaviors and comprise the majority of neurons in the central nervous system.

What is myelin?

Myelin is used for the protection of your nervous system by forming an insulating wrapping around large nerve axons. In the peripheral nervous system, myelin is formed by Schwann cells (a type of supporting cell) that wrap repeatedly around the axon. In the central nervous system, myelin is formed by repeated wrappings of processes of oligodendrocytes (a different type of supporting cell). The process of each cell forms part of the myelin sheath. The space between the myelin from individual Schwann cells is a bare region of the axon called the node of Ranvier. Overall, nerve conduction is faster in myelinated fibers because it jumps from one node of Ranvier to the next (a process called salutatory, or jumping, conduction).

What is demyelination?

Demyelination is the term used for a loss of myelin, a substance in the brain's white matter that insulates nerve endings. Myelin helps the nerves receive and interpret messages from the brain at maximum speed. When nerve endings lose this substance, they cannot function properly, leading to patches of scarring, or "sclerosis." The result may be multiple areas of sclerosis. The damage slows or blocks muscle coordination, visual sensation, and other functions that rely on nerve signals.

What is Guillain-Barrè syndrome?

In the autoimmune disorder known as Guillain-Barrè syndrome, the body's immune system attacks part of the peripheral nervous system. The immune system starts to destroy the myelin sheath that surrounds the axons of many peripheral nerves, or even the axons themselves. The myelin sheath surrounding the axons speeds up the transmission of nerve signals and allows the transmission of signals over long distances. But in diseases such as Guillain-Barrè, in which the peripheral nerves' myelin sheaths are injured or degraded, the nerves cannot transmit signals efficiently. Consequently, muscles begin to lose their ability to respond to the brain's commands, commands that must be carried through the nerve network. The brain also receives fewer sensory signals from the rest of the body, resulting in an inability to feel textures, heat, pain, and other sensations. Alternately, the brain may receive inappropriate signals that result in tingling, "crawling" skin or painful sensations. Because the signals to and from the arms and legs must travel the longest distances, these extremities are most vulnerable to interruption. Although painful in many respects, most patients recover from even the most severe cases of Guillain-Barrè syndrome, although some continue to have a certain degree of weakness.

What is the peripheral nervous system in vertebrates?

The peripheral nervous system has two divisions: the sensory division and the motor division. The sensory division has two sets of neurons: one set (from the eyes, ears, and other external sense organs) brings in information about the outside environment; the other set supplies the central nervous system with information about the body itself, such as the acidity of the blood.

The motor division includes the somatic nervous system (which carries signals to skeletal muscles and skin, mostly in response to external stimuli, and controls voluntary actions); and the neurons of the autonomic nervous system, which are involuntary. This latter system is further divided into the sympathetic division (which prepares the body for intense activities and is responsible for the "fight-or-flight" response) and the parasympathetic division, or "housekeeper system," which is involved in all responses associated with a relaxed state such as digestion.

How are vertebrates' brains organized?

The vertebrate brain is divided into three regions: the hindbrain, the midbrain, and the forebrain, with the size of each region of the brain varying from species to species. The hindbrain may be considered an extension of the spinal cord and coordinates motor reflexes; thus, it is often described as the most primitive portion of the brain. The midbrain is responsible for processing visual information, and finally, the forebrain is the center for processing sensory information in fish, amphibians, reptiles, birds, and mammals.

What is a reflex arc?

A reflex is an involuntary response to a specific stimulus; the most simple nerve response is called a reflex arc—an inborn, automatic, and protective response of humans (and some other organisms). The most well-known example is the knee-jerk response, when a doctor taps gently around your kneecap with a hammer. This reflex is due to

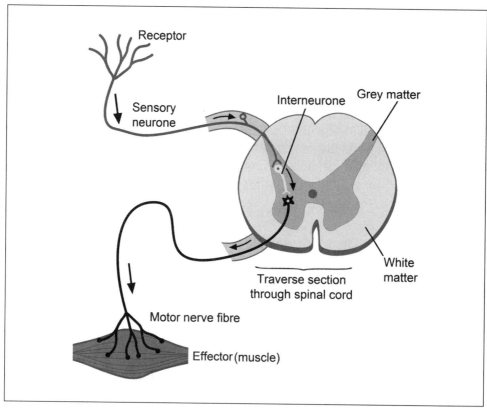

In a reflex arc, a receptor nerve that reacts to some outside stimulation sends a signal to the spinal cord, which is then relayed directly to a motor nerve, which then stimulates a muscle, bypassing the brain completely.

only two neurons—the sensory neuron that, after the tap, sends an impulse to the motor neuron that causes the thigh muscle to contract. An even more complex reflex, for example, when you jerk your hand away from a hot stove, involves three neurons—a sensory neuron transmits an impulse to the interneuron in the spinal cord; that sends an impulse to the brain for processing and one to the motor neuron to the muscle to make a change—in other words, pull your hand away quickly.

Who proposed that the left and right sides of the brain differ in function?

American neuropsychologist and neurobiologist Roger Sperry (1913–1994) conducted the pioneering research on the different functions of the left side and right side of the brain. The left side of the brain controls language, logic, and mathematical abilities. In contrast, the right side of the brain is associated with imagination, spatial perception, artistic and musical abilities, and emotions. Sperry received the Nobel Prize in Physiology or Medicine in 1981 for his work.

What are some diseases that affect the nervous system?

Epilepsy, multiple sclerosis, and Parkinson's disease are only a few of the diseases of the nervous system. The following lists some of the reasons for these diseases:

- *Epilepsy* is a nervous system disorder in which clusters of neurons in the brain sometimes signal abnormally. In epilepsy the normal pattern of neuronal activity becomes disturbed, causing strange sensations, emotions, and behavior or sometimes convulsions, muscle spasms, and loss of consciousness. It is thought that many causes of epilepsy exist. In particular, anything that disturbs the normal pattern of neuron activity—from illness to brain damage to abnormal brain development—can lead to seizures. It may also develop because of an abnormality in brain wiring, an imbalance of nerve signaling chemicals called neurotransmitters, or some combination of these factors.

- *Multiple sclerosis* (MS) is a chronic, potentially debilitating disease that affects the myelin sheath of the central nervous system and is thought to be an autoimmune disease. In MS, the body directs antibodies and white blood cells against proteins in the myelin sheath surrounding nerves in the brain and spinal cord. This causes inflammation and injury to the myelin sheath.

- *Parkinson's disease* is a progressive neurological disorder that results from degeneration of neurons in a region of the brain that controls movement. This degeneration creates a shortage of the brain-signaling chemical (neurotransmitter) known as dopamine, causing the movement impairments—most often shaking of various parts of the body, including the hands and head—that characterize the disease.

What are two of the most common forms of dementia?

The term "dementia" describes a group of symptoms that are caused by changes in brain function. The two most common forms of dementia in older people are Alzheimer's disease and multi-infarct dementia (sometimes called vascular dementia). These types of dementia are irreversible, which means they cannot be cured. In Alzheimer's disease, nerve cell changes in certain parts of the brain result in the death of a large number of cells, with symptoms ranging from mild forgetfulness to serious impairments in thinking, judgment, and the ability to perform daily activities.

In multi-infarct dementia, a series of small strokes or changes in the brain's blood supply may result in the death of brain tissue. The location in the brain where the small strokes occur determines the seriousness of the problem and the symptoms that arise. Symptoms that begin suddenly may be a sign of this kind of dementia. People with multi-infarct dementia are likely to show signs of improvement or remain stable for long periods of time, then quickly develop new symptoms if more strokes occur. In many people with multi-infarct dementia, high blood pressure is to blame.

Do all brain cells work the same way?

Two basic types of cells are in the nervous system: neurons that carry messages and the support cells that both feed and protect them. In the central nervous system—comprised of the brain and spinal cord—these support cells are known as neuroglial or glial cells. These cells perform a variety of functions in maintaining the health of the neurons of the brain. It is estimated that glial cells actually comprise about 90 percent of the cells in the nervous system.

How do neurons carry messages?

Imagine a long trail of dominoes running from your finger to the spinal cord in your back. When you touch a hot object, the messages this action produces in neurons is similar to the reaction that would happen if you knocked over the first domino of the set. After the first domino falls, each domino that follows it would be knocked over in turn. This process is similar to what happens between neurons. The message is carried from the fingertip touching the hot object all the way to the spinal cord. Neurons maintain a concentration of sodium and potassium ions; the openings present in membranes allow these ions to pass through. As the sodium and potassium gates open and then close, the electrical charge of the nerve cell changes and messages are transmitted along the neuron membrane.

How do brain cells store memories?

The part of your brain responsible for processing memory is the hippocampus. It is believed that memories are formed at the level of individual nerve cells. The synapse is the point at which adjoining nerve cells touch, and it is this juncture that is the building block of memory systems. Information moves across the synapse and the information signal is carried inside a cell by a second messenger (known as cyclic AMP), which then activates other cell machinery. The end result is the switching on of a gene that regulates memory. The product of the gene, a protein, promotes synaptic growth and can convert short-term memory to long-term memory.

How is sensory information transmitted to the central nervous system?

Sensory information is transmitted to the central nervous system through a process that includes stimulation, transduction, and transmission. A physical stimulus (for example, light or sound pressure) is converted into nerve cell electrical activity in a process called transduction; the electrical activity is then transmitted as action potentials to the central nervous system.

What are the main types of receptors?

Receptor cells are cells that receive stimuli, with each type of receptor responding to a particular stimulus. The five main types of receptors are as follows:

- *Pain receptors* are probably found in all animals. However, it is difficult to understand nonhuman perception of pain. Pain often indicates danger, and the animal or individual retreats to safety.

- *Thermoreceptors* in the skin are sensitive to changes in temperature. Thermoreceptors in the brain monitor the temperature of the blood to maintain proper body temperature.

- *Mechanoreceptors* are sensitive to touch, pressure, sound waves, and gravity. The sense of hearing relies on mechanoreceptors.

- *Chemoreceptors* are responsible for taste and smell.

- *Electromagnetic receptors* are sensitive to energy of various wavelengths, including electricity, magnetism, and light. The most common types of electromagnetic receptors are photoreceptors that detect light and control vision.

IMMUNE SYSTEM

How does the immune system work?

The immune system has two main components: white blood cells and antibodies circulating in the blood. The antigen-antibody reaction forms the basis for this immunity: When an antigen (antibody generator)—a harmful bacterium, virus, fungus, parasite, or other foreign substance—invades the body, a specific antibody is generated to attack the antigen. The antibody is produced by B lymphocytes (B cells) in the spleen or lymph nodes. An antibody may either destroy the antigen directly or it may "label" it so that a white blood cell (called a macrophage or scavenger cell) can engulf the foreign intruder. After a human has been exposed to an antigen, a later exposure to the same antigen will produce a faster immune system reaction, and thus, the necessary antibodies will be produced more rapidly and in larger amounts. Artificial immunization (such as the vaccine for polio) uses this antigen-antibody reaction to protect the human body from certain diseases by exposing the body to a safe dose of antigen to produce effective antibodies as well as a "readiness" for any future attacks of the harmful antigen.

What are nonspecific defenses?

Nonspecific defenses do not differentiate between various invaders, and depending on the organism, include barriers such as skin, hide, and the mucous membrane lining the respiratory and digestive tracts, phagocytic white blood cells, and certain chemicals. The nonspecific defenses are the first to respond when a foreign substance enters the body.

What are allergies?

Allergies, autoimmune diseases, and immunodeficiency diseases are different kinds of disorders of the immune system. Allergies are abnormal sensitivities to a substance that is harmless to many other people. Common allergens include pollen, certain foods, cosmetics, medications, fungal spores, and insect venom. The antibody immunoglobulin E (IgE) is responsible for most allergic reactions. When exposed to an allergen, IgE antibodies attach themselves to mast cells or basophils. Mast cells are normal body cells that produce histamines and other chemicals. When exposed to the same allergen at a later time, the individual may experience an allergic response when the allergen binds to the antibodies attached to mast cells, causing the cells to release histamine and other inflammatory chemicals. While most allergic reactions are expressed as a runny nose, difficulty in breathing, skin rashes and eruptions, or intestinal discomfort, a severe allergic reaction results in anaphylactic shock.

What are autoimmune diseases?

Autoimmune diseases occur when the immune system rejects the body's own molecules. Insulin-dependent diabetes, rheumatoid arthritis, systemic lupus, and rheumatic fever are autoimmune diseases. In contrast, in immunodeficiency diseases such as AIDS, the immune system is too weak to fight disease.

ANIMAL SENSES

How do animals and people identify smells?

The sense of smell allows animals and humans, as well as other organisms, to identify food, mates, and predators. This sense also provides sensory pleasure (for example, of flowers) and warnings of danger (for example, of chemical dangers). Specialized receptor cells in the nose have proteins that bind chemical odorants and cause the receptor cells to send electrical signals to the olfactory bulb of the brain; from there, the cells in the olfactory bulb relay this information to olfactory areas of the forebrain to generate perception of smells.

And when it comes to the ability to smell, humans are not as good as other animals. For example, a bloodhound has about 300 million scent receptors in its nose; a human has a mere five million. To compare, it is often said the human has the scent receptors the size of a postage stamp, whereas the bloodhound has one the size of a handkerchief.

What animal, insect, and fish have the best sense of smell?

In the animal category, the bear wins in terms of the land animal with the best sense of smell—mainly because its brain is a third of the size of ours, yet the part devoted to smell is five times larger. Bears also have large noses with special interior folds that carry thousands of smell receptors. The giant male silk moths (*Bombyx mori*) are thought to have the best sense of smell in the insect world. Their antennae are covered with about 65,000 tiny bristles, most of which are chemoreceptors. And the winner in the fish category is the shark—a fish that has two-thirds of its brain dedicated to smell—allowing it to detect a drop of blood from more than a mile away.

In the insect world, the giant male silk moth is believed to have the most acute sense of smell.

How are taste and smell related?

By convention, air-breathing vertebrates, including humans, associate taste with materials that come in direct contact with the animal, usually through the mouth. By contrast, smell is associated with substances that reach the animal from a distance, usually through the nose. However, the distinction between the two becomes blurred when considering animals that live in water. Although fish have well-developed chemoreceptors, scientists generally do not refer to the senses of "taste" and "smell" in fish.

What are the three types of photoreceptors among invertebrates?

The three different types of eyes are represented by different types of photoreceptors among invertebrates. They are: 1) eye cup, 2) compound eye, and 3) single-lens eye. The eye cup is a cluster of photoreceptor cells that partially shield adjacent photoreceptor cells. The compound eye consists of many tiny light detectors (photoreceptors). Crayfish, crabs, and nearly all insects have compound eyes. Single-lens eyes, found in cephalopods such as squids and octopi, are similar to cameras. They have a small opening, the pupil, through which light enters.

Do all animals see the same colors?

No, all animals do not see the same colors. The list of animals that can or can't see color is long, but here are some general observations:

- Most reptiles, fish, insects, and birds appear to have a well-developed color sense—and some can see even more colors in the spectrum than other animals. For example, bees and butterflies can see a color range that extends into the

Why do humans have certain-colored eyes?

In 2008, scientists discovered that humans with blue eyes have one single, common ancestor. They believe a genetic mutation took place about 6,000 to 10,000 years ago and caused the blue-eyed people we know today. Originally, all humans had brown eyes, but a genetic mutation of what is called the OCA2 gene in our chromosomes that "turned off" the ability to create brown eyes—not turning it off entirely, but essentially diluting brown eyes to blue. The variation in color is actually due to the amount of melanin (the pigment that gives color to our hair, eyes, and skin) in the eye's iris, with blue-eyed people having a small degree of variation in the amount of melanin in their eyes.

ultraviolet—which allows them to see the special patterns of leaves that guide the insects to the flowers. Birds can see five to seven colors.

- Most mammals are color-blind. Primates are unusual in the mammal world—for example, humans have three different types of visual pigments (called trichromacy). But overall, most mammals have only one visual pigment (such as a few nocturnal animals) or two visual pigments (called dichromats).

- Dogs and cats can see two colors, but weakly; rabbits can see blue and green; and squirrels can see blues and yellows.

Can animals hear different sound frequencies than humans?

The frequency of a sound is the pitch. Frequency is expressed in Hertz (Hz). Sounds are classified as infrasounds (below the range of human hearing), sonic range (within the range of human hearing), and ultrasound (above the range of human hearing).

Animal	Frequency Range Heard (Hz)
Dog	15–50,000
Human	20–20,000
Cat	60–65,000
Dolphin	150–150,000
Bat	1,000–120,000

How do some animals locate the source of a sound?

Locating a source of a sound for many animals is a major necessity—not only for hunting food, but also to avoid becoming prey themselves. For example, cats (and many other animals) have extremely keen hearing, not only in terms of frequency range, but the way they locate a source of sound. One way they determine a sound source is called the interaural time difference (ITD)—the difference in arrival time of a sound wave between a cat's two ears. The cat uses ITD to localize the origin of the sound, but because they have symmetrically placed ears, determining a source may be more difficult, since each ear is about the same height and distance from the nose as the other. To compensate, cats will often tilt their head, altering the position of the ears relative to the origin of the sound, which helps to better localize the sound's origin as the sound waves reach the ears at different times.

Of course, one of the top hunters of the animal world is the owl. They are also one of the best "listeners" and are extremely accurate when locating the source of sound (mostly prey). The asymmetrical location of their ears help them to hear better, with one ear further forward on the head than the other ear. In addition, owls have good eyesight—one of the largest binocular fields of vision in the animal world—with the tilting and bobbing often seen in an owl allowing it to better see, judge the position of, and hunt its prey.

The owl's unique anatomy greatly enhances its ability to quickly locate prey.

Do animals suffer from allergies?

Yes, some animals seem to have allergies. In particular, we know more about domestic than wild animal allergies (although some researchers claim that wild animals would have fewer allergies because of their "wild," more natural diets). For example, veterinarians report that dogs, cats, and even pet birds suffer from allergies. They may be allergic to food, insect bites, dust, household chemicals, or pollen. Instead of having runny noses and watery eyes, most domestic animals experience itchy skin conditions, difficulty in breathing, or disruptions in the digestive tract.

REPRODUCTION

How does asexual reproduction differ from sexual reproduction?

Asexual reproduction produces offspring with the exact genetic material of the parent. Only one individual is needed to produce offspring via asexual reproduction. Sexual reproduction produces offspring by the fusion of two gametes (haploid cells) to form one zygote (diploid cell). The male gamete is the sperm, and the female gamete is the egg.

What are some methods of asexual reproduction?

Budding, fission, and fragmentation are methods of asexual reproduction. In budding a new individual begins as an outgrowth, or bud, of the parent. Eventually, the bud de-

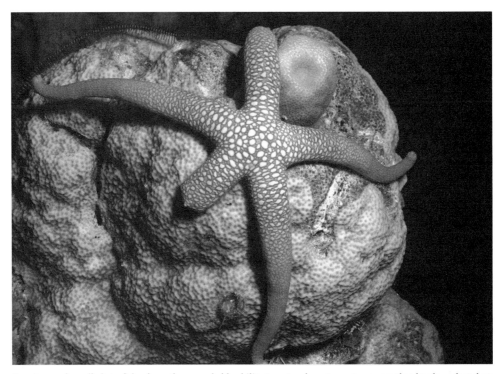

Sea stars—also called starfish—have the remarkable ability to not only regenerate an arm that has been lost, but also to reproduce asexually. A lost arm, if not eaten by a predator, can form an entirely new sea star!

taches from the parent and develops into a new individual. Budding is common among sponges and coelenterates such as hydras and anemones. Fission is the division of one individual into two or more individuals of almost equal size. Each new individual develops into a mature adult. Some corals reproduce by dividing longitudinally into two smaller but complete individuals. Fragmentation is the breaking of the parent into several pieces. It is accompanied by regeneration when each piece develops into a mature individual. Sea stars are well known for reproducing by fragmentation and regeneration. (For more about reproduction processes, see the chapter "Cellular Basics.")

What names are used for juvenile animals?

Along with the names for adult animals—male and female—the young animals also have names. The following lists some of the animal names for their young:

Animal	Name for Young
Ant	Antling
Antelope	Calf, fawn, kid, or yearling
Bear	Cub
Beaver	Kit or kitten
Bird	Nestling

Animal	Name for Young
Bobcat	Kitten or cub
Buffalo	Calf, yearling, or spike-bull
Camel	Calf or colt
Canary	Chick
Caribou	Calf or fawn
Cat	Kit, kitling, kitty, or pussy
Chicken	Chick, chicken, poult, cockerel, or pullet
Chimpanzee	Infant
Cicada	Nymph
Clam	Littleneck
Cod	Codling, scrod, or sprag
Condor	Chick
Cougar	Kitten or cub
Cow	Calf (m. bullcalf; f. heifer)
Coyote	Cub, pup, or puppy
Deer	Fawn
Dog	Whelp
Dove	Pigeon or squab
Duck	Duckling or flapper
Eagle	Eaglet
Eel	Fry or elver
Elephant	Calf
Elk	Calf
Fish	Fry, fingerling, minnow, or spawn
Fly	Grub or maggot
Frog	Polliwog or tadpole
Giraffe	Calf
Goat	Kid
Goose	Gosling
Grouse	Chick, poult, squealer, or cheeper
Horse	Colt, foal, stot, stag, filly, hog-colt, youngster, yearling, or hogget
Kangaroo	Joey
Leopard	Cub
Lion	Shelp, cub, or lionet
Louse	Nit
Mink	Kit or cub
Monkey	Suckling, yearling, or infant
Mosquito	Larva, flapper, wriggler, or wiggler
Muskrat	Kit
Ostrich	Chick
Otter	Pup, kitten, whelp, or cub
Owl	Owlet or howlet
Oyster	Set seed, spat, or brood
Partridge	Cheeper
Pelican	Chick or nestling
Penguin	Fledgling or chick

Animal	Name for Young
Pheasant	Chick or poult
Pigeon	Squab, nestling, or squealer
Quail	Cheeper, chick, or squealer
Rabbit	Kitten or bunny
Raccoon	Kit or cub
Reindeer	Fawn
Rhinoceros	Calf
Sea Lion	Pup
Seal	Whelp, pup, cub, or bachelor
Shark	Cub
Sheep	Lamb, lambkin, shearling, or yearling
Skunk	Kitten
Squirrel	Dray
Swan	Cygnet
Swine	Shoat, trotter, pig, or piglet
Termite	Nymph
Tiger	Whelp or cub
Toad	Tadpole
Turkey	Chick or poult
Turtle	Chicken
Walrus	Cub
Weasel	Kit
Whale	Calf
Wolf	Cub or pup
Woodchuck	Kit or cub
Zebra	Colt or foal

Can certain animals regenerate parts of their bodies?

Regeneration occurs in a wide variety of animals; however, it progressively declines as the animal species becomes more complex. Regeneration frequently occurs among primitive invertebrates. For example, a planarian (flatworm) can split symmetrically, with the two sides turning into clones of one other. In higher invertebrates regeneration occurs in echinoderms such as starfish and arthropods such as insects and crustaceans. Starfish are known for their ability to develop into complete individuals from one cut-off arm. Regeneration of appendages (limbs, wings, and antennae) occurs in insects such as cockroaches, fruit flies, and locusts and in crustaceans such as lobsters, crabs, and crayfish. For example, regeneration of a crayfish's missing claw occurs at its next molt (shedding of its hard cuticle exterior shell/skin in order to grow and the subsequent hardening of a new cuticle exterior). Sometimes the regenerated claw does not achieve the same size of the missing claw. However, after every molt (a process that occurs two to three times a year) the regenerated claw grows and will eventually become nearly as large as the original claw. On a very limited basis, some amphibians and reptiles can replace a lost leg or tail.

What is hermaphroditism?

Hermaphroditic animals have both male and female reproductive systems. Hermaphroditism provides a means for animals to reproduce sexually without finding mates. For example, individuals in many species of tapeworms fertilize their own eggs. In other species such as earthworms, each individual serves as a male and female during mating, both donating and receiving sperm.

What are the differences between external and internal fertilization?

External fertilization is common among aquatic animals, including fish, amphibians, and aquatic invertebrates. Following an elaborate ritual of mating behavior to synchronize the release of eggs and sperm, both males and females deposit their gametes in the water at approximately the same time in close proximity to each other. The water protects the sperm and eggs from drying out. Fertilization occurs when the sperm reach the eggs. Internal fertilization requires that sperm be deposited in or close to the female reproductive tract. It is most common among terrestrial animals that either lay a shelled egg, such as reptiles and birds, or when the embryo develops for a period of time within the female body. For example, certain sharks, skates, and rays have internal fertilization. The pelvic fins are specialized to pass sperm to the female. In most of these species, the embryos develop internally and are born alive.

What animal has the longest gestation period?

Gestation is the period of time between fertilization and birth. The animal with the longest gestation period is not a mammal; it is the viviparous (give birth to live young) amphibian, the alpine black salamander, which can have a gestation period of up to thirty-eight months. It lives at altitudes above 4,600 feet (1,402 meters) in the Swiss Alps and most often has two fully metamorphosed young.

What mammals have the shortest and longest gestation periods?

The shortest gestation period known among mammals is twelve to thirteen days, shared by three marsupials: the American or Virginian opossum (*Didelphis marsupialis*); the rare water opossum, or yapok (*Chironectes minimus*) of central and northern South America; and the eastern native cat (*Dasyurus viverrinus*) of Australia. The young of each of these marsupials are born while still immature and complete their development in the ventral pouch of their mother. While twelve to thirteen days is the average, the gestation period is sometimes as short as eight days. The longest gestation period for a mammal is that of the African elephant (*Loxodonta africana*), with an average of 660 days and a maximum of 760 days.

Why are some animals so brightly colored while others are so dull?

Bright coloration has two general functions. The individual with such coloration is trying to advertise either to members of its own species or to those of other species. Within species, communication revolves around reproductive behavior. For example, male redwinged blackbirds use their brightly colored red-and-white shoulder patches to advertise their territory ownership to potential mates and rivals. An experiment with zebra finches found that females were more likely to choose mates if the identification bands on their legs were red rather than some other color. Across species, communication is usually threatening: Warning (or aposematic) coloration is a method used by animals with stings or poison to circumvent attacks. For example, stinging insects like wasps and bees use similar color patterns of yellow and black to advertise their arsenal. Poison dart frogs (family Dendrobatidae) also use bright coloration in this way. Dull or camouflage coloration (as in a flounder) is an alternative strategy. By hiding, these species hope to avoid predation.

What are examples of courting behavior?

Male stickleback fish swim in a stereotypical manner as they court potential mates. Male bowerbirds build elaborate towers of vegetation to entice females. Female moths release pheromones that attract males from up to a mile away. Male African elephants use low-frequency sounds to find females who are sexually receptive. And, of course, male birds and frogs use vocalization to attract mates.

How do animals show sexual readiness?

Unlike female humans, most female mammals have an estrous cycle instead of a menstrual one. Around the time of ovulation, these females experience estrus, a period of sexual receptivity. Sexual readiness can be advertised in a variety of ways, including physical and behavioral changes. Among females in species with estrus, it is common to see a swelling of the external genitalia. Female chimpanzees and other primates demonstrate this phenomenon. Male mandrills exhibit sexual maturity and dominance by the vibrancy of the coloration on their facial ridges and posterior flesh. When male elephants reach sexual maturity, they experience a condition typified by leakage of fluid from the penis as well as oozing of "tears" from the face (called musth). Of course, members of both sexes are likely to also exhibit changes in behavior, actively seeking out the opposite sex, for example. Cats and dogs in heat will go to extraordinary lengths to meet a potential mate.

What is delayed implantation?

Delayed implantation is a phenomenon that lengthens the gestation period of many mammals. The blastocyst (a stage in embryo development a little beyond fertilization) remains dormant, and its implantation in the uterine wall is postponed for a period of time lasting from a few weeks to several months. Many mammals (including bears, seals, weasels, badgers, bats, and some deer) use this to extend their gestation period through delayed implantation so they give birth at the time of year that offers the best chance of survival for their young.

How many eggs does a spider lay?

The number of eggs varies according to the spider species. Some larger spiders lay more than 2,000 eggs, but many tiny spiders lay one or two and perhaps no more than a dozen during their lifetime. Spiders of average size probably lay a hundred or so. Most spiders lay all their eggs at one time and enclose them in a single egg sac; others lay eggs over a period of time and enclose them in a number of egg sacs.

How many eggs are produced by sea urchins?

The number of eggs produced by sea urchins is enormous. It has been estimated that a female of the genus *Arbacia* contains about eight million eggs. In the much-larger genus *Echinus*, the number reaches twenty million.

Do humans use pheromones in reproduction?

This is a debated question—especially since it has been difficult to determine the existence of pheromones in humans. Scientists have detected that human eggs release a chemical signal that allows them to "communicate" with sperm; it is thought that some human females respond to pheromones with the signals actually coordinating with the menstrual cycle: one increasing the likelihood of ovulation and the second suppressing ovulation. But human pheromones are evasive. One study suggests that over time, humans have lost some of our pheromones because of our ability to see in color. They suggest that when early "humans" developed color vision about twenty-three million years ago, they stopped producing pheromones to attract the opposite sex. In fact, most primates have a gene called TRP2—thought to signal pheromones—but in humans, it is no longer active.

Do animals "marry"?

Although they don't perform ceremonies and have receptions, scientists equate "marrying" with an animal being monogamous, or having only one mate. It is estimated that 90 percent of bird species are monogamous—that is, one male mates with one female to produce offspring. Some of these pair bonds may extend beyond a single mating season, so this could be considered a form of "marriage." The type of pair bonds a species will form is dependent on their ecological niche and is heavily influenced by the needs of their offspring. Altricial offspring require large amounts of parental care for survival (like

---

How can you tell male and female lobsters apart?

The differences between male and female lobsters can only be seen when they are turned on their backs. In the male lobster the two swimmerets (forked appendages used for swimming) nearest the carapace (the solid shell) are hard, sharp, and bony; in the female the same swimmerets are soft and feathery. The female also has a receptacle that appears as a shield wedged between the third pair of walking legs. During mating the male deposits sperm into this receptacle, where it remains for as long as several months until the female uses it to fertilize her eggs as they are laid.

humans) and demand the efforts of two parents; therefore, they are more likely to be found in monogamous species.

How is temperature related to the gender of alligator embryos?

The gender of an alligator is determined by the temperature at which the eggs are incubated. High temperatures of 90 to 93°F (32–34°C) result in males; low temperatures of 82 to 86°F (28–30°C) yield females. This determination takes place during the second and third week of the two-month incubation. Further temperature fluctuations before or after this time do not alter the gender of the young. The heat from the decaying matter on top of the nest incubates the eggs.

The temperature of alligator eggs influences the gender of the hatchlings. Warmer temperatures result in more males being born, and lower temperatures causes more females to develop in the embryonic stage.

What is a mermaid's purse?

Mermaid's purses are the protective cases in which the eggs of dogfish, skates, and rays are released into the environment. The rectangular purse is leathery and has long tendrils streaming from each corner. The tendrils anchor the case to seaweed or rocks, where the case is protected during the six to nine months it takes for the embryos to hatch. Empty cases often wash up on beaches.

What is unique about egg incubation in some amphibians?

Unlike most toads and frogs, the female Surinam toad (*Pipa pipa*) carries her eggs in special pockets in the skin on her back. Each egg develops in its own pocket in the female's skin. The tadpoles' tails are "plugged in" to the mother's system, similar to the placenta of mammals, exchanging nutrients and gases. The tadpoles develop quickly, undergoing metamorphosis while still in the pockets. Upon transformation into miniature frogs, they break free of their pocket walls to begin independent lives.

What is the importance of an external egg in reproduction?

Species that have an external egg usually produce a greater number of zygotes because mating between males and females is not required for successful reproduction. The external egg of most species has a leathery outer covering to prevent desiccation.

Which birds lay the largest and smallest eggs?

The elephant bird (*Aepyornis maximus*), an extinct flightless bird of Madagascar also known as the giant bird or roc, laid the largest known bird eggs. Some of these eggs measured as much as 13.5 inches (34 centimeters) in length and 9.5 inches (24 centimeters) in diameter. The largest egg produced by any living bird is that of the North African ostrich (*Struthio camelus*). The average size is 6 to 8 inches (15–20.5 centimeters) in length and 4 to 6 inches (5–15 centimeters) in diameter. The smallest mature egg, measuring less than 0.39 inch (1 centimeter) in length, is that of the vervain hummingbird (*Mellisuga minima*) of Jamaica. Obviously, the larger the bird, the larger the egg. But when compared with the bird's body size, the ostrich egg is one of the smallest eggs, while the hummingbird's egg is one of the largest. The Kiwi bird of New Zealand lays the largest egg relative to body size of any living bird, with an egg that weighs up to 1 pound (0.5 kilograms).

What is unusual about the way the emperor penguin's eggs are incubated?

Each female emperor penguin (*Aptenodytes forsteri*) lays one large egg. Initially, both sexes share in incubating the egg by carrying it on his or her feet covered with a fold of skin. After a few days of passing the egg back and forth, the female leaves to feed in the open water of the Arctic Ocean. Balancing their eggs on their feet, the male penguins shuffle about the rookery, periodically huddling together for warmth during blizzards and frigid weather. If an egg is inadvertently orphaned, a male with no egg will quickly adopt it. Two months after the female's departure, the chick hatches. The male feeds it with a milky substance he regurgitates until the female returns. Now padded with blubber, the females take over feeding the chicks with fish they have stored in their crops. The females do not return to their mate (and own offspring) but wander from male to male until one allows her to take his chick. It is then the males' turn to feed in open water and restore the fat layer they lost while incubating.

What are some animals with pouches, which are used to carry their young?

Marsupials (meaning "pouched" animals) differ from all other living mammals in their anatomical and physiological features of reproduction. Most female marsupials—kangaroos, bandicoots, wombats, banded anteaters, koalas, opossums, wallabies, Tasmanian devils, etc.—have an abdominal pouch called a marsupium, in which their young are carried. In some small terrestrial marsupials, however, the marsupium is not a true pouch but merely a fold of skin around the mammae (milk nipples).

The short gestation period in marsupials (in comparison to other similarly sized mammals) allows their young to be born in an "undeveloped" state. Consequently, these animals have been viewed as "primitive" or second-class mammals. However, some scientists now see that the reproductive process of marsupials has an advantage over that of placental mammals. A female marsupial invests relatively few resources during the brief gestation period, more so during the lactation (nursing period) when the young are in the marsupium. If the female marsupial loses its young, it can conceive again sooner than a placental mammal in a comparable situation.

Which mammals lay eggs and suckle their young?

The duck-billed platypus (*Ornithorhynchus anatinus*), the short-nosed echidna or spiny anteater (*Tachyglossus aculeatus*), and the long-nosed echidna (*Zaglossus bruijni*)—indigenous to Australia, Tasmania, and New Guinea, respectively—are the only three species of mammals that lay eggs (a non-mammalian feature) but suckle their young (a mammalian feature). These mammals (order Monotremata) resemble reptiles in that they lay rubbery shell-covered eggs that are incubated and hatched outside the mother's body. In addition, they resemble reptiles in their digestive, reproductive, and excretory systems and in a number of anatomical details (eye structure, presence of certain skull bones, pectoral [shoulder] girdle, and rib and vertebral structures). They are, however,

The remarkable platypus, which looks like a combination between a duck and a beaver, is one of only three mammal species that lays eggs.

classed as mammals because they have fur and a four-chambered heart, nurse their young from gland milk, are warm-blooded, and have some mammalian skeletal features.

What names are used for male and female animals?

Numerous names are attached to male and female animals—from domesticated to wild. The following chart lists some of the male and female names for specific animal groups:

Animal	Male name	Female name
Alligator	Bull	Cow
Ant	Drone	Queen
Ass	Jack, jackass	Jenny
Bear	Boar or he-bear	Sow or she-bear
Bee	Drone	queen or queen bee (the worker bees are also females)
Camel	Bull	Cow
Caribou	Bull, stag, or hart	Cow or doe
Cat	Tom, tomcat, gib, pussy, or queen	Tabby, grimalkin, malkin, boarcat, ramcat, or gibeat
Chicken	Rooster, cock, stag,	Hen, partlet, or biddy or chanticleer
Cougar	Tom or lion	Lioness, she-lion, or pantheress
Coyote	Dog	Bitch
Deer	Buck or stag	Doe
Dog	Dog	Bitch
Duck	Drake or stag	Duck
Fox	Fox, dog-fox, stag,	Vixen, bitch, or she-fox reynard, or renard
Giraffe	Bull	Cow
Goat	Buck, billy, billie, or	She-goat, nanny, nannie, billie-goat or nannie-goat
Goose	Gander or stag	Goose or dame
Guinea pig	Boar	Sow

260

Animal	Male name	Female name
Horse	Stallion, stag, horse,	Mare or dam stable horse, sire, or rig
Impala	Ram	Ewe
Kangaroo	Buck	Doe
Leopard	Leopard	Leopardess
Lion	Lion or tom	Lioness or she-lion
Lobster	Cock	Hen
Manatee	Bull	Cow
Mink	Boar	Sow
Moose	Bull	Cow
Mule	Stallion or jackass	She-ass or mare
Ostrich	Cock	Hen
Otter	Dog	Bitch
Ox	Ox, beef, steer, or bullock	Cow or beef
Partridge	Cock	Hen
Peacock	Peacock	Peahen
Pigeon	Cock	Hen
Quail	Cock	Hen
Rabbit	Buck	Doe
Reindeer	Buck	Doe
Robin	Cock	Hen
Seal	Bull	Cow
Sheep	Buck, ram, male-sheep, or mutton	Ewe or dam
Swan	Cob	Pen
Termite	King	Queen
Tiger	Tiger	Tigress
Turkey	Gobbler or tom	Hen
Walrus	Bull	Cow
Whale	Bull	Cow
Woodchuck	He-chuck	She-chuck
Zebra	Stallion	Mare

ANIMAL BEHAVIOR

BEHAVIOR BASICS

What is the definition of animal behavior?

In its broadest sense, animal behavior covers all kinds of movement and responses to environmental changes. It is the term used to describe what an animal does—in particular, it refers to behavior critical for its survival and for the successful production of the animal's offspring.

Who were the first scientists thought to have studied animal behavior?

Aristotle (384–322 B.C.E.) wrote ten volumes on the *Natural History of Animals*. The Roman naturalist Pliny (23–79 C.E.; also called Pliny the Elder) also extensively observed and recorded organisms for his book *Natural History*. In more recent times, naturalist Charles Darwin (1809–1882) recorded (in his journal) the behavior of the marine iguanas of the Galapagos Islands. Darwin also published a book, *The Expression of the Emotions of Man and Animals* (1872), in which he showed how natural selection would favor specialized behavioral patterns for survival.

What is ethology?

Ethology is the study of organism behavior and its relationship to its evolutionary origins. It emerged in Europe in the mid-1930s and was first recognized as a subdiscipline of biology. It differed from traditional biological studies of animals in that scientific principles were applied to the study of animal behavior, with practitioners using both field observations and laboratory experiments.

Who were three important scientists in early ethology studies?

Three scientists who helped establish the field of ethology (and who all shared the Nobel Prize in 1973) were Dutch ethologist (scientist who studies animals' behavior in their natural habitat) and ornithologist Niko Tinbergen (1907–1988), Austrian ethologist and zoologist Konrad Lorenz (1903–1989), and Austrian zoologist, entomologist, and ethnologist Karl von Frisch (1886–1982)—all who lay the four cornerstones of ethological study: causation, development, evolution, and function of behavior. Lorenz studied imprinting behavior of birds, von Frisch worked on the "dance" of the honeybees, and Tinbergen studied aggressive behavior of stickleback fish.

What were three basic concepts in early animal behavior studies?

The three basic concepts in animal behavior were developed by the first three scientists who were foremost in ethological studies: Niko Tinbergen (1907–1988), Konrad Lorenz (1903–1989), and Karl von Frisch (1886–1982).

Fixed Action Pattern—Niko Tinbergen's studies were based on the aggressive threat displays of adult male three-spined stickleback fish (*Gasterosteus aculeatus*). From his work, he developed what is called fixed action pattern (FAP), or an innate behavior pattern that is genetic and thus independent of individual learning and that once begun is carried to completion no matter how useless. It is initiated by an external stimuli (called a sign stimuli), and when the stimuli are exchanged between members of the same species, they are called releasers. In the case of the stickleback fish, the releaser for a territorial attack between two males is the red belly of the intruder—with the fish even attacking a fake fish (in the lab) painted with the color red.

Imprinting—Konrad Lorenz became famous for his work in the field of avian ethology, particularly in his studies of imprinting. By raising goslings from the time they were hatched, Lorenz was able to make the goslings follow him rather than their own mother. This work led to the theory that the goslings were genetically programmed to exhibit a certain behavior with regard to any large organism that was near them during a critical early period of their life.

Communication—Karl von Frisch carried out extensive work on honeybee communication—in particular, the "waggle dance" in bees. He determined the precise pattern of movements done by the worker honeybees when they returned to the hive. From this "dancing communication," the other bees would know the direction and distance of a food source.

Do animals use imprinting?

Imprinting is a prime example of the relative influence of nature (genes) and nurture (environment) on behavior. The organism is born with a sketchy outline of an object (parent, reproductive partner) drawn from its genetic component, and this is then filled in by its experiences in the environment. Therefore, while not all species have been found to meet the scientific definition of imprinting, it is highly likely that all animals will exhibit at least some behaviors that combine genetic and environmental influences.

What does the term umwelt refer to with regard to animal behavior?

The term "umwelt" was first used by Baltic German biologist Jacob von Uexküll (1864–1944) to describe the part of the environment that an animal can actually perceive with its sense organs and nervous system. Animals display this concept in the wild, but the same behavior can be seen with domestic animals. For example, if a stranger comes into your house and tries to pet your dog, this may be perceived as a threat within the dog's umwelt. The practical side of this is that you should be aware of your pet's umwelt when introducing it to new experiences and people.

What is anthropomorphism?

Many humans are guilty of anthropomorphism—attributing human characteristics and feelings to nonhumans. For example, *Bambi* is a well-known story written by writer Felix Salten (1869–1945) in 1923; the inspiration for the story came from the wildlife he saw while vacationing in the Alps. When the story was eventually made into a Disney movie, Bambi had become a talking animal, complete with human feelings and emotions. Anthropomorphism can obscure the true motivation for an animal's behavior—especially if a human actually thinks a wild Kodiak bear cub is just another "cute animal" that acts like the bear cubs portrayed by Hollywood movies or even "faked videos" over the Internet. (In other words, mother bears don't act kindly to anyone getting near their cubs!)

What is sociobiology?

Sociobiology—considered by some as a subdiscipline of behavioral ecology (see below for more about behavioral ecology)—is the study of the social organization of a species. Sociobiology attempts to develop rules that explain the evolution of certain social systems.

Who is Jane Goodall?

British primatologist Jane Goodall (1934–) is famous for her studies of chimpanzees in Tanzania. She began her career as a secretary in Nairobi, Kenya, for anthropologist Louis B. Leakey (1903–1972); she later spent her time studying chimpanzees. After more than forty years of research, Goodall showed that chimpanzees could make and use tools (a behavior previously attributed only to humans). She was also able to distinguish the individual personalities among the chimpanzees she studied. She currently continues her work through the Jane Goodall Institute.

Who was Dian Fossey?

American zoologist Dian Fossey (1932–1985) was an occupational therapist who, inspired by the writings of the American naturalist George Schaller (1933–), decided to study the endangered mountain gorilla of Africa. She was trained in field work by Jane Goodall and went on to watch and record the behavior of mountain gorillas in Zaire and Rwanda. She eventually obtained a Ph.D. in zoology from Cambridge University and in 1983 published a book on her studies, *Gorillas in the Mist*. In 1985, she was found murdered in her cabin in Rwanda; her death is still unsolved. The movie based on Fossey's work, *Gorillas in the Mist*, was released in 1988, with actress Sigourney Weaver playing the role of Dian Fossey. The movie was filmed in Rwanda and Kenya and galvanized support for the plight of the gorillas.

American zoologist Dian Fossey worked to study and protect gorillas in Africa. She was murdered in 1985, and many theorize she died at the hands of poachers who felt she was interfering in their business.

How is animal behavior studied in the field?

Animal behavior is studied by construction of an ethogram—a listing and description of all naturally observed behaviors. Behavior can also be studied through the use of manipulative investigations, both in the field and in the laboratory. In order to be objective, all observers must record behavior patterns in exactly the same way; observations can then be statistically analyzed. For example, animal behavior can be sorted into broad categories (such as courtship and feeding) or into more specific patterns (such as an attack, chase, and/or aggressiveness).

How does an animal's energy level affect its behavior?

Every animal has a finite amount of energy available for use in a unit of time. This energy usage is its metabolic rate (measured in calories or kilocalories) plus the energy required for life activities. The energy budget places limitations on an animal's behavior. For example, an ectothermic (cold-blooded) lizard uses less energy because it does not maintain a constant body temperature; thus, ectotherms such as amphibians and reptiles control body temperature by behavior. Endothermic (warm-blooded) animals like birds and mammals have a higher energy budget, most of which is used to maintain internal body temperature. This all means that endotherms require more energy than ectotherms of a similar size, so they may spend a greater portion of their day searching for food. Other factors that influence energy requirements are age, sex, size, type of diet, activity level, hormonal balance, and time of day. For example, in general, the energy expenditures (in kilocalories/kilogram weight per day) vary depending on the animal—a deer mouse (small mammal) expends 438, a penguin (larger bird) expends 233, a human (large mammal) expends 36.5, and a python (reptile) expends 5.5.

What kinds of behavior do protozoa exhibit?

Protozoa react to changes in their immediate environment, but evidence has not been found that this is learned behavior. For example, paramecia will avoid a strong chemical or physical stimulus by turning to locate an escape route; for instance, cool water cause paramecia to swim away since they prefer warmer temperatures. (For more about protozoa, see the chapter "Bacteria, Viruses, and Protists.")

ANIMAL INSTINCT, LEARNING, AND EMOTIONS

What is learning in the animal world?

Learning in the animal world is considered to be a sophisticated process in which a response or responses become modified from experiences. Learning is actually based on the life span and the complexity of an organism's brain, and if the organism's life span is short, even if they have the brain capacity to learn, learning will not occur. (In such a case, other types of learning take over, such as fixed action pattern [FAP]; see above for more about FAP.)

What is cognition?

Cognition is the highest form of learning and consists of the perception, storage, and processing of information gathered by sensory receptors. Whether or not animals have such mental abilities has long been studied, and is called cognitive ethnology. Long ago, scientists believed that it was the human's use of tools that set us apart, thinking-wise, from other animals. Today, the definition is a bit more detailed and involves complex observations of animals, such as whether animals have memory, how do animals make decisions based on their surroundings, or even how do they "remember," such as a squirrel remembering where he hides nuts in the autumn.

What is associative learning?

Associative learning is a method of learning in an organism in which one stimulus becomes linked to another based on an experience (or experiences). It includes classical conditioning and operant conditioning, or what is also called trial and error learning.

What is an example of classical conditioning?

Russian physiologist Ivan Pavlov's experiment on dogs is one of the most famous examples of classical conditioning. In these well-known investigations, minor surgery was performed on a dog so that its saliva could be measured. From there, the dog was deprived of food, a bell was sounded, and meat powder was placed in the dog's mouth. The meat powder caused the hungry dog to salivate—an example of an unconditioned reflex. However, eventually, after many trials, the dog would salivate at the sound of the bell without meat

Ivan Pavlov (1849–1936) was a Russian physiologist who became famous for his experiments with dogs, in which the animals performed a specific behavior in response to a certain stimulus (see above)—an example of classical conditioning. Although he never thought much of the then fledgling science of psychology, Pavlov's work on conditioned reflexes has been far reaching, from elementary education to adult training programs. Pavlov was awarded the Nobel Prize in Physiology or Medicine (1904) for his study of the physiology of digestion.

powder being offered. This is classical conditioning, also called a conditioned reflex or classical Pavlovian conditioning. In fact, such conditioning is evident every day in many humans' lives, such as advertising campaigns that link an unrelated stimulus with a desired behavior. For example, a beautiful woman in a beer commercial entices men to buy beer, or a commercial of a handsome man influences women to buy a certain perfume.

What scientist started the idea of operant conditioning?

American psychologist B. F. Skinner (1904–1990) extensively studied trial and error learning in animals (later known as operant conditioning). A standard setup for his research involved the following: an animal is placed in a cage (known as a Skinner box; for example, Skinner often used a rat cage) that has a bar or pedal that yields a food reward when pressed. Once the animal has practiced the behavior, it will continue to press the bar repeatedly, having learned to associate this activity with food. By releasing food only when the animal completes some task, the observer can train the subject to perform complex behaviors on demand. These operant conditioning techniques have been used to teach behaviors to various animals, such as training pigeons to play table tennis with their beaks.

What is an example of operant conditioning in the wild?

A good example of the associative learning called operant conditioning in the wild is the interaction between monarch butterflies and certain birds. For example, as part of their life cycle, female monarch butterflies lay their eggs exclusively on milkweed plants. After a few days the eggs hatch and a yellow-, black-, and white-striped caterpillar emerges from each egg. These caterpillars are totally dependent on milkweed plants. Although the plants contain toxic substances (cardenolides) that are poisonous to other animals, the toxin is harmless to the monarch. Blue jays spend much of their day searching for food and will often eat insects such as adult monarch butterflies to supplement their otherwise-vegetarian diet. However, if the food tastes bad, the blue jays will vomit up the food and will then learn individually to avoid the food in the future. Thus, wild monarch butterflies with high levels of cardenolide concentrations are less susceptible to natural predation by birds—in other words, a good example of operant conditioning in the wild.

Does genetics have anything to do with behavior?

Yes, probably in some ways. The first scientific theory is that some scientists believe that all behavior is genetically programmed, and if a behavior is under genetic control, then it should follow a sequence of events—the stimulus, releasing mechanism, and a fixed pattern, all genetically controlled. However, the mainstream scientific view is that most behaviors are a result of a combination of genetic programming and environmental learning—the so-called "nature and nurture."

Who once believed that animal studies could predict human behavior?

Canadian-born, English evolutionary biologist and physiologist George Romanes (1848–1894) was one of the first scientists to investigate the comparative psychology of intelligence. He believed that by studying animal behavior, one could gain insights into human behavior. However, his theory was based on inferences rather than direct observations of comparable behavior.

Are humans the only animals who can think?

Before we can answer this question, we must define what is meant by thought. Thought can be defined in several ways; in particular, it is philosophical rumination and/or the processing of perceptions of the natural world. Because we are still trying to translate animal communication into human language, it is difficult to provide definitive proof of philosophical thought processes. Studies by animal behaviorists suggest that animals with a varied social life (such as chimpanzees) perceive the world in ways similar to hu-

---

What animal's brain was recently studied with an MRI to see what it was thinking?

Trying to determine how an animal thinks has been futile in most cases—especially efforts focused on how to translate their barks, chirps, and sundry other noises. Scientists realized that it would involve scanning the brain of the animal with instruments such as a functional Magnetic Resonance Imaging (fMRI) while performing a cognitive task. And while human brain studies using an MRI are common, such experiments on other animals are not as common. One main reason was because putting an animal under sedation defeats the purpose of checking out their brain in response to certain stimuli. In 2012, scientists reasoned that dogs that underwent extensive training—such as for the Navy Seals—would not be as "wild" as dogs that were not trained as rigorously. They were able to get two well-trained dogs into an MRI and monitor the animals' reactions. They did find a human-canine "connection": After training for specific signals to get a treat, when the dogs saw such a treat signal, the part of the brain called the caudate region showed activity—and in humans, this region is associated with rewards.

mans. However, since we do not share a common verbal language with animals, it is impossible to know what they are truly thinking.

Can animals—and infant humans—recognize different languages?

In the past decade, scientists have tried to understand how and why human babies and other animals develop language. For example, in one experiment, scientists compared language discrimination in human newborns and cotton-top tamarin monkeys. Each group was presented with twenty sentences in Japanese and twenty sentences in Dutch. Infant reactions were gauged by their interest in sucking on a pacifier; when infants first heard sentences in Dutch, they sucked rapidly on their pacifiers, but after a while they grew bored with the Dutch sentences and the rate of sucking slowed. When someone started speaking in Japanese, they showed increased interest by increasing their rate of sucking.

Language discrimination was studied in tamarin monkeys by changes in the facial orientation toward or away from the loudspeaker. Similar to the infant reactions, the monkeys looked at a loudspeaker broadcasting Dutch sentences and looked away when bored. When someone started speaking Japanese sentences, they looked back at the loudspeaker. Results indicate shared sensitivities between monkeys and humans in the ability to discriminate between languages.

Among the invertebrates, which are the most intelligent?

Cephalopods (squid, octopi, and nautilus) are unique among the invertebrates because of their intelligence. For example, octopi can be taught to associate geometric shapes with either punishment (a mild electric shock) or reward (food). This can then be used to train them to avoid one type of food and reach for another. Research has indicated that octopi are also tool users; with their flexible arms and suckers, octopi are able to manipulate their environment, as in building a simple home. After an octopus has selected a home site, it will narrow the entrance size by moving small rocks.

Besides humans, what vertebrates are the most intelligent?

The answer to this question is highly debated—even saying "besides humans" is bound to result in a heated discussion. Many lists of "the most intelligent" animals exist, and some of them include such vertebrates as rats and sheep. But according to the famous American behavioral biologist Edward O. Wilson (1929–), the ten most intelligent animals are as follows: 1) chimpanzee (two species); 2) gorilla; 3) orangutan; 4) baboon (seven species, in-

Cephalopods like this octopus are extremely intelligent. In one experiment, an octopus was able to figure out how to unscrew a lid off a glass jar in order to retrieve a piece of food inside.

cluding drill and mandrill); 5) gibbon (seven species); 6) monkey (many species, especially macaques, the patas, and the Celebes black ape); 7) smaller toothed whale (several species, especially killer whale); 8) dolphin (many of the approximately eighty species); 9) elephant (two species); 10) pig. Still another list based on a scientific study includes 1) dolphins; 2) chimps and orangutans; 3) elephants; 4) parrots; and 5) crows. Like all similar scientific studies, the list of animals not only changes as new information is learned, but also varies depending on the researcher's criteria for "intelligent."

How and when do songbirds learn to sing?

Through analysis of many bird species, ethnologists have found two major types of song development: by imitating the songs of others, particularly adults of the same species, and the invention or improvisation of unique songs. For example, observations of male song sparrows, particularly during their first month of life, show that when the birds arrive at a new habitat, they memorize the songs of the male song sparrows in that neighborhood.

As for when birds begin to sing, it varies depending on the type of bird. Taking the same example of the song sparrow, the males generally learn to sing during a critical period between ten and fifty days after hatching. In some birds such as the mouse wren, the learning period for song development is influenced by the photoperiod (amount of daylight) and social interactions with other adult birds.

Do animals play?

Many animals (mammals and some birds, particularly) have been observed at play during different stages of development. Although play would seem to be random, scientists have described three patterns: 1) social, for establishing relationships with other animals; 2) exercise, for development of muscles; and 3) exploration of an object. All of these occur during the wrestling, chasing, and tumbling activities of the young of many species. Some juvenile play, such as lion cubs capturing mice, may be practice for adult activities such as hunting.

Do animals have emotions?

Many pet owners say that they know when their animal is happy or sad, and evidence has been found to show that animals do exhibit emotion. Researchers have found that emotions are accompanied by biochemical changes in the brain that can be measured. When scientists examine the physiological changes found in humans that correlate with certain emotional states (for example, anger, fear, or lust), they find that these changes can also be observed in certain animal species. For example, a study of stress among African baboons showed that social behavior, personality, and rank within the troop can influence the levels of stress hormones. In addition, field observations have recorded expressions that correlate with pleasure, play, grief, and depression. In one good example, primate ethologist Jane Goodall (1934–), watching the reaction of a young chimp after the death of his mother, maintained that the animal "died of grief." Even with this evidence, it is impossible to truly know how another organism "feels."

271

Who was the first primate taught to use sign language?

Samuel Pepys (1633–1703), famous for his seventeenth-century diary, wrote about what he called a "baboone" and suggested that it might be taught to speak or make signs. And although it was long known that primates use a number of methods of communication in the wild, early attempts (1900–1930s) to teach primates simple words were failures. A 1925 scientific article suggested sign language as an alternative to verbal language in communicating with primates. In the 1960s, researchers tried to teach chimps and gorillas a modified form of sign language. It began with Washoe the chimpanzee, followed by the gorillas Michael (now dead) and Koko. Washoe learned a little over one hundred signs, but Koko has a working vocabulary of over 1,000 signs and understands about 2,000 words of spoken English.

Why do they say elephants have emotions?

Elephants—both African and Asian—are the largest land animals on the planet and are also one of the most intelligent. According to research, they are able to express many of the emotions we usually associate with humans, such as anger, self-awareness, play, joy, grief, and love. For example, the animals have a deep emotional attachment to their family members, grieve for the loss of family members, and are almost "psychologically affected" when a family member is killed, for example, by poachers. One of the main reasons for these emotions—and intelligence—has to do with their brains: Elephants have a large and complex neocortex (part of the brain involved in higher functions, such as sensory perception, spatial reasoning, and conscious thought) similar to other intelligent beings, such as dolphins, apes, and humans. They also have one of the largest hippocampuses (when compared to other animals, such as primates and humans)—the part of the brain linked to emotions through the processing of certain types of memory.

What is habituation?

Habituation is the decreased response to a stimulus that is repeated without reinforcement, which can be very important to an animal in its natural surroundings. For example, young ducklings run for cover when a shadow (a possible predator) passes overhead; gradually, however, the ducklings learn the types of shadows that are dangerous and hopefully which are harmless. This, of course, only applies to ducklings in the wild, not the ones that are fed in parks or your backyard. Because they are used to humans, they may not, unfortunately, respond to a predator's shadow in time.

Why were humans the only primates who learned to speak?

Scientists used to think that apes were not intelligent enough to speak; however, it is now thought that an ape's vocal cords are not "built" for speech. After many years of observation, it is now known that apes do use vocal communication, but it is usually in the form of hoots and grunts, with accompanying gestures.

But apes are not the only primates, and researchers have discovered that most primates lack the vocal anatomy necessary to make more "sophisticated" sounds. But they still wanted to know how humans could have developed speech. In 2012, researchers began to look closer at the sounds made by other primates. One study suggested that human speech may have begun similar to a behavior many primates have called lipsmacking, in which the animals move their jaws, tongues, and lips in much the same way humans do when they talk. In addition, in 2013, researchers studying the gelada, a monkey found in the highlands of Ethiopia and closely related to the baboon, noticed that the animal made a gurgling noise. They believe that the noises made by the monkeys have some speechlike properties—a sound that they describe as a cross between a yodel and a baby's gurgle.

What is cooperation in animals?

Cooperation in animals is a social behavior that allows individuals to carry out a behavior—such as hunting—which will make the group more successful than if they carried out the behavior separately. For example, lions, wild dogs, and coyotes will hunt in a pack; their cooperation allows the prey to be caught more efficiently than if just one animal attacked.

Do animals have friends?

Animals often do form social attachments (friends) among their peers and other animals. For example, among savanna baboons, bonds between males and females are a central feature of a society. Other "friendlike" behaviors have been seen, too: allogrooming is when

Lions are cooperative hunters, taking down large prey by coordinating their efforts. Lionesses do much of the work, but male lions also participate in hunting for their prides.

the animal grooms another animal, usually of the same species. Also, reciprocal altruism is a phenomenon in which an animal helps another unrelated animal. These enduring bonds range from a moose "in love" with a horse to a cat who snuggles with a pet mouse.

What is agonistic behavior in animals?

Agonistic behavior in animals is actually another name for aggressive behavior. It often includes threats or actual combat to settle disputes among individual animals, whether it is about food, mating, shelter, or territory. Animals show aggression through sound (such as growls, barks, trumpeting), sight (changing coloration or inflating body structures), and even scent. They can change the way they move, where they perch, or how much tooth enamel they display. For example, yawning among male mandrills is often not an expression of boredom but rather an opportunity to display their well-honed canine teeth. Or chest-beating in gorillas is part of an aggressive behavior display usually presented by a silverback (male) against unrelated silverbacks. A chest-beating display, accompanied by hoots and barks, may also be used to impress females. Also, in ritualistic or symbolic aggression, the behavior often prevents either animal from serious harm. For example, when a dog shows aggression by baring its teeth, growling, and its back hairs stand on end, they are merely ways of making the dog look "bigger" and more menacing. Most dogs meeting with such a behavior will turn with its tail between its legs, run off, and usually won't approach the other dog again.

Why do animals defend territories?

Animals defend territories to protect their assets such as food, water, or mates. Animals will also defend the area where their offspring live because in evolutionary terms, offspring are the ultimate asset. Animals use pheromones—natural airborne or glandular chemicals that the animal releases—to mark their territory. This can be accomplished by leaving bits of fur, sweat, and dander on a visual object or by scent-marking with small amounts of urine. For example, outdoor cats rub pheromones from their face, paws, and leg glands against trees and other brush to leave a message to other outdoor cats that the area is their territory.

Can animals behave altruistically?

Altruism is the performance of a behavior that will benefit the recipient at a cost to the donor, such as risking one's life to save another. Numerous examples exist of animals exhibiting altruistic behavior. For instance, adult crows may act as "nannies" for other crows, instead of increasing their fitness by producing their own offspring. Another example is a ground squirrel that will warn others of the presence of a predator, even though making such a call may draw the attention of the predator to itself. In studying social insects, American biologist Edward O. Wilson (1929–) found that in many species of social insects, workers forego reproduction entirely (they are sterile) in order to help raise their sisters. This behavior can be explained in two possible ways. The donor is performing the act either in hope that the recipient will someday return the favor (reciprocal altruism) or because the recipient is a family member. In the game of evolution, winners are those who leave the greatest number of copies of their genes in the subsequent generations, so this "kin selection" form of altruism may not be so altruistic, after all.

BEHAVIORAL ECOLOGY

What is behavioral ecology?

Behavioral ecology investigates the relationship between the environment and animal behavior. It emphasizes the evolutionary roots of the behavior, in contrast to the classical studies involving animals in laboratory settings. American biologist George C. Williams (1926–2010), in his book *Adaptation and Natural Selection* (1966), first posed the question as to how behavior affects evolutionary fitness. By showing that behavior is responsive to the environmental forces that drive natural selection (and evolutionary fitness), researchers have demonstrated that the environment plays a crucial role in determining which behaviors are exhibited in natural settings.

Why do animals migrate?

Animals migrate for a variety of reasons, including climate (too hot or too cold during part of the year), food availability (seasonal), and breeding (some animals need a specific environment to lay eggs or give birth). Animals move between locations to take advantage of optimal environments; for example, migrating birds move twice a year, with most species flying great distances to and from their ancestral breeding grounds—most often from cold to warm areas. In the spring in the Northern Hemisphere, birds return to the north to mate, lay eggs, and raise their young; in the fall, shortening days and colder weather trigger hormonal changes that signal to the birds to return to the warmer climates in the south, where food is more plentiful.

When do the swallows return to Capistrano?

As part of their annual migration, thousands of swallows arrive each year around March 19th at the San Juan Capistrano mission in San Juan, California. They depart around October 23rd for their 6,000-mile (9,656-kilometer) flight to Goya, Argentina. March 19th and October 23rd are the feast days for Saints Joseph and San Juan, respectively.

How does a homing pigeon find its way home?

Scientists have several hypotheses to explain the homing flight of pigeons, and both are still debated. One hypothesis involves an "odor map," in which young pigeons learn how to return to their original point of departure by smelling different odors that reach their home in winds from varying directions. They would, for example, learn that a certain odor is carried on winds blowing from the east. If a pigeon were transported eastward, the odor would tell it to fly westward to return home. Another hypothesis proposes that a bird may be able to extract its home's latitude and longitude from the Earth's magnetic field.

People often think of migration in terms of bird behavior, but sea creatures and land animals migrate, too. For example, the wildebeest of East Africa travel each year between the Masai Mara in Kenya and the Serengeti in Tanzania in what is called the Great Migration.

Still another suggestion is that the birds are able to fly home by using learned landmarks. It may be proven in the future that none of these theories explains the pigeon's navigational abilities—that some combination of the theories is the actual mechanism.

What animals make some of the longest migrations?

According to researchers (who actually outfitted a tern with a geolocator tracking device), the arctic tern (*Sterna paradisaea*) has the longest known migration distance, flying from its Greenland breeding grounds to the Weddell Sea on the shores of Antarctic—a round trip that can total as much as 44,000 miles (71,000 kilometers). Running a close second appears to be the Sooty shearwater (*Puffinus griseus*), a bird that migrates nearly 40,000 miles (64,000 kilometers) a year, flying from New Zealand to the North Pacific Ocean every summer in search of food. Other animals whose migration distance varies depending on the species are as follows: For example, gray whales (*Eschrichtius robustus*) migrate the longest of any whale, migrating about 12,500 miles (20,117 kilometers); the caribou (*Rangifer tarandus caribou*) migrates the longest for a land animal—about 700 miles (1,127 kilometers); and the desert locust (*Schistocerca gregaria*) migrates the longest for any insect—about 2,800 miles (4,506 kilometers).

Why do birds flock?

The old adage about "safety in numbers" certainly holds true for flocking birds and other social groups. Flocks, herds, schools, and others provide a refuge for the young and an antipredator advantage for all. In addition, the simultaneous movements of a large number of prey are apt to confuse predators and decrease the likelihood that they will capture any one individual.

Why do birds fly in formation?

Some birds, like Canadian geese and whooping cranes, often fly in V-formation because of the phenomenon called wingtip vortices. The lead bird in the V-formation "breaks up" the wall of air that the flock flies into; the swirling air then helps push along the birds behind. However, being the

Migratory birds like these geese fly in a V-shaped formation in order to reduce drag and conserve energy.

lead bird is hard work, and the leader may then drop back to allow another bird to take on the work of the lead bird. The V-formation also gives these birds the ability to watch each other and communicate (via honking) about likely landing spots.

How do animals know which direction to travel?

The ability to navigate from one place to another is found among a diverse group of species from bats, salmon, locusts, and frogs to even bacteria. Animals use a variety of cues to find their way, including the position of the Sun, Moon, and stars, topographic features of the landscape, or even meteorological cues (for example, prevailing winds). But one idea in particular has gained more verification: animals using magnetic fields. For example, in 2012, scientists found that pigeons have "magnetic compasses" (also called biological compasses) in their brain cells that allow them to detect the Earth's magnetic field and thus determine direction.

Do females of all species take care of their offspring?

While most females of a species are the most common primary caregivers, in some species (for example, the seahorses) males are the primary caregivers for their offspring. Male parental care may be as simple as defending the nest against potential predators, or as time-consuming as providing food and the shelter of their bodies for young hatchlings. Other species have no caregivers—for example, when it comes to guppies, no parental care by either sex is provided.

What is meant by the "alpha male" with animals?

In animal groups with multiple individuals, those at the top of the dominance hierarchy are designated as the alpha male (and alpha female); next in line would be the beta individuals. Alpha individuals control the behavior of the other animals and may be the only individuals that mate within the group. They may also be the decision makers, as in determining which direction the group travels, where the group sleeps, and so forth. For example, wolves and gorillas usually have an alpha male in their group.

What is meant by the phrase "pecking order"?

Pecking order refers to the dominance hierarchy (or the relative ranking) of animals within the same species. For example, flocks of chickens actually do have a pecking order, which is where this term originated. The chickens establish their pecking order by using their beaks, and in most cases, a group of chickens quickly reach an agreement, and thus each hen will know her status within the group. In the case of chickens, "status" determines who gets the best food and treats, and who gets to sleep in the prime spot on the roosts. The phrase has also been adapted in human figures of speech—for example, in business, it often refers to one's position within the company.

Do animals commit murder?

If murder is defined in human terms—the killing of members of the same species—then some species do indeed commit murder. This could be the result of an altercation to determine dominance within the group or a battle over a resource like food or a mate. Animals, including lions and langur monkeys, have also been known to commit infanticide—the killing of infants of their own species; in most of these cases, infanticide has been linked to the arrival of a new alpha male in the group. Scientists surmise that by killing the infants in the group (who were fathered by some other male), the new alpha can bring their mothers into sexual receptivity faster and thus ensure a chance for reproductive success.

How do elephants find each other across the savannah?

Although elephants are well known for the trumpeting calls that they make when angry or disturbed, they are also capable of using ultrasound (sounds above the range of human hearing) and infrasound (sounds below the range of human hearing) to communicate with one another. Researchers have concluded that elephants may be able to hear ultrasonic calls from as far away as 2.5 miles (4 kilometers). In contrast, it is estimated that an infrasonic call by a male elephant could in fact cover an area of 11.6 square miles (30 square kilometers).

Do whales really talk to one another?

Whales produce low-frequency sounds that allow them to communicate across long distances. Research has found that among the fin whales, only males produce these calls. The long, low frequency sounds of male fin whales attract females to patches of food, where mating can then occur. Scientists now realize that this means that the increasing amounts of sonar activity from ships in the ocean may interfere with the ability of these males (and other ocean species) to find mates, which may in turn threaten many species' survival.

Humpback whales, which can grow thirty to forty-five feet in length, migrate between the Caribbean Ocean and off the coasts of Canada and New England in the Atlantic. Like other whales, they communicate using low-frequency sound that can travel for miles.

What is parasitism?

Parasitism is an interaction in which one organism (the parasite) co-opts the resources of another organism (the host). By definition, the host is hurt by the association while the parasite benefits. Parasitism can be physical, like the parasitic worms found in the internal organs of animals, or social, like the brood or nest parasitism found in some birds. In these species, resident birds are tricked into incubating the eggs (and raising the chicks) of interlopers; one good example in the bird world is the cowbird.

What is nest parasitism?

From an evolutionary perspective, if breaking the rules is to your advantage, you should do so—as long as you don't get caught. Taking advantage of another's hard work has been documented in a number of species, most notably the brown-headed cowbird. Cowbirds find food by following large mammals, such as cows, across open country and eating the bugs disturbed by their hooves. Additionally, when it comes time to reproduce, female cowbirds simply fly off into nearby woods and lay their eggs in the nests of other species, which then raise the cowbird offspring with (or instead of) their own.

What is the difference between a territory and a home range?

A territory is a defended area which can be as small as the space around a female red-winged blackbird's nest or as large as the backyard that your dog defends. A home range, in contrast, is simply the area where an animal spends its time. Home ranges may be shared with members of the same species and may overlap those of other species. For example, the

Pufferfish are any of a number of species found in warm seas that use a special adaptation of the gullet to inflate their bodies to nearly twice the normal size. Pufferfish, blowfish, and similar animals do this in response to a perceived threat. The increased size and unpalatable-looking spines make the potential prey look quite unappetizing to predators. One of these fish, the *Fugu rubripes*, or the Japanese pufferfish, is a specialty in sushi restaurants—but only specially trained chefs can safely prepare the fish for consumption by minimizing the presence of the fish's deadly toxin, called tetrodotoxin, a compound 1,000 times deadlier than cyanide.

home range of an adult polar bear may cover an area of 20,000 square miles (50,000 square kilometers), or an area about the size of Nova Scotia in Canada. The home range of each polar bear varies due to food availability and condition of the ice—a factor that is currently changing rapidly because of global climate change and the disappearance of Arctic ice.

Why do animals pretend to be hurt?

Among bird watchers, the female killdeer is well known for pretending to be hurt, especially when a potential predator appears within the vicinity of her nest. In an act worthy of the human stage, the killdeer will adopt a posture of wing-dragging, making it appear that she is injured and an easy catch. The female will gradually lead the predator away from her nest—usually filled with eggs or hatchlings—eventually flying off when the predator is at a safe distance from the nest. Another pretender is the common or Virginia opossum (*Didelphis virginiana*)—the only marsupial in the southern and eastern United States. When threatened or frightened, the opossum will lie quite still, with stiffened limbs and a fixed gaze, appearing as if dead. When the perceived threat is over, the opossum will return to its normal, nocturnal activities.

Who studied fly behavior?

American physiologist and entomologist Vincent Dethier (1914–1993) spent most of his life researching chemical responses in insects. He wrote numerous articles and books, both for the general public and specifically for children. *To Know a Fly*, perhaps his most famous book, is considered one of the classic books in entomology. The book details physiological research on flies, particularly with regard to chemoreception (how an organism responds to a chemical stimulus, most often through taste or smell)—all with a touch of humor.

Do animal societies have a culture?

Culture can be defined as the set of societal rules that are passed from one generation to the next, with the parents or caregivers teaching juveniles what they need to know in order to participate in their society. Animals sometimes exhibit "culture"; for example, elephant families rely on memory and the knowledge of the matriarch (oldest female) to respond to social cues. She controls the direction in which the herd moves and where and for how long feeding occurs, and when danger strikes, the other herd members cluster around the matriarch. Perhaps most striking is the tender loving care lavished on young elephants. The bond between mother and daughter elephants lasts up to fifty years. (For more about elephants and emotions, see this chapter.)

How does an animal's expression reflect behavior?

Although only primates have the facial musculature to be truly expressive, many species use their appearance to convey information. For example, pufferfish expand their bodies to look more threatening, as do dogs and cats when they raise their hackles and puff up their fur. Cats, horses, and others also use their ears to relay information about their intentions. And certainly, many a pet owner has learned to read his dog's "mood" by the movement of the animal's tail.

Why do some animals hunt only at night?

In the arms race between predator and prey, a change in the behavior of one species can drive adaptation in another (coevolution). Many rodent species, for example, are adapted to life at night, but their predators have gained adaptations to night work as well. Owls and foxes have special adaptations that allow them to hunt at night for nocturnally active rodents, such as field mice, moles, and voles.

Do animals laugh?

So far, no example of animal laughter comparable to that of humans has been documented. However, researchers have reported that under certain conditions some species emit special vocalizations that could be considered akin to laughter. In one specific case, a researcher has been able to identify a huffing sound unique to dogs at play. Scientists studying the behavior of nonhuman primates and rats have also reported similar observations.

Do animals cry?

Crying in response to emotional distress has been documented only in humans. However, many animals, particularly the young, demonstrate their response to distress by changes in vocalizations and movement.

Do animals have emotions? Many pet owners will swear that when their dogs play they are smiling and experiencing joy. Researchers believe that the "huffing" sound dogs make when playing is a canine version of laughter.

For example, many of us have heard a newborn kitten calling for its mother, or a hatchling bird calling to its parent when it wants food. Whether these are actual "crying babies" is a matter of interpretation—they may be calls for food or warmth. While tear production (the lacrimal response) is found in a number of animals (but not crocodiles!), the tears are used to maintain the cleanliness and moisture of the eyes and not to display emotion.

Why do some animals only have a few offspring?

Females have a finite amount of energy that they can allocate to the growth and maintenance of offspring. Therefore, two basic strategies have evolved for successful reproduction: The animal produces many small offspring, only some of whom may survive; or they produce just a few larger offspring, each with a higher likelihood of survival.

What is polygyny?

Polygyny is a type of mating system that involves pair bonds between one male and several females. While this may at first seem unfair to females, in a polygynous system every female can usually have a mate, while only the most desirable males will find a mate. In human societies, polygyny is actually the most common type of mating system. This is reinforced by genetic data—it suggests that the Y chromosome sequences vary little within human populations. Thus, relatively few men have contributed a larger fraction of their chromosome pool in every generation.

What is polyandry?

Polyandry is the opposite of polygyny; that is, one female having several male mates. It is relatively rare in nature, but one notable example would be bees, where one queen in a hive mates with several males.

What is the role of the queen bee in a honeybee society?

A honeybee society is made up of a single queen, a few male drones, and up to 80,000 female workers. The queen is the only female in the hive capable of reproduction, and

How do bats hunt?

Bats hunt a variety of prey. Some bats are insectivores, capable of catching their prey on the wing. Vampire bats hunt for large, slow-moving organisms to use for sustenance. Fruit-eating (frugivorous) bats are adapted to finding ripe fruit distributed throughout the forest. Depending on the food type, bats may rely on their sight and/or echolocation to find their prey. In echolocation, bats emit ultrasonic sounds that bounce off the objects around them. These signals are fielded by the specialized folds of flesh on the face and around the ears, allowing the bat to judge the direction and distance of objects. (For more about bats, see the chapter "Aquatic and Land Animal Diversity.")

she can lay as many as 1,000 eggs in a single day. While the queen's job is solely to lay eggs, the workers' jobs change over time. Young workers serve as nurses to the larvae that develop from the eggs; older workers serve first as hive builders and then as foragers. And the males? Their job is to fertilize the new queens.

How do animals know what prey to eat?

Once an animal has located a potential food, it must decide whether to consume or ignore it. In field studies, behavioral ecologists have observed that animals usually select food that will yield the highest rate of energy return for the energy spent "capturing" the food. Very few animals actually eat all of the food they are capable of consuming. This is known as optimal foraging strategy. As an example, crows that live in the Pacific Northwest often find littleneck clams, which they drop on rocks to crack the clams and then eat them. However, the crows do not eat all the clams they locate; they only eat those clams that are larger and thus contain more energy.

How is an ant colony organized?

Ants are the dominant social insects and, numerically, the most abundant. At any one time, at least 10^{15} (one quadrillion) ants are alive on the planet! In general, ant colonies contain a variety of castes (groups of individuals with a common job) like workers or soldiers, plus a queen who is in charge of egg-laying. For example, the largest and most aggressive workers make up the soldier caste, whose job it is to protect the colony against dangerous invaders. Because of the way that sex is determined among ants, males are haploid (contain one set of chromosomes) while females are diploid (contain two sets of chromosomes). This causes a close interrelatedness among members of a colony and is theorized as one of the reasons for the evolution of social colonies.

How does social deprivation affect animals?

The effect of a lack of parental care on the social development of young monkeys was studied by American psychologist Harry Harlow (1905–1981) and his colleagues beginning in the 1950s. In a now classic experiment, Harlow was able to show that the mother-infant bond was so important to young rhesus monkeys that the infants preferred a soft cuddly fake mother to a fake mother built from wire, even if it had a nursing bottle attached. Depending on the age of the monkey and the duration of the treatment (total isolation, isolation with fake mother, and so on), monkeys in

Even though each individual ant is far from intelligent, they are all very good at their individually assigned tasks. Thousands of ants in a colony function together as one living organism in a highly efficient system.

these studies later exhibited a range of behavioral deficits, including rocking and swaying, poor maternal behavior, and a failure in understanding communication signals from other monkeys.

How does an animal's niche affect its behavior?

A niche can be defined as the natural history of a species, where it lives, what it eats, and so forth. Those constraints in turn have an effect on behavior. For example, lizard species found in deserts from different parts of the world use a very similar behavior to harvest water from the moist air in the early morning. They use the bumps or ridges on their heads as condensing spots and then orient their bodies so that the water rolls toward the snout, where it can be licked off by the tongue.

Why do so many round animals seem to show only simple behaviors?

Examples of "round" animals include members of the phylum Cnidaria (hydras, jellyfish, corals) and phylum Echinodermata (starfish, sea urchins, sand dollars). An animal with radial symmetry usually has a nerve net that only allows very simple types of behavior. Animals that are round are usually sessile (nonmoving). This is in contrast to animals that display bilateral symmetry, which have a distinct head/tail and in which the animals can be divided into different planes. Bilaterally symmetrical animals usually move in a specific direction. (For more about animal symmetry, see the chapter "Aquatic and Land Animal Diversity.")

What is a biological clock?

A biological clock controls a biological rhythm; it involves an internal pacemaker with external (usually environmental) cues. An environmental signal that cues the clock for animals is called a *zeitgeber*, a German term meaning "time-giver." Examples of zeitgebers include light and dark cycles, high and low tides, temperature, and food availability. A biological rhythm is a biological event or function that is repeated over time in the same order and with a specific interval. Biological rhythms are evident when an animal's behavior can be directly correlated to certain environmental features that occur at a distinct frequency. Biological clocks control animal behaviors such as when migration, mating, sleep, hibernation, and eating occur. Some examples of biological rhythms: A tidal rhythm would include the oysters (feeding) and fiddler crab (mating and egg laying)—12.4 hours; circadian rhythm would include the fruit fly (an adult emerging from a pupa)—twenty-four hours; a circannual rhythm, such as a woodchuck coming out of hibernation or a robin migrating or mating—twelve months; or an intermittent rhythm, such as a lion that needs to feed because it's hungry, or a river fish called a shiner that reproduces when the river floods—going from days to several years.

How do animals recognize each other?

We know that animals can use scent, color, and sound to recognize individuals, and they may also be able to recognize other attributes as well. A recent study on sheep intelligence indicates that easily herded animals may be smarter than originally thought. The sheep were shown pictures of other sheep and were subsequently rewarded if they moved toward a selected picture—in other words, the sheep learned which face produced a reward. Ultimately, it was shown that sheep were able to pick a selected picture 80 percent of the time and could remember up to fifty images for two years.

Can animals use tools?

A tool can be defined as any object used by an animal to perform a specific task. Chimpanzees carefully select twigs that they then prepare as probes to fish out termites from mounds. Sea otters use rocks to crack open clamshells. Birds will drop clams onto rocks to crack their shells, and crows will use sticks to pry out insects from a hole. Japanese macaques use the sea to wash sand off food items.

What is hibernation?

Hibernation (from the Latin term *hiberna*, meaning "winter") is a period of dormancy practiced by animals to overcome wintry environmental conditions. Hibernation involves a decrease in metabolic rate (the rate of burning calories), heart rate, respiration, and other functions (such as urine production and rate of digestion). These rates dive so low that the animal's body temperature approaches that of its surroundings. Small animals whose increased metabolic rate forces them to find an alternative to starving during the winter months are more likely to hibernate than larger animals.

Alpine marmots, which live in Europe, hibernate up to eight months every year!

285

What are some "different" hibernating animals?

Several animals hibernate, or become inactive in a cold climate—when animals become inactive in a hot climate, it is called estivation (see below). It is interesting to note, too, that hibernation in reptiles (such as the garter snake below) is called brumation, and differs from hibernation in mammals (like the alpine marmots below) because the reptile is not living off its fat reserves like mammals. The following lists some interesting examples of hibernators:

- Alpine marmots (*Marmota marmota*)—The mammals called alpine marmots are found in the mountain regions of central and southern Europe and usually hibernate for up to eight months (and are mating and having babies the other four months). During their hibernation, their heartbeats slow from about 120 per minute to around five beats per minute—and they take a mere one to three breaths per minute (the rates may increase if the temperatures reach below freezing).

- The Common Poorwill (*Phalaenoptilus nuttallii*) is the only known hibernating bird; it lowers its body temperature and stays in logs or a rock overhang for up to five months. But it only takes around seven hours for the bird to come back to its normal temperature.

- The garter snake—a collection of many different types, for example, the Eastern Garter Snake (*Thamnophis sirtalis sirtalis*)—are found in colder northern regions of North America. In the winter, their metabolism slows down so the snakes hardly use any energy over the winter; the snakes have relatively nothing to eat, so they have to conserve their energy. The majority of the time garter snakes brumate not as solitary creatures, but in groups—with some of the best sites (usually below the frost line) containing hundreds, if not thousands, of snakes.

Why is brown fat important to some animals that hibernate?

Many mammals have both brown and white adipose tissues (cells that store a large droplet of fat that swells when fat is stored and shrinks when fat is used for energy). They are both called triglyceride lipids, but brown fat tissue has the ability to generate heat. Although called brown, brown adipose tissue varies in color from dark red to tan, reflecting its lipid content. It is most commonly found in newborn animals and in most species, disappears by adulthood. But it does have some use, especially to mammalian hibernators: These animals have exceptionally well-developed brown fat (in fact, some scientists even refer to it as "the hibernation gland"). The supply of brown fat built up during the spring and summer and is then used during the winter months when the animal hibernates; thus, brown fat becomes an important tissue in the rewarming process that the animal undergoes after hibernation.

Do bears in nature—or in a zoo—hibernate?

As for bears, while they are certainly less active during the winter, they do not truly hibernate. Instead, they take very long naps known as "winter sleep." In winter conditions

What primate was recently discovered hibernating?

Most primates are not known for their hibernation habits—except for the western fat-tailed lemur that hibernates for seven months in a tree hole. But in 2013, scientists discovered two new lemurs in Madagascar that did hibernate—Crossley's dwarf lemur and the Sibree's dwarf lemur. Unlike more hibernators that lower their body temperatures and hide in special hidden spots in the cold winter, the western fat-tailed lemurs actually hibernate from the cold *and* the heat, with temperatures that can reach 85°F (29.4°C) over a long, dry season—when food and water are in short supply.

But the two new lemurs live in the high-altitude forests, where it does go below freezing. The scientists found that although the animals burrow underground and breathe once every several minutes, their temperature does not drop, remaining constant. This may indicate that primates as hibernators are more prevalent than we think.

they find protected areas like a cave or hollow log and conserve energy by taking extended naps. This is why it is always dangerous to disturb a wintering bear, because it is merely napping, not hibernating. In zoos, because temperatures in cages and enclosures remain warm throughout the year and the bears have access to a food supply continually replenished by keepers, bears remain active year-round.

Why do animals pace in zoos?

Pacing is an indication of lack of stimulation. A recent doctoral study found that larger animals, which have a larger home range in the wild, are particularly prone to pacing in captivity, a behavior known as cage stereotypy. Most reputable zoos attempt to assuage pacing by providing what is known as "animal enrichment." Enrichment may include constantly changing objects, odors, and sounds to mimic the types of stimulations that would be found in the animal's natural habitat. As an example, in order to keep life interesting for poison dart frogs from South America, keepers will hide their crickets inside a coconut with holes drilled in it. The frogs then have to seek out the food inside the chirping coconut.

What is estivation?

Estivation (from the Latin term *aestas*, meaning "summer") is a process by which animals become dormant during the summer rather than the winter. Estivation may be used as a survival strategy against intense heat (ground squirrels), drought (snails), or both. The Columbian ground squirrel both hibernates and estivates, beginning its dormant period in the late summer and continuing it until the next May.

What is torpor?

Torpor is a short-term decrease in body temperature and metabolic rate. Animals such as hummingbirds and bats go through daily periods of torpor that allow them to reduce their energy requirements at night or when hunting is poor. Torpor can be considered as a type of brief sleep, but it is distinct from a state of hibernation.

BEHAVIOR OF ANIMALS IN MOTION

What are the problems an animal must overcome to move?

In contrast to other organisms, animals are able to move. The two forces an animal overcomes to move are gravity and friction. Aquatic animals do not have much difficulty overcoming gravity, since they are buoyant in water. However, because water is dense, the problem of resistance (friction) is greater for these animals. Many of them have sleek shapes to help them swim. Terrestrial animals tend to have fewer problems with friction since air has less resistance than water. However, terrestrial animals must work harder to overcome gravity.

Which animals can run faster than a human?

The cheetah, the fastest mammal, can accelerate from 0 to 45 miles per hour (mph; 0–64 kilometers per hour [kph]) in 2 seconds; it has been timed at speeds of 70 mph (112 kph) over short distances. In most chases, cheetahs average around 40 mph (63 kph); although they have been clocked at up to about 68 mph (110 kph). Humans can run very short distances at almost 28 mph (45 kph) maximum. Most of the speeds given in

The cheetah has been clocked running over 70 mph (112 kph) over short distances.

the table below are for distances of .25 miles (0.4 kilometers). The following lists animals and their estimated maximum speeds in miles and kilometers per hour:

Animal	Maximum speed (miles per hour)	Maximum speed (kilometers per hour)
Cheetah	68	110
Pronghorn antelope	61	98.1
Wildebeest	50	80.5
Lion	50	80.5
Thomson's gazelle	50	80.5
Quarter horse	47.5	76.4
Elk	45	72.4
Cape hunting dog	45	72.4
Coyote	43	69.2
Gray fox	42	67.6
Hyena	40	64.4
Zebra	40	64.4
Mongolian wild ass	40	64.4
Greyhound	39.4	63.3
Whippet	35.5	57.1
Rabbit (domestic)	35	56.3
Mule deer	35	56.3
Jackal	35	56.3
Reindeer	32	51.3
Giraffe	32	51.3
White-tailed deer	30	48.3
Wart hog	30	48.3
Grizzly bear	30	48.3
Cat (domestic)	30	48.3
Human	27.9	44.9

How do fleas jump so far?

The jumping power of fleas comes both from strong leg muscles and from pads of a rubberlike protein called resilin. The resilin is located above the flea's hind legs. To jump, the flea crouches, squeezing the resilin, then it relaxes certain muscles. Stored energy from the resilin works like a spring, launching the flea. A flea can jump well both vertically and horizontally, which is why they are able to attach themselves so readily to not only our pets, but to us. In fact, some species can jump 150 times their own length! To match that

The flea's remarkable hind legs can produce a powerful burst of energy by using a rubber-like protein called resilin.

record, a human would have to spring across the length of two and a quarter football fields—the height of a one hundred-story building—in a single bound. The common flea (*Pulex irritans*) has been known to jump 13 inches (33 centimeters) in length and 7.25 inches (18.4 centimeters) in height.

What causes the Mexican jumping bean to move?

The bean moth (*Carpocapa saltitans*) lays its eggs in the flower or in the seed pod of the spurge, a bush known as *Euphorbia sebastiana*. The egg hatches inside the seed pod, producing a larva or caterpillar. The jumping of the bean is caused by the active shifting of weight inside the shell as the caterpillar moves. The jumps of the bean are stimulated by sunshine or by heat from the palm of the hand.

How fast do fish swim?

The maximum swimming speed of a fish is somewhat determined by the shape of its body and tail and by its internal temperature. The cosmopolitan sailfish (*Istiophorus platypterus*) is considered to be the fastest fish species, at least for short distances, swimming at greater than 60 miles per hour (95 kilometers per hour). Some American fishermen, however, believe that the bluefin tuna (*Thunnus thynnus*) is the fastest, but the fastest speed recorded for this species so far is 43.4 miles per hour (69.8 kilometers per hour). Data is extremely difficult to secure because of the practical difficulties in measuring the speeds. The yellowfin tuna (*Thunnus albacares*) and the wahoe (*Acanthocybium solandri*) are also fast, timed at 46.35 miles per hour (74.5 kilometers per hour) and 47.88 miles per hour (77 kilometers per hour) respectively during 10- to 20-second sprints. Flying fish swim at 40-plus miles per hour (64-plus kilometers per hour), dolphins at 37 miles per hour (60 kilometers per hour), trout at 15 miles per hour (24 kilometers per hour)—and humans can swim 5.19 miles per hour (8.3 kilometers per hour).

What is the fastest snake on land?

The black mamba (*Dendroaspis polylepis*), a deadly poisonous African snake that can grow up to 13 feet (4 meters) in length, has been recorded reaching a speed of 7 miles per hour (11 kilometers per hour). A particularly aggressive snake, it chases animals at high speeds, holding the front of its body above the ground.

Do all birds fly?

No, and among the flightless birds, the penguins and the ratites are the best known. Ratites include emus, kiwis, ostriches, rheas, and cassowaries—they are called ratites because they lack a keel on the breastbone. On the other hand, although penguins cannot fly, they really don't need to, because some types, such as the Adelie, can reach the remarkable speeds of 40 to 100 miles (64–161 kilometers) per hour in the oceans, outswimming some smaller and larger fish. All of these birds have wings but lost their power to fly millions of years ago. Many birds that live isolated on oceanic islands, such as the great auk, apparently became flightless in the absence of predators because they never needed their wings to escape.

How fast do some birds fly?

Hummingbirds fly at speeds up to 71 mph (114 kph). Small species beat their wings fifty to eighty times per second, and higher in courtship displays. The following lists some birds and their fastest flights:

Bird	Speed (miles per hour)	Speed (kilometers per hour)
Peregrine falcon	168–217	270.3–349.1
Swift	105.6	169.9
Merganser	65	104.6
Golden plover	50–70	80.5–112.6
Mallard	40.6	65.3
Wandering albatross	33.6	54.1
Carrion crow	31.3	50.4
Herring gull	22.3–24.6	35.9–39.6
House sparrow	17.9–31.3	28.8–50.4
Woodcock	5	8

How far do some types of hummingbird migrate?

The longest migratory flight of a hummingbird documented to date is the flight of a rufous hummingbird from Ramsey Canyon, Arizona, to near Mt. Saint Helens, Washington, a distance of 1,414 miles (2,277 kilometers). Hummingbird studies, however, are difficult to complete because so few banded birds are recovered.

How fast does a hummingbird's wings move?

Hummingbirds are the only family of birds that can truly hover in still air for any length of time (although such birds as Kestrels do hover, it is not for as long). They need to do so in order to hang in front of a flower while they perform the delicate task of inserting their slim sharp bills into the flower's depths to drink nectar. Their thin wings are not contoured into the shape of aero-foils, so they do not generate lift in this way. Instead, their pad-

The remarkable hummingbird can hover in place as it flaps its wings dozens of times per second. The ruby-throated hummingbird (pictured here) flaps its wings 50 times per second, while the bee hummingbird can beat its wings an astonishing 200 times per second.

dle-shaped wings are, in effect, hands that swivel at the shoulder. They beat them in such a way that the tip of each wing follows the line of a figure-eight lying on its side. The wing moves forward and downward into the front loop of the eight, creating lift. As it begins to come up and goes back, the wing twists through 180° so that once again it creates a downward thrust. The hummingbird's method of flying does have a major limitation: The smaller a wing, the faster it has to beat in order to produce sufficient downward thrust. An average-sized hummingbird beats its wings twenty-five times a second. The bee hummingbird, native to Cuba, is only 2 inches (5 centimeters) long—it beats its wings at an astonishing 200 times a second.

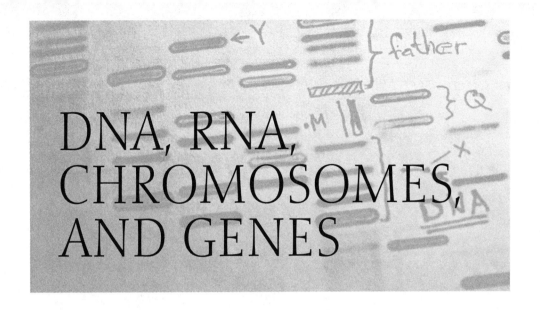

DNA, RNA, CHROMOSOMES, AND GENES

HISTORY OF NUCLEIC ACIDS

What are nucleic acids?

DNA (deoxyribonucleic acid) and RNA (ribonucleic acid) are nucleic acids. They are molecules comprised of monomers (structural unit of a polymer) known as nucleotides. These molecules may be relatively small (as in the case of certain kinds of RNA) or quite large (a single DNA strand may have millions of monomer units). Individual nucleotides and their derivatives are important in living organisms. For example, ATP, the molecule that transfers energy in cells, is built from a nucleotide, as are a number of other molecules crucial to metabolism.

Which came first—DNA or RNA?

The first molecule had to be able to reproduce itself and carry out tasks similar to those done by proteins. However, proteins, even though bigger and more complicated than DNA, can't make copies of themselves without the help of DNA and RNA. Therefore, RNA was the likely candidate as the first "information" molecule—mainly because scientists have found that RNA, unlike DNA, can replicate and then self-edit.

What is the difference between DNA and RNA?

DNA and RNA are both nucleic acids formed from a repetition of the simple building blocks called nucleotides. A nucleotide consists of a phosphate (PO_4), sugar, and a nitrogen base, of which five types exist: adenine (A), thymine (T), guanine (G), cytosine (C), and uracil (U). In a DNA molecule, this basic unit is repeated in a double helix structure made from two chains of nucleotides linked between the bases; these links are either between A and T or between G and C. (The structure of the bases does not allow other kinds of links.)

293

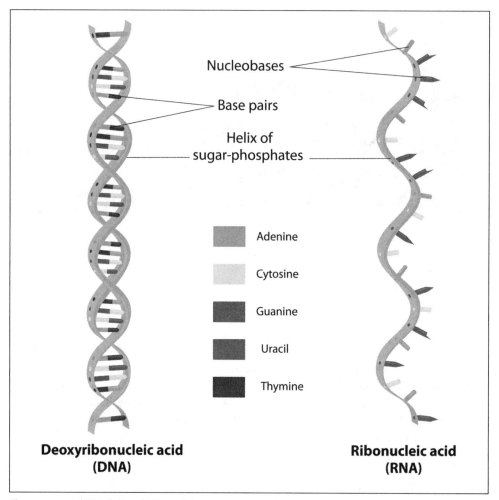

Nucleobases

Base pairs

Helix of
sugar-phosphates

Adenine

Cytosine

Guanine

Uracil

Thymine

**Deoxyribonucleic acid
(DNA)**

**Ribonucleic acid
(RNA)**

The structures of DNA (left) and RNA (right) are shown here. The arrangement of adenine, thymine, guanine, cytosine, and uracil nucleotydes are used to send instructions to cells to make proteins.

RNA is also a nucleic acid, but it consists of a single chain instead of a double; the sugar is ribose rather than deoxyribose; the bases are the same as in DNA, except that the thymine (T) is replaced by another base called uracil (U), which, like the thymine in DNA, links to adenine (A); and the RNA exists in three different forms (for more about these forms, see ahead). All RNA is formed in the nucleus (eukaryotic cells) or in the nucleoid region (prokaryotic cells).

Other than in DNA, what are some examples of how the body uses nucleotides?

Other than in DNA, nucleotides can also act as messengers or modulators in the body. For example, the nucleotide adenosine (adenine plus ribose without the phosphate) may be the most important type of immunomodulator (a compound that increases or decreases neuron, or nerve, activity). In fact, adenosine is used clinically to stop a person's heart when it is beating erratically; the natural heart pacemaker cells then return the heart to its normal rhythm. Adenosine is also thought to play a role when you feel fatigued or tired; in fact, neuroscientists currently theorize that caffeine works to maintain alertness by interfering with the reception of adenosine on a cell's surface.

What term was originally used for DNA?

DNA was originally called nuclein because it was first isolated from the nuclei of cells. In the 1860s, Swiss biochemist Friedrich Miescher (1844–1895), while working in Germany at the University of Tübingen lab of German biochemist and molecular biologist Felix Hoppe-Seyler (1825–1895), was given the task of researching the composition of white blood cells. Washing off the pus from used bandages he obtained from a nearby hospital, he studied the white blood cells by isolating a new molecule from the cell nucleus (white blood cells have very large nuclei). He called the substance nuclein (which we now call DNA). The substance was rich in nitrogen and phosphorus and also contained carbon, hydrogen, and oxygen. From there, Hoppe-Seyler checked and verified the important work of his student Miescher.

What did scientists once think made up genes?

Before the 1940s, scientists knew about genes and other inherited materials, but many believed that proteins, not DNA, were the molecules responsible. Their reasoning seemed logical at that time: Proteins are a major constituent of all cells and come in a wide array of functions and varieties, so it would make sense that this was one more protein controlling yet another function: our genetic makeup. In addition, a great deal more was known about the structure and function of proteins than about DNA. It was

What experiments transformed nonlethal bacteria into lethal bacteria—all thanks to DNA?

In 1928, an army medical officer, Frederick Griffith (1878–1939), was trying to find a vaccine against *Streptococcus pneumoniae*. In the course of his work, he found that two strains of the bacteria existed: one had a smooth coat S but was lethal; the other form had a rough coat R, which was nonlethal when injected into mice. He decided to investigate what would happen if he injected both heat-killed S bacteria and live R bacteria into mice. To his surprise, the mice injected with this combination died. Upon closer examination of the blood, living S bacteria were found. Something had occurred that transformed the nonlethal R bacteria into S bacteria. Subsequent experiments throughout the 1940s attempted to find the identity of the transforming factor—and it was eventually found to be DNA. (For more about bacteria, see the chapter "Bacteria, Viruses, and Protists.")

not until scientists began to look deeper into the DNA molecule that its true nature, so to speak, was found.

How was DNA shown to be the genetic material for all cellular organisms?

The proof that the material basis for a gene is DNA came from the work of Canadian-born American physician and medical researcher Oswald T. Avery (1877–1955), Canadian-born American geneticist Colin M. MacLeod (1909–1972), and American geneticist Maclyn McCarty (1911–2005) in a paper published in 1944. This group of scientists followed the work of Frederick Griffith (see sidebar) in order to discover what causes non-lethal bacteria to transform into a lethal strain. Using specific enzymes, all parts of the S (lethal) bacteria were degraded, including the sugarlike coat, the proteins, and the RNA. The degradation of these substances by enzymes did not affect the transformation process. Finally, when the lethal bacteria were exposed to DNase, an enzyme that destroys DNA, all transformation activity ceased. They discovered that the transforming factor was DNA, thus proving that DNA was an agent that carried genetic characteristics, not proteins.

Who were the scientists to further support the idea of DNA as genetic material?

In 1952, American geneticist and bacteriologist Alfred D. Hershey (1908–1997) and American geneticist Martha C. Chase (also known as Martha Epstein; 1927–2003) further carried out experiments to prove that DNA was the true carrier of genetic material—not protein. By tagging certain bacteriophages (a virus that infects and replicates within a bacteria) with radioactive isotopes, they were able to follow the effects, noting when certain radioactive elements entered or did not enter the bacteria. This showed that protein was not infecting the bacteria, but DNA from the virus's nucleus.

When was RNA discovered?

By the 1940s, another kind of nucleic acid other than DNA was discovered, this one called RNA. Phoebus Levene (1869–1940), a Russian-born chemist, further refined the work of German biochemist Albrecht Kossel (1853–1927), who was awarded the 1910 Nobel Prize for determining the composition of nuclein. At the time of Kossel's work, it was not clear that DNA and RNA were different substances. In 1909, Levene isolated the carbohydrate portion of nucleic acid from yeast and identified it as the pentose sugar ribose. In 1929, he succeeded in identifying the carbohydrate portion of the nucleic acid by isolating it from the thymus of an animal. It was also a pentose sugar, but it differed from ribose in that it lacked one oxygen atom. Levene called the new substance deoxyribose. Thus, these studies defined the chemical differences between DNA and RNA by their sugar molecules.

What is the "RNA Tie Club"?

In 1953, shortly after the publication of James Watson (1920–1958) and Francis Crick's (1916-2004) paper on the structure of DNA, George Gamow (1904–1968), a Russian physicist, wrote to Watson and Crick and suggested a mathematical link between DNA structure and the structure of twenty amino acids. The "RNA Tie Club" was an outgrowth of this proposal; members included Watson, Crick, and seventeen others, each taking a nickname from one of the twenty amino acids. Each member of the club was presented with a specially designed tie corresponding to the appropriate amino acid.

What was the role of Maurice Wilkins in early DNA research?

Maurice Wilkins (1916–2004) was trained as a physicist and worked briefly on the Manhattan Project, but when he became disillusioned with working on the nuclear project, he turned to the field of biophysics. He worked at Kings College, London, with John Randall (1905–1984), where together they began to use X-ray crystallography to study DNA. Both Wilkins and

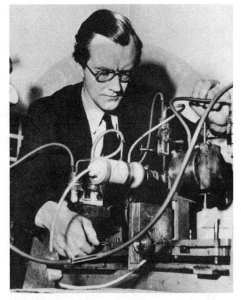

After becoming disillusioned working on the Manhattan Project, physicist Maurice Wilkins switched to biophysics and helped develop ways to study DNA using X-ray crystallography.

Rosalind Franklin (1920–1958) worked in the same laboratory, but their relationship was not one of cooperation—a factor that ultimately slowed the progress of their work.

Who was Rosalind Franklin?

Rosalind Franklin (1920–1958) was a chemist by training; in 1951, she worked at King's College, London, in the lab of English physicist and biophysicist John Randall (1905–1984). Both Franklin and Randall were working on the structure of DNA using the relatively new field of X-ray crystallography. Through meticulous research on the DNA molecule, Franklin took photographs that indicated a helical structure. Randall presented Franklin's work at a seminar where it was then (without Franklin's knowledge) provided to the competitors (Watson and Crick) at Cambridge University. This research was crucial to the detailed description of DNA that was published in 1953. Because the Nobel Prize is only awarded to the living, Franklin, who died of cancer in 1958, did not share the award—it was given to Watson, Crick, and Wilkins in 1962.

What is Chargaff's rule?

Chargaff's rule is based on the data generated by Austro-Hungarian-born American biochemist Erwin Chargaff (1905–2002) in his study of the base composition of DNA from various organisms. By comparing the nitrogen base composition of various organisms, he found that in all double-stranded DNA, the amount of adenine equals the amount of thymine, and the amount of guanine equals the amount of cytosine (A=T, G=C). The "law of complementary base pairing" refers to the pairing of nitrogenous bases in a specific manner: in particular, purines pair with pyrimidines. More specifically, adenine must always pair with thymine, and guanine must always pair with cytosine. The basis of this law came from the data from Chargaff's studies and is known as Chargaff's law or rule. (For more about DNA base pairing, see this chapter.)

DNA AND RNA

What is DNA?

Deoxyribonucleic acid (DNA) is the genetic material for all cellular organisms, with the "nucleic" part coming from the location of DNA in the nuclei of eukaryotic cells. The discovery of DNA is considered the most important molecular discovery of the twentieth century. It is actually a polymer (long strand) of nucleotides. A nucleotide has three component parts: a phosphate group, a five-carbon sugar (deoxyribose), and a nitrogen base. If you visualize DNA as a ladder, the sides of the ladder are made of the phosphate and deoxyribose molecules, and the rungs are made of two different nitrogen bases. The nitrogen bases are the crucial part of the molecule with regard to genes. Specific sequences of nitrogen bases make up a gene.

How is a DNA molecule held together?

Although DNA is held together by several different kinds of chemical interactions, it is still a rather fragile molecule. The nitrogen bases that constitute the "rungs" of the ladder are held together by hydrogen bonds. The "sides" of the ladder (the phosphate and deoxyribose molecules) are held together by a type of covalent bond called a phosphodiester bond. Because part of the DNA molecule is polar (the outside of the ladder) and the rungs (nitrogen bases) are nonpolar, other interactions—called hydrostatic interactions—occur between the hydrogen and oxygen atoms of DNA and water. The internal part of the DNA tends to repel water, while the external sugar-phosphate molecules tend to attract water. This creates a kind of molecular pressure that glues the helix together.

What genome will help scientists understand modern humans?

In 2010, scientists at the Max Planck Institute for Evolutionary Anthropology used a toe bone excavated in Denisova Cave in southern Siberia to generate a high-quality genome from one of the most elusive primates in evolution—a Neanderthal (also seen as Neandertal) individual. The subject of Neanderthals (*Homo neanderthalensis*) has been highly debated since the first fossil discovery in 1857 in Germany. They have long been regarded as an ancestor of modern humans—albeit bigger, with larger brains—that split from *Homo sapiens sapiens*, while others believe they were an evolutionary dead end. Other fossils have since been found—including those in Siberia and Croatia. In this modern study, the researchers used about 0.00134 ounces (0.038 grams) of the toe bone to sequence the Neanderthal genome, then compared it to other partial genome sequences from other Neanderthals. The entire sequence is now available for the scientific community to examine and interpret—with hopes that more will be known about the Neanderthals—and the genetic changes that occurred in the genomes of modern humans after they parted with the Neanderthals.

What are the nitrogenous bases of DNA?

The four nitrogenous bases in DNA are adenine (A), thymine (T), cytosine (C), and guanine (G). These are further divided into two types based on their structure: thymine and cytosine (called pyrimidines) have single-ring structures, while adenine and guanine (called purines) have double-ring structures. A double-ring base is always paired with a single-ring base on the opposite strand.

What recent discovery includes a "quadruple helix" DNA?

In 2013, researchers at Cambridge University in England uncovered a four-stranded structure inside human cells—what they call a quadruple helix or G-quadruplex. The research showed that the four-stranded DNA appeared more frequently during the so-called "s-phase," when the cell copies its DNA just prior to dividing; it also seems to form where substantial amounts of the base guanine (G) exist. This, they believe, may be important to the study of cancers, which are usually driven by genes that mutate to increase DNA replication. If so, this may be yet another step in helping eradiate certain cancers.

What is DNA supercoiling?

When DNA is not being replicated or specific genes are not transcribed, its normal form is two strands twisted around a helical axis, much like a spiral staircase turning clockwise. However, during DNA replication or transcription, enzymes alter the structure of DNA such that additional twists are added (positive supercoiling) or subtracted (negative supercoiling). Either type of supercoiling makes DNA even more compact. A group

What was a recent connection between computers and human DNA?

In 2013, researchers developed a biological computer—all with the help of human DNA. Essentially making a biological transistor—called a transcriptor—from DNA, they created a living computer. The transcriptor used a group of natural proteins that controlled how the enzyme (RNA polymerase) flowed along a DNA molecule, similar to how electrons are controlled inside an electrical wire. From there, they used something similar to a computer's language—of 0s and 1s—to open and close logic gates, but biologically. They hope to eventually develop this living computer to take on such tasks as sensing when a cell has been exposed to a material, such as sugar—or even to tell cells to start or stop dividing depending on the stimuli in their surrounding environment.

of enzymes, the topoisomerases, are able to disentangle DNA strands; they are called topoisomerases because they control the topology of DNA.

How is DNA unzipped?

DNA is unzipped during its replication process; the two strands of the double helix are separated and a new complementary DNA strand is synthesized from the parent strands. Also, during DNA transcription, one DNA strand, known as the template strand, is transcribed (copied) into an mRNA strand. In order for the two strands of DNA to separate, the hydrogen bonds between the nitrogen bases must be broken. DNA helicase (an enzyme) breaks the bonds. However, the enzyme does not actually unwind the DNA; special proteins first separate the DNA strands at a specific site on the chromosome, which are called initiator proteins.

What is needed for DNA replication?

DNA replication is a complex process requiring more than a dozen enzymes, nucleotides, and energy. Eukaryotic cells have multiple sites called origins of replication; at these sites, enzymes unwind the helix by breaking the double bonds between the nitrogen bases. Once the molecule is opened, separate strands are kept from rejoining by DNA stabilizing proteins. DNA polymerase molecules read the sequences in the strands being copied and catalyze the addition of complementary bases to form new strands.

Is DNA always double-stranded?

No, DNA is not always double-stranded. Certain viruses have only a single strand of DNA. And at temperatures greater than 176°F (80°C), eukaryotic DNA will become single-stranded. This strand will not always have the characteristic structure and can even form a hairpin, stem, or a cross shape.

Can DNA show what our ancient ancestors looked like?

Yes, to a point: Thanks to advances in DNA analysis of the human genome, researchers can determine the eye and hair color of ancestors dead for up to around 800 years. After comparing genomes across thousands of people, scientists have uncovered genetic variations at twenty-four different points in the human genome—all linked to eye and hair colors. Not only is this useful in knowing what your ancestors looked like, but it will no doubt solve plenty of historical controversies and mysteries in which color photos, paintings, or other records are missing.

What is a base pair?

The phrase "base pair" is a way of describing the length of DNA. Since most DNA is double-stranded, every nitrogen base containing nucleotide on one strand of the molecule is paired with a complementary base on the other strand. Thus, "ten base pairs" refers to a segment of DNA that is ten nucleotides in length and two strands in width.

How did DNA help "discover" King Richard III?

The question "Where was the body of King Richard III buried?" has long been a curiosity in England. Historians knew that Richard rode out from Leicester to meet his death on the Bosworth battlefield in August of 1485, ending the Plantagenet reign—but no one knew where his body ended up. Legends told of tossing the body in the river or bringing the body back to town. Still others say the body was claimed by Franciscans and buried hastily near the high altar of their church—and it was at that church where the alleged remains of King Richard were found.

In 2013, several tests were carried out—including radio carbon dating of bone samples and several forensic pathologists determining the cause of death based on the bones. But one clue was the key: comparing the DNA from the leg bone of Canadian Michael Ibsen, believed to be a direct descendant of Richard's sister Anne. When all the tests came in, the researchers concluded that the bones found at the church site were truly those of King Richard. Alas, King Richard's lineage apparently stops there, as both Ibsen and his sister (mitochondrial DNA is passed on through the women) have no children.

How fast is DNA copied in both prokaryotes and eukaryotes?

In prokaryotes, about 1,000 nucleotides can be copied per second; for example, the bacteria *E. coli* can be copied in about forty minutes. Since the eukaryotic genome is immense compared to the prokaryotic genome, one might think that the eukaryotic DNA replication would take a very long time. However, actual measurements show that the chromosomes in eukaryotes have multiple replication sites per chromosome. Eukaryotic cells can replicate about 500 to 5,000 bases per minute; the actual time to copy the entire genome (collection of genes) would depend on the size of their genome. (For more about prokaryotes and eukaryotes, see the chapter "Cellular Basics.")

How does DNA correct its own errors?

Spontaneous damage to DNA occurs at a rate of one event per billion nucleotide pairs per minute. Assuming this rate in a human cell, DNA is damaged every twenty-four hours at 10,000 different sites in the body. DNA has a number of quality control mechanisms. DNA polymerase (the enzyme that catalyzes DNA replication) has a proofreading function that immediately corrects 99 percent of these errors during replication. Those errors that pass through are corrected by a mismatch repair system. When a mismatch base (such as A-G) is detected, the incorrect strand is cut and the mismatch is removed. The gap is then filled in with the correct base, and the DNA is resealed.

Where is DNA found in a cell?

In addition to the nuclear DNA of eukaryotic cells, mitochondria (an organelle found in both plant and animal cells) and chloroplasts (found in plant and algal cells) both contain DNA. Mitochondrial DNA contains genes necessary to cellular metabolism. Chloroplast DNA contains genetic information essential to photosynthesis. (For more about cells, see the chapter "Cellular Basics.")

Why is DNA such a stable molecule?

DNA is a stable molecule due to what is called the "stacking interaction" between the adjacent base pairs in the helix, along with the strong hydrogen bonding between the bases. In fact, this stability has allowed researchers to gather DNA from extinct species.

How has DNA become commercialized?

DNA has been extensively commercialized in its applications to plant biotechnology, genetically modified organisms, gene therapy, gene patents, and applications to forensic science. DNA jewelry, artwork, and apparel can also be purchased—even music CDs have been created based on DNA sequences!

How does smoking affect the DNA of lung cells?

Smoking appears to alter gene expression in lung cells. The proteins produced by the bronchial cells of smokers show increased synthesis of genes, which can eventually evolve into cancer (become carcinogenic). It appears that depending on how these genes are configured, they can control cancer development, cancer suppression, or airway inflammation and varies according to the number of years a person spent smoking. The good news is that two years after a person stops smoking, his/her gene expression levels begin to resemble those of people who have never smoked.

What are the three types of RNA?

The three types of RNA are mRNA (messenger RNA), tRNA (transfer RNA), and rRNA (ribosomal RNA). The following lists the characteristics of each:

- *Messenger RNA*, or mRNA, "reads" and carries the genetic code from DNA to the ribosome. A DNA strand provides the pattern [template] for the formation of the mRNA; the mRNA "reads" the DNA code, allowing the mRNA to be synthesized as a strand complementary to the DNA strand.

- *Transfer RNA*, or tRNA, is the translation molecule. It recognizes the nucleic acid message and converts it into polypeptide language. On one part of the molecule is a section of three nucleotides known as the "anticodon" that matches the pair rules for a specific codon. The other end of the molecule, which looks something like an inverted three-leaf clover, is the amino acid attachment site. A tRNA with a specific anticodon will bind to only one kind of amino acid, ensuring the accuracy of translation from mRNA to polypeptide.

- *Ribosomal RNA*, or rRNA, is associated with the mRNA at the ribosome; along with proteins, it makes up the ribosome (formed in the nucleolus).

What is a codon?

A codon is the three-unit sequence (AUG, AGC, etc.) of mRNA nucleotides that codes for a specific amino acid. Since only twenty amino acids are commonly used and sixty-four (4 × 4 × 4) codon sequences are possible, the genetic code is described as "both degenerate and unambiguous." Each codon codes for only one amino acid, but each amino acid may have more than one matching codon.

What is the genetic code?

The genetic code is a chart depicting the relationship between each of the possible mRNA codons and their associated amino acids. The codons are grouped according to the amino acid they code for. Present in the code as well are the "start" and "stop" codons. The start codon actually codes for methionine, which is always the first amino acid in the polypeptide sequence. (Methionine may appear elsewhere in the polypeptide as well.) A stop codon signals the end of coding. Instead of a tRNA, with its amino acid in tow, a release factor matches the stop codon during translation, causing the polypeptide to be released from the ribosome. It is interesting to note that the genetic code contains only one start codon—AUG (and also codes for methionine)—but three stop codons—UAA, UGA, and UAG.

What is transcription and translation?

Transcription is how DNA makes RNA; it comes in three stages. Simply put, the first stage is the initiation—a step that begins when an enzyme, RNA polymerase, recognizes and binds to the DNA in a certain spot called the promoter region. Once the RNA polymerase is attached, the DNA transcription of the DNA template begins. The next step is the elongation—this means the strand continues as the RNA polymerase adds more nucleotides to the end of a growing chain. It does this by prying the two strands of DNA apart and attaching the RNA nucleotides based on the base pairing rules—C with G and A with U. Finally, in the termination stage, the RNA polymerase transcribes what is called a termination sequence (AAUAAA) and is cut free from the DNA template. Translation is the

What is cDNA?

Complementary DNA (cDNA) is single-stranded DNA that is complementary to a certain sequence of messenger RNA. It is usually formed in a laboratory by the action of the enzyme reverse transcriptase on a messenger RNA template. Complementary DNA is a popular tool for molecular hybridization or cloning studies. For example, if scientists are cloning a human gene in a bacterium, the long intervening, noncoding sequences called introns get in the way. Thus the scientists must insert a gene with no introns, so they take mRNA from cells and use the enzyme called reverse transcriptase to make DNA transcripts of this RNA. The resulting DNA molecule—cDNA—has the complete coding sequence the researchers need without the introns.

process by which the ribosomes synthesize proteins using a mature mRNA transcript that is produced during transcription. Simply put, genetic information stored in DNA, through transcription, changes to mRNA; then through translation, it changes to an amino acid.

Are all forms of RNA molecules unstable?

Not all forms of RNA are unstable. Messenger RNA (mRNA) molecules, however, can vary in their stability, depending on their rate of degradation and synthesis in cells and the amount of a particular protein needed.

How is RNA made from DNA—and DNA from RNA?

In eukaryotes, first the DNA of the specific gene unwinds. Then enzymes, known as RNA polymerases, use the DNA sequence, the pairing rules (U-A, G-C, C-G, A-T), and available RNA nucleotides to efficiently copy the DNA sequence into RNA. Thus, the DNA nucleotide adenine matches the RNA nucleotide uracil; thymine matches adenine; cytosine matches guanine; and guanine matches cytosine.

An RNA sequence can go to DNA, too. A process known as reverse transcription (see above) can convert an RNA sequence into DNA by using an enzyme known as reverse transcriptase. First observed in retroviruses like HIV, reverse transcriptase has also been identified as playing a role in the copying of DNA segments from one site to another in the genome.

Why are ribozymes important to RNA?

Ribozymes are often referred to as "molecular scissors" that cut RNA. They were discovered in the early 1980s by Canadian-American molecular biologist Sidney Altman (1939–) and American chemist Thomas Cech (1947–), who both won the Nobel Prize in Chemistry for their work in 1989. The ability of ribozymes to recognize and cut specific sequences of RNA allows certain genes to be turned off. The importance of ribozymes has become even more obvious, as they are often used in human genetic studies.

What is meant by the term "RNA world"?

"RNA world" refers to a hypothetical stage in life's origin on Earth. The term was coined in 1986 by American molecular biologist Walter Gilbert (1932–) to explain two tasks carried out by the RNA: to store information and to act as an enzyme. Gilbert proposed that these early "RNA" molecules randomly assembled together in the primordial soup, carrying out simple metabolic activities. (For more about early life on Earth, see the chapter "Basics of Biology.")

How does a ribosome participate in protein synthesis?

Ribosomes are different than the ribozymes mentioned in the question above—they actually serve as the site of translation (see above). This combination of RNA and protein is a meeting place for mRNA and tRNA. Structurally, a ribosome is composed of two parts known as the large and small subunits. Each of these is a combination of protein and a type of RNA: rRNA. At the beginning of translation, the two subunits form a structure around the mRNA molecule as the first tRNA (the one matching the first methionine—also called fMet) arrives. The completed ribosome has niches that hold up to three tRNAs at a time. Because a cell has so many ribosomes at any one time, rRNA is the most common type of RNA found in cells.

How many human RNA genes exist?

About 250 genes code for short RNA sequences instead of protein. These RNA strands appear to regulate the activity of other genes, particularly those involved in embryonic development. (For more about genes, see below and the chapter "Heredity, Natural Selection, and Evolution.")

CHROMOSOMES

What is a chromosome?

A chromosome is the threadlike part of a cell that contains DNA and carries the genetic material of a cell. In prokaryotic cells, chromosomes consist entirely of DNA and are not enclosed in a nuclear membrane. In eukaryotic cells, the chromosomes are found within the nucleus and contain both DNA and RNA.

Who first observed chromosomes?

Chromosomes were observed as early as 1872, when Baltic German biologist Edmund August Friedrich Russow (1841–1897) described seeing items that resembled small rods during cell division; he named the rods "Stäbchen." Belgian cytologist Edouard van Beneden (1846–1910) used the term "bâtonnet" in 1875 to describe nuclear duplication. The following year, French embryologist Edouard-Gérard Balbiani (1825–1899) described that at the time of cell division the nucleus dissolved into a collection of *bâtonnets étroits* (French for "narrow little rods"). German biologist Walther Flemming (1843–1905) discovered that the chromosomal "threads" (or "Fäden") split longitudinally during mitosis (Flemming is also called the founder of cytogenetics).

What organisms have the least and most number of chromosomes—and some in between?

Ophioglossum reticulatum, a species of fern also known as Adders-tongue, has the largest number of chromosomes with more than 1,260 (630 pairs). To date, the organism with the least number of chromosomes is the male Australian ant, *Myrmecia pilosula*, with one chromosome per cell (male ants are generally haploid—that is, they have half the number of normal chromosomes while the female ant has two chromosomes per cell). Another contender is bacteria that have one circular chromosome consisting of DNA and associated proteins. Some of the more common plants and animals also have different numbers of chromosomes: Humans (*Homo sapiens sapiens*) have forty-six; dogs (*Canis lupus familiaris*) have seventy-eight; cats (*Felis catus*) have thirty-eight; a mouse (*Mus musculus*) has forty; a fruit fly (*Drosophila melanogaster*) has eight; a mosquito (*Aedes aegypti*) has six; a potato (*Solanum tuberosum*) has forty-eight; and yeast (*Saccharomyces cerevisiae*) has thirty-two.

What are telomeres?

Unique structures known as telomeres—protective structures composed of DNA—are found at the end of eukaryotic chromosomes (eukaryotes are organisms with cells that contain complex structures). Experiments have determined that without telomeres, the chromosome structure could be damaged; the DNA of the chromosome tends to stick to other pieces of DNA, and enzymes (deoxyribonucleases) are more likely to degrade or digest the ends of the chromosomes.

How are chromosomes assembled?

Chromosomes are assembled on a scaffold of proteins (histones) that allow DNA to be tightly packed. Five major types of histones exist, all of which have a positive charge; the positive charges of the histones attract the negative charges on the phosphates of DNA, thus holding the DNA in contact with the histones. These thicker strands of DNA and proteins are called chromatin. Chromatin is then packed to form the familiar structure of a chromosome. During mitosis, chromosomes acquire characteristic shapes that allow them to be counted and identified.

Can people have missing or extra chromosomes?

Yes, people can live with this chromosomal abnormality, depending on which chromosomes are copied or missing; for example, Down syndrome (an extra copy of chromosome 21) and Turner syndrome (a female with only one X chromosome). Conversely, almost 1 percent of all conceptions are triploid (three copies of each chromosome), but over 99 percent of these die before birth.

What are the shortest and longest chromosomes?

Among human chromosomes, the length ranges from fifty to 250 million base pairs. The longest chromosome is chromosome 1, with 300 million bases (approximately 10 percent of the human genome), and the shortest is chromosome 21, with 50 million bases.

What are homologous chromosomes?

In diploid organisms, chromosomes occur in matching pairs, with the exception of the sex chromosomes. Not all organisms are diploid; for example, bacteria have only one circular chromosome, and some insects may have an odd number of chromosomes. Every human somatic cell (excluding egg or sperm cells) has forty-six chromosomes—twenty-two somatic chromosome pairs and one sex chromosome pair. The somatic chromosome pairs, called homologous chromosomes, carry the same sequence of genes for the same inherited characteristics.

How can organisms with just one set of chromosomes reproduce?

An organism that is haploid (one set of chromosomes) can reproduce by mitosis to produce more haploid cells or a multicellular haploid organism. This is typical of some algae and fungi. (For more about haploid cells and mitosis, see the chapter "Cellular Basics.")

What are autosomes?

Autosomes are chromosomes that contain information (gene sequences) available to both sexes. For example, these chromosomes contain genes controlling most "body" traits.

What are sex chromosomes?

Sex (X and Y) chromosomes are found in mammals. For example, in human beings, the number of chromosomes is forty-six (twenty-three pairs); of these, twenty-two pairs are autosomes and the remaining pair of chromosomes are known as the sex chromosomes, or the X and Y chromosomes. Each human cell normally contains two of these chromosomes: the XX leading to the female, and the XY leading to a male. The combination YY cannot occur.

The Y chromosome found only in males has the sex-determining gene, SRY. The SRY determines whether or not an individual will develop testes and produce appreciable quantities of testosterone (the hormone that generates male characteristics). No correlate gene on the X chromosome for the formation of ovaries is known; therefore, it is the absence of the Y chromosome that determines a female.

What recent chromosome discovery may tell something about the human family tree?

A DNA sample from an African-American living in South Carolina revealed that the human Y chromosome may be much older than previously thought—pushing back the most recent common ancestor for the Y chromosome lineage tree by almost 70 percent to around 338,000 years ago. The Y chromosome is the hereditary factor determining the male sex; because it does not exchange genetic material with other chromosomes, it is easier to trace the ancestral relationships between lineages. After analysis, it was shown that this may be the oldest known branch of the human Y chromosome—a new divergent lineage that apparently branched from the Y chromosome tree even before the first appearance of anatomically modern humans in the fossil records.

When is the sex of an organism determined?

The sex of a new organism is determined at the instant the egg is fertilized by the sperm. For example, in humans, the female's egg cell contains a single X chromosome; the sperm cell may contain either an X or Y chromosome (all, of course, with the other twenty-two autosomes in the egg and sperm cells).

What is the National Geographic Genographic Project?

The National Geographic Genographic Project was launched in 2005—a multiyear study that uses genetics, especially in terms of DNA, as a tool to understand how humans have migrated around the world over their history. The first phase included around 500,000 participants from over 130 countries; the second phase, which started in late 2012, is involving citizens in the project even more, offering an updated participation kit that allows members of the public to learn more about their ancestral makeup and at the same time add to the scientific database that will help scientists trace early routes of human migration.

Can an organism have more than two complete sets of chromosomes?

Having one or more extra sets of chromosomes is known as polyploidy. In humans, it is virtually impossible to survive to adulthood as a polyploidy. Such a condition is more common among plants (in fact, this is a common method by which new plant species arise) and some groups of fish and amphibians; for example, some salamanders, frogs, and leeches are polyploids. But note: polyploids that arise within a species are different from those that arise due to the hybridization of two distinct species; the former are known as autopolyploids and the latter are referred to as allopolyploids.

What are "mitochondrial Eve" and "Y chromosome Adam"?

Until recently, mitochondrial Eve and Y chromosome Adam were proposed by some researchers to be the one pair of humans that lived at a certain point in human evolution, from whom all humans descended. The main reason the mitochondria are useful for such evolutionary studies is that they are passed intact from the mother to the children, with no mixing of the father's genes; plus, the DNA within the mitochondria quickly picks up mutations and keeps them, while the Y chromosome does not exchange genetic material with other chromosomes. With more studies,

---

Can I see a gene?

No, a gene cannot be seen because it is submicroscopic. We can see a chromosome, which contains genes, and geneticists can pinpoint the location of a gene on that chromosome, but the actual gene cannot be seen.

such as the Human Genome Project and others, scientists now believe that humans did not evolve from a single genetic region from these two humans only. Instead, they believe that pockets of genetically isolated communities eventually would produce our human diversity.

GENES

What is a gene?

A gene is one of the complex protein molecules that are associated with chromosomes. They are responsible for—as a unit or in certain biochemical combinations—the transmission of certain inherited characteristics from the parent to the offspring. The terms gene, from the Greek term *genos*, meaning "to give birth to," and genotype were first used in 1909 by a Danish botanist, Wilhelm Johannsen (1857–1927), who is considered to be one of the architects of modern genetics. The average size of a vertebrate gene is about 30,000 base pairs (for more about base pairs, see this chapter). Bacteria, because their sequences containing only coding material, have smaller genes of about 1,000 base pairs each. Human genes are in the 20,000 to 25,000 base pair range, although sizes greater than 100,000 base pairs are in some organisms.

How are genes connected to RNA?

A gene has a section of DNA that will be used as a template to build a strand of RNA or protein. In addition to this information, each gene also contains a promoter region, which indicates where the coding information actually begins, and a terminator, which delineates the end of the gene.

How are genes controlled?

Genes are controlled by certain mechanisms that vary depending on whether the organism is a prokaryote or a eukaryote. Bacteria (prokaryote) genes can be regulated by DNA binding proteins that influence the rate of transcription or by global regulatory mechanisms that refer to an organism's response to specific environmental stimuli such as heat shock. This is particularly important in bacteria. Gene control in eukaryotes depends on a complex set of regulatory elements that turn genes off and on at specific times. Among these regulatory elements are DNA binding proteins as well as proteins that, in turn, control the activity of the DNA binding proteins.

What is a centimorgan?

A centimorgan (cM) is a way of measuring the distances between genes on a chromosome. This unit was named in honor of American biologist Thomas H. Morgan (1866–1945), who was one of the first to map genes onto chromosomes and who formulated the chromosome theory of heredity. Typically, a distance of a cM corresponds to a distance of about one million base pairs.

What is a chromosome map?

A chromosome map lists the sequence of genes found on a given chromosome. Chromosome maps are usually determined by breeding experiments in which the ratio of the offspring with certain combinations of traits indicates how far apart those traits are on the chromosome.

What is a pseudogene?

As genes are copied and transposed, mutations may occur that make the copies nonfunctional. This is thought to be the origin of pseudogenes, which resemble actual genes but which, while stable, do not produce a polypeptide.

What is gene redundancy?

Gene redundancy refers to having multiple copies of the same gene. This assures that essential genes that are required in large amounts, like those for ribosomal RNA (rRNA), can be transcribed at multiple sites. In addition, gene redundancy provides for a source of gene sequences that can be modified by natural selection without immediately destroying the organism.

Can one gene control another?

No, one gene cannot actually control another gene, but one gene can mask the effect of another gene, which is a process called epistasis. For example, a gene in Labrador dogs controls the amount of melanin—in particular the dominant allele B results in large amounts of melanin; a recessive allele b causes less amounts. When bred, a BB or Bb dog is black, while a bb dog is brown. Yet another gene controls whether or not melanin is there at all—the gene E allows for more amounts of melanin while the recessive form e does not. When bred, an ee dog is yellow, while an Ee or EE dog is not yellow and has melanin. Thus, interaction of two genes, the B and E genes, controls coloration in Labrador dogs.

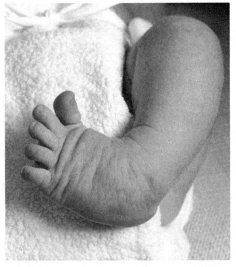

Some birth defects, such as clubfoot, are the result of polygenic inheritance in which both the mother and father must donate a gene that results in the malformation.

What are some examples of polygenic inheritance?

Phenotypes of a given trait that are controlled by more than one gene are described as having polygenic inheritance; for example, in humans, it includes height, weight, skin color, and intelligence. Some congenital malformations (birth defects), like clubfoot, cleft palate, or neural tube defects, are also the result of multiple gene interactions.

What is gene expression?

Gene expression refers to the molecular product of a particular gene and is another way of describing the phenotype. The phenotype of an organism is a result of the chemical interactions that are guided by the DNA (gene blueprint) for a specific polypeptide.

What is meant by a genetic fingerprint?

Just as a real fingerprint is used to identify people individually, a genetic (or DNA) fingerprint is a unique pattern of DNA sequences for each individual. (For more about how genetic fingerprinting is used, see the chapter "Biology in the Laboratory.")

Are all genes in a genome used by an organism?

No, not all genes are used by an organism all the time. Since protein synthesis is an energy intensive cellular process, proteins are not produced unless they are needed for a specific cell function. For example, before a human has reached adult height, cells are continually producing human growth hormone, a protein that encourages bone and muscle growth. However, at a certain age (it varies by individual) the gene will become dormant and will no longer produce growth hormone.

What is the difference between a gene and a chromosome?

The human genome contains twenty-four distinct, physically separate units called chromosomes. Arranged linearly along the chromosomes are tens of thousands of genes. The term "gene" refers to a particular part of a DNA molecule defined by a specific sequence of nucleotides. It is the specific sequence of the nitrogen bases that encodes a gene. The human genome contains about three billion base pairs, and the length of genes varies widely.

What is an allele?

An allele is an alternative form of a gene; usually each gene has two alleles, although the number may vary from one trait to another. Each individual inherits one allele from

Can the environment affect genes?

Yes, the environment can affect genes. For example, some genes are sensitive to temperature—like the enzyme that controls the synthesis of black pigment melanin in Siamese cats, which is only active at cool temperatures. The cooler the body part, the darker the pigment; since the body is warmer than the extremities, it remains lighter in color. During the winter if your Siamese cat goes outdoors, the fur may become darker, while Siamese kittens are usually white due to the warmth of the mother's body.

the mother and one from the father. Alleles for a trait are located on corresponding spot on each chromosome.

GENETICS AND THE HUMAN GENOME

What is the relationship between probability and genetics?

Probability is a branch of mathematics used to predict the likelihood of an event occurring. It is also an important tool for understanding inheritance patterns of specific traits. Two rules are applied to genetic inheritance: 1) The rule of multiplication is used when determining the probability of any two events happening simultaneously. For example, if the probability of having dimples is ¼, and the probability of having a male child is ½, then the probability of having a boy with dimples is ¼ × ½, or ⅛) The rule of addition is used when determining the outcome of an event that can occur in two or more ways. For example, if we were to consider the probability of having a male child or of having a child with dimples, then the probability would be ½ + ¼, or ¾. In other words, chances are 3 out of 4 that a given birth will produce a male child or a child with dimples.

What is meant by the modern era of genetics?

Gregor Mendel's work (see the chapter "Heredity, Natural Selection, and Evolution") was really not appreciated until advances in cytology enabled scientists to better study cells. In 1900, Dutch botanist Hugo de Vries (1848–1935), German botanist and geneticist Carl Correns (1864–1933), and Austrian botanist Erich von Tschermak (1871–1962) examined Mendel's original 1866 paper and repeated the experiments. In the following years, chromosomes were discovered as discrete structures within the nucleus of a cell. In 1917, American geneticist Thomas Hunt Morgan (1866–1945), while at Columbia University, extended Mendel's findings to the structure and function of chromosomes. This and subsequent findings in the 1950s were the beginning of the modern era of genetics.

What are a Barr body—and the Lyon hypothesis?

In 1949, Canadian physician and medical researcher Murray Barr (1908–1995) noticed a dark body in the neurons of female cats. It was later identified as a structure found only in the nucleus of females. It was named a Barr body in honor of its discoverer. The Lyon hypothesis refers directly to a Barr body. It was proposed by English geneticist Mary Frances Lyon (1925–) in 1961 that a Barr body is actually an inactivated X chromosome. According to this hypothesis, female mammals sequester one X chromosome in each of their cells during the early stages of development. This folded chromosome becomes the dark body of Barr's observation. This means that both males and females rely on the information from only a single X chromosome. Therefore, it is only one X chromosome that provides genetic information in both males and females.

Who first discovered the link between genetics and metabolic disorders?

Archibald Garrod (1857–1936) was a British physician who became interested in a rare but harmless disease called alkaptonuria. A unique characteristic of this condition is that the patient's urine turns black when exposed to air. Previously, physicians thought the disorder was due to a bacterial infection; however, Garrod observed that the disorder was more common in children of first-cousin marriages and followed the Mendelian description of an autosomal recessive disease. Garrod, a follower of the then fledgling science of biochemistry, suspected that the urine turned dark because patients lacked the enzyme to break down the protein that caused the urine to darken. He referred to alkaptonuria as an inborn error of metabolism. Unfortunately, very few people understood metabolic pathways at that time, so his contributions remained largely unnoticed. By the 1950s his work was verified, earning Garrod the title of "father of chemical genetics."

What is Thomas Hunt Morgan's contribution to genetics?

American evolutionary biologist, geneticist, and embryologist Thomas Hunt Morgan (1866–1945) won the Nobel Prize in Physiology or Medicine in 1933 for his discovery of the role played by chromosomes in heredity. He is perhaps most noted for his "fly lab" at Columbia University in New York, where he collected Drosophila (fruit fly) mutants. Morgan studied fruit flies much in the same way that Mendel studied peas. He found that the inheritance of certain characteristics, such as eye color, was affected by the sex of the offspring.

Why were Beadle and Tatum important to genetics?

The work of American geneticist George Beadle (1903–1989) and American geneticist Edward Tatum (1909–1975) in the 1930s and 1940s demonstrated the relationship between genes and proteins. They grew the orange bread mold *Neurospora* on a specific growth medium. After exposing the mold to ultraviolet (UV) radiation, the mold was unable to grow on a medium unless it was supplemented with specific amino acids. The UV radiation caused a single gene mutation that led to the production of a mutant enzyme. This enzyme, in turn, caused the mold to exhibit a mutant phenotype. Beadle and Tatum's work was important in demonstrating that genes control phenotypes through the action of proteins in metabolic pathways. In other words, one gene affects one enzyme. Since Beadle and Tatum were awarded their Nobel Prize in 1958, scientists have discovered that the relationship between genes and their proteins is much more complex.

What is meant by genetic determinism?

Genetic determinism is a theory proposing that behavior and character are entirely shaped by genes. Since only genetic material passes from one generation to the next, it seemed logical at one time to assume that all of our traits (physical, behavioral, psychological) were the result of gene sequences. More recent and extensive research has demonstrated the importance of environmental factors in determining how or which genes are actually involved in the development of the individual.

313

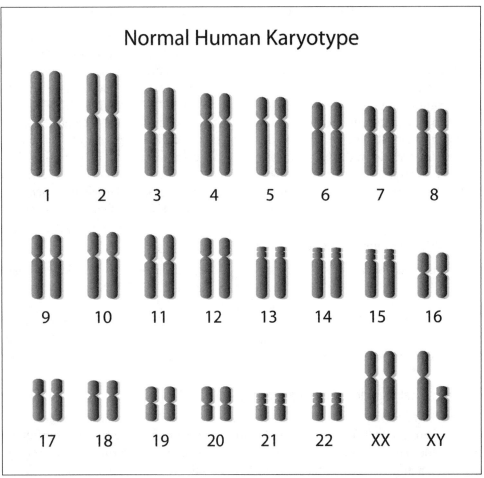

Normal Human Karyotype

1 2 3 4 5 6 7 8

9 10 11 12 13 14 15 16

17 18 19 20 21 22 XX XY

The normal human karyotype contains twenty-three chromosome pairs, including either an XX pair for women or an XY pair for men.

What is a karyotype?

A karyotype is a snapshot of the genome; it can be used to detect extra or missing chromosomes, chromosomal rearrangements, or chromosomal breaks. Any cell that contains a nucleus can be used to make a karyotype. However, white blood cells seem to work best for human karyotypes. After the cells are cultured, they are killed by using a drug that halts mitosis—the chromosomes are then stained, observed, and a size order chart is produced.

Why are some species more commonly used for genetic studies than others?

Species with a relatively small genome, with a short generation time from seed to seed, and that are adaptable to living in captivity are appealing as experimental organisms. Even though many of these species bear little physical resemblance to humans, they do share part of our genome, so they can answer some of the questions we have about genetic inheritance and gene expression. For example, some commonly used species in genetic research include the plant *Arabidopsis thaliana*, with a genome size of 120 million base pairs; the orange bread mold *Neurospora*, a fungus with forty million base pairs; *Drosophila melanogaster*, a fruit fly with 170 million base pairs; and a *Caenorhabditis elegans*, or roundworm, with ninety-seven million base pairs.

Why is the human genome such a small size?

The process of translation uses an mRNA pattern to produce a string of amino acids technically known as a polypeptide. To turn a polypeptide into a protein with the shape that is crucial to its function requires several more steps. Sugars, lipids, phosphate groups, or other molecules may be added to the amino acids in the chain. The polypeptide may be "trimmed" as well; the first few amino acids may be removed. Some proteins, like insulin, only become functional after cleavage by an enzyme. Because of this ability to create variations from the basic polypeptide chain, a given gene may code for several different functional structures. This helps to explain the unexpectedly small size of the human genome. Our genes actually code for several hundred thousand different proteins. While most proteins fold spontaneously into their specific conformation, some require the help of other molecules to achieve this. Chaperone proteins guide newly synthesized polypeptides into their functional three-dimensional shape.

What are examples of sex-linked traits?

Sex-linked traits are most commonly found on the X chromosome and include color blindness (both red and green types), hemophilia (types A and B), icthyosis (a skin disorder causing large dark scales), and Duchenne muscular dystrophy. Very few traits are Y-linked; for example, although relatively rare, hairy ears are a Y-linked trait.

What is epigenetics?

Just as the term "epidermis" refers to the layer of the skin above (or beyond) the dermis, the term "epigenetics" describes nongenetic causes of a phenotype. For example, in genetic imprinting, a different expression of an allele is produced depending on the parent who transmits it. In humans, an example comes from two medical problems—both that result from a deletion on chromosome 15: In one called the Prader-Willi syndrome—originating from a deletion on the paternal chromosome 15—affected children have small hands and feet, short stature, mental retardation, and are obese. In the Angelman syndrome—originating from a deletion on the maternal chromosome 15—affected children have a large mouth and tongue, severe mental and motor retardation, and a happy disposition, accompanied by excessive laughter.

What is the Human Genome Project (HGP)?

The HGP was begun in 1990 and finally completed in 2003. According to the official HGP website (http://www.ornl.gov/sci/techresources/Human_Genome/home.shtml), the goals are as follows: To identify the number and sequence of genes in human DNA; to store this information in public databases for scientists and the public; to improve the tools used to analyze the genome data and possibly transfer these technologies to the private sector; and finally, and many scientists believe most importantly, to address and understand the ethical, legal, and social issues that may stem from the project.

Overall, the project has been very successful, especially in the analysis of the data. And it's easy to see why the data will be used for many more years to come: According to the Human Genome Project, humans have about 20,000 to 25,000 genes in their DNA—the goal of the project has been to identify all those genes. In addition, the project goals included determining the sequences of the three billion chemical base pairs that make up human DNA.

How have scientists used the data from the Human Genome Project so far?

The data from the Human Genome Project is being used in many different ways, albeit many of them still in their infancy. For example, certain medical treatments are being uncovered through genomic analysis—it has already fueled the discovery of more than 1,800 disease genes; more than 2,000 genetic tests are now available for human conditions; the first human-made life forms have been created (which may help in the future of energy production or help enhance agriculture through bioengineering); even a sequence for the Neanderthal genome (for more about the Neanderthal genome, or Neandertal, see this chapter)—and other advancements in the new field of paleogenomics—will help to shed light on human prehistory.

What are ELSI and HUGO?

ELSI began in 1990 as a part of the Human Genome Project and is thus part of the National Human Genome Research Institute that deals with the ethical (E), legal (L), and social (S) issues (I)—also called the Ethical, Legal, and Social Implications Research Program. It started because of the increasing availability of genetic information and represents the largest bioethics study in the world. The program funds and manages studies and supports workshops, research and policy conferences, and many more events related to these topics. In addition, HUGO, also known as the Human Genome Organization, was conceived in early 1988 at the first meeting on genome mapping and sequencing in Cold Spring Harbor, New York. It is an international consortium created to coordinate the work of human geneticists around the world and is incorporated in Geneva, Switzerland.

Who made the first gene map?

Although gene maps are a relatively recent (past thirty years) means of locating genes, geneticists of the early twentieth century had prototype gene maps. American zoologist

What is the 1000 Genomes Project?

According to the Wellcome Trust in the United Kingdom that runs the project, the 1000 Genomes Project is a scientific effort to sequence the genomes of at least 1,000 people to understand and create the most detailed and medically useful catalog of human genetic variation to date—hopefully to be used in future studies of people with particular diseases. They can do this for a good reason: any two humans are more than 99 percent the same at the genetic level—but the very small percent that is not the same can tell a great deal about such things as susceptibility to disease or even reactions to environmental factors. Although the goal started with 1,000 people, by 2011, the project increased its sequencing to include 2,500 genomes, sourced from about twenty different populations around the world. For example, populations being sequenced include Chinese in the metropolitan Denver, Colorado area; Japanese in Tokyo; Maasai in Kinyawa, Kenya; people of Mexican ancestry in Los Angeles; Utah residents with ancestry from northern and western Europe; and people of African ancestry in the southwestern United States.

and geneticist Edmund Beecher Wilson (1856–1939) and his colleagues were the first to demonstrate that the genetic differences between males and females were due to a special pair of chromosomes in the cell. American evolutionary biologist, geneticist, and embryologist Thomas Hunt Morgan (1866–1945), American geneticist Calvin Bridges (1889–1938), and their colleagues were able to place a gene known to be inherited differently by males and females onto one of the sex chromosomes. This was the beginning of the first gene map.

What is eugenics?

Sir Francis Galton (1822–1911), a cousin of Charles Darwin, founded eugenics. After reading Darwin's work on natural selection, Galton thought that the human species could be improved by artificial selection—the selective breeding for desirable traits (which was already a method used for domesticated animals). In Galton's plan, those with desirable traits would be encouraged to have large families, while those with undesirable traits would be kept from breeding. However, Galton's theory overlooked two important points: the importance of environmental factors and the difficulty of removing recessive traits from the gene pool. Recessive alleles can be passed from one individual to the next as part of a (heterozygous) genotype, thereby escaping detection for generations. Galton's work was enthusiastically adopted in both the United States and Europe: In the United States between 1900 and 1930, eugenics gave rise to changes in federal immigration laws and the passage of state laws requiring the sterilization of "genetic defectives" and certain types of criminals, while unfortunately, in Europe, eugenics became a cornerstone of the Nazi movement.

GENETIC MUTATIONS

What is a mutation?

A mutation is an alteration in the DNA sequence of a gene. Mutations are a source of variation to a population, but they can have harmful effects in that they may cause diseases and disorders. One example of a disease caused by a mutation is sickle cell disease, in which a change occurs in the amino acid sequence (valine is substituted for glutamic acid) of two of the four polypeptide chains that make up the oxygen-carrying protein known as hemoglobin, or our blood.

What are the different kinds of mutations?

Any rearrangement within a gene is a mutation, and such rearrangement is usually due to a random genetic accident. The mutation can be harmless and simply add to the variation within an organism's genome, or it can be harmful with dire consequences.

Mutations may occur within the gametic cells (germinal mutation) or within any non-sex cell (somatic mutation). The difference is that a germinal mutation will affect all cells of an individual, while a somatic mutation will only affect those cells produced by mitosis from the original mutated cell. The following are various categories of mutations:

- A point mutation is a change in a single DNA base
- A deletion occurs when information is removed from a gene
- An insertion occurs when extra DNA is added to a gene
- Frameshift mutations occur when one or two bases are either added or deleted

How can chromosomes become damaged?

Chromosomes can become damaged physically (in their appearance) and molecularly (in the specific DNA sequences they contain). Chromosomes can break randomly or due to exposure to ionizing radiation, which acts like a miniature cannonball, blasting the strands of DNA. Other factors such as physical trauma or chemical insult can also cause breaks. When a chromosome breaks, the broken part may then reattach to another chro-

Most diamond doves are all white, but this dove has fawn-colored feathers due to a genetic mutation.

How do some cats have stripes and others spots?

Researchers recently discovered why certain cats have stripes and others spots—and it all has to do with mutations. In particular, tabby house cats and king cheetahs share a gene responsible for the cats' stripes and the cheetahs' spots. But when a mutation in the single gene (called Taqpep) occurs, the tabby will develop patches, not stripes, and the cheetah's spots become wide stripes. The biggest question remains: What is the biological and evolutionary significance of this mutation?

mosome, a process called translocation. Sometimes the broken chromosomes may attach to each other, forming a ring. Any type of chemical change is known as a mutation.

Could a mutation have shaped modern humans?

Scientists are currently trying to determine if a small mutation of a gene known as EDAR helped to shape human evolution. About 30,000 years ago, this small change may have helped humans in Asia survive excessive heat and humidity by endowing them with extra sweat glands. But the researchers believe this was not the only change from mutations—and point to mutation of genes involved in bone density, skin color, hair changes, and immune system function—all necessary for humans to spread to new environments around the world. But the interpretations are not easy—after all, "reading" how mutations have been passed down and tweaked over thousands of years of human evolution still remains difficult.

How often is DNA damaged in various organisms?

Depending on the animal or plant, DNA can be damaged, causing mutations; in particular, the rate of mutation can vary according to different genes in the various organisms. For example, considering how many cells are in the human body and how often it occurs, DNA replication is fairly accurate. And in bacteria, spontaneous damage to DNA is low, occurring at the rate of one to one hundred mutations per ten billion cells.

Are all mutations bad?

No, not all mutations are bad. Although people may use the term "mutant" in a disparaging manner, mutations are important to a population's gene pool because of the variation they contribute. Variations cannot occur without mutations and, following along Darwin's thinking, no natural selection as a result.

What is pleiotropy?

Pleiotropy refers to the ability of one single gene to influence several other body characteristics. This gene interaction in chickens can cause the frizzle trait—a gene that

319

causes malformed feathers. This mutation makes it so the chicken cannot keep warm, causing several changes in various organ systems. In Siamese cats, the allele responsible for the animal's color patterns of light fur on the body and dark at the extremities is the same one responsible for many Siamese cats being cross-eyed.

In humans, an example involves sickle cell anemia, a disorder in which a single point mutation in the amino acid sequence for hemoglobin (red blood cells) results in many different medical problems. In this case, red blood cells produce abnormal hemoglobin molecules which, because of their odd shape, tend to stick together and crystallize. This causes the normal disk shape of red blood cells to change to a sickle shape (hence the name of the disorder). Sickle-shaped red blood cells can clog small vessels, causing pain and the possibility of brain damage and heart failure. In addition, since some of the hemoglobin in these cells is abnormal, less oxygen is available, leading to physical weakness and anemia. If left untreated, the anemia can impair mental function.

HEREDITY, NATURAL SELECTION, AND EVOLUTION

EARLY STUDIES IN HEREDITY

Who was Mendel?

Austrian monk Gregor Mendel (1822–1884) was the founding father of modern genetics. His work with the garden pea, *Pisum sativuum*, was not consistent with the nineteenth-century ideas of inheritance. In his world, scientists believed that inheritance was essentially a mix of fluid that blended and passed from parents to children (see "blending theory" below).

Mendel was the first to demonstrate that distinct physical characteristics could be passed from generation to generation. In particular, Mendel studied peas of distinct and recognizable plant varieties; his experiments included the characteristics of height, flower color, pea color, pea shape, pod color, and the position of flowers on the stem. His theory was one called particulate inheritance, in which certain inherited characteristics were carried by what he called elementes, which eventually became the now well-known name "genes."

What is the homunculus theory?

The homunculus theory had its roots in the late seventeenth century, when illustrations depicted sperm containing a miniaturized adult—a preformed human called a homunculus. It was thought at that time that the sperm (and its homunculus) contained all of the traits for building a baby and that the egg and womb served simply as incubators.

What is the blending theory?

In the nineteenth century, around Gregor Mendel's time, the blending theory was the commonly held belief that characteristics were mixed in each generation. For example, breeding two horses, one with a light-colored coat and the other dark, would result in offspring that would have an intermediate coat color. If this held true, then eventually all organisms would become more alike in each generation. Although this theory persisted for many years, it was eventually supplanted by the work of Mendel and other modern geneticists.

Why was Mendel successful in his work while others were not?

Using a simple organism like the garden pea, Mendel was able to control pollination among his experimental plants, and most importantly, he used true breeding plants

Austrian monk Gregor Mendel is considered the father of modern genetics.

with easily observable characteristics, such as flower color and height. He also kept meticulous records and discovered consistent data that involved thousands of plant-breeding experiments over eleven years. Interestingly, he studied physics and math under Christian Doppler (1803–1853; the eventual discoverer of the Doppler Effect) at the University of Vienna, which helped Mendel understand statistics and experimentation and eventually led him to determine the laws for his famous heredity works.

Why was Mendel's work disputed?

Some scientists feel that Mendel's work was too perfect to be accurate, even though the validity of his conclusions were (and are) not in doubt. A closer look at Mendel's work suggests that he may not have reported the inheritance of traits that did not show independent assortment, and thus he may have "skewed" some of his data. A British statistician and population geneticist, R. A. Fisher (1890–1962), pointed out in 1936 that Mendel's data fit the expected ratios much closer than chance alone would indicate. However, Mendel's published data is comparable to the work of others in his field, and his conclusions are still accepted as part of the core of genetics.

Who developed the Punnett square?

Geneticist Reginald Punnett (1875–1967) invented a table called the Punnett square for displaying the genetics cross, which graphically represented Mendel's laws of segregation and independent assortment. He was the founding member of the Genetical Society and, along with English geneticist William Bateson (1861–1926), created the *Journal of Genetics*. Punnett was particularly interested in the genetics of chickens, discovering a sex-linked characteristic that was used to separate male and female chicks after hatching—this was especially important in improving food production after World War I.

What was the principle of dominance?

Gregor Mendel discovered this principle (also often called the law of dominance) in his classic studies of inheritance patterns. When genes of certain allelic pairs have contrasting effects on the same trait, only one can be expressed, while the other one will remain "masked." The gene that is expressed is called the dominant allele; the one that is masked is the recessive allele. In order to understand this, geneticists use symbols to represent the genes—usually the dominant genes are represented by capital (italicized or not) letters, while the recessive genes are represented by lowercase (italicized or not) letters.

For example, in the garden pea, the allele for "tall" plants is T (tall plants are dominant) and for "short" plants t (short plants are recessive). If a plant is pure tall (TT) and is crossed with a pure small plant (tt), the offspring will be a hybrid, but will only exhibit the dominant trait (the recessive trait remains "hidden"). The following shows this example:

TT × tt = offspring.

	T	T
t	Tt	Tt
t	Tt	Tt

All tall offspring result in the above example.

What is incomplete dominance and codominance?

Both incomplete dominance and codominance are also forms of inheritance—how certain genetic traits are expressed in offspring. Incomplete dominance is characterized by blending where neither trait is dominant. For example, if a long watermelon (*LL*) is crossed with a round watermelon (*RR*), it produces an oval squash (*LR*); a black animal (*BB*) crossed with a white animal (*WW*) results in a gray (*BW*) animal. Codominance is the instance in which both traits show—in other words, both alleles are dominant and both are expressed in the offspring. For example, the pink petal color in four-o'clock flowers and sickle-cell anemia in humans are all examples of codominance.

What is meant by the terms "true breeding," homozygous, and heterozygous?

Individuals when bred to others of the same genotype produce only offspring of that genotype; this is called true breeding. When two inherited alleles are alike, they are said to be homozygous individuals (AA, aa), an example of true breeding; if the two members of the allelic pair are different, the combination is said to be heterozygous individuals (Aa; also called hybrid).

What is the law of independent assortment?

This Mendelian principle deals with the prediction of the outcome of dihybrid (two-trait) crosses—in other words, two individuals hybrid for two or more traits that are not on the same chromosome (called a dihybrid cross; see below). The law states that during gamete formation, the alleles of a gene for one trait, such as plant height, segregate independently from those for another trait, such as seed color. From this, Mendel concluded that traits are transmitted to offspring independently of one another. In other words, the separation of alleles for one trait does not influence or control the distribution of alleles from the second trait. This law holds true as long as the two traits in question are located on separate chromosomes or are so distant from each other on the same chromosome that they sort independently.

What is an example of a dihybrid cross?

A good example of a dihybrid cross are the plant traits for height and seed color (for example, if Tt = height and Cc = color). If two true breeding plants are crossed, then the offspring would be as follows:

	TC	Tc	tC	tc
TC	TTCC	TTCc	TtCC	TtCc
Tc	TTCc	TTcc	TtCc	Ttcc
tC	TtCC	TtCc	ttCC	ttCc
tc	TtCc	Ttcc	ttCc	ttcc

What are phenotypes and genotypes?

The phenotype is the physical manifestation of the genotype and the polypeptides it codes for—for example, the rose. The color and shape of the petals (the phenotype) are the result of chemical reactions within the cells of each petal. Polypeptides synthesized from the directions encoded within the cell's genes (the genotype) are part of those reactions. Different versions of a gene will produce different polypeptides, which in turn will cause different molecular interactions and ultimately a different phenotype.

What is a hybrid?

A hybrid is produced as the offspring of two true breeding organisms of different strains (AA × aa). If the hybrids (Aa) are then mated, the result is a hybrid cross.

How does meiosis relate to Mendel's laws?

Mendel's law of segregation directly refers to meiosis. He had suggested that gametes contain "characters" that control specific traits, and these characters separate when gametes form. Gametes are formed by meiosis, in which the chromosome number is reduced by one half (from diploid to haploid). (For more about diploid, haploid, and meiosis, see the chapter "Cellular Basics.")

NATURAL SELECTION

Who was Charles Robert Darwin?

English naturalist Charles Robert Darwin (1809–1882) was the first to propose the theory of natural selection—an idea that revolutionized all aspects of natural science. Darwin was born into a family of physicians and planned to follow his father and grandfather in that profession. Unable to stand the sight of blood, he studied divinity at Cambridge and received a degree from the university in 1830.

Who was Erasmus Darwin?

Erasmus Darwin (1731–1802) was the grandfather of British naturalist Charles Robert Darwin (1809–1882). Erasmus was a physician, inventor, and natural scientist who published a book on his ideas (*Zoonomia*) between 1794 and 1796. The book contained poetic couplets describing Erasmus Darwin's ideas about science and, in particular, the evolution of life. Erasmus Darwin's hypothesis was that all the animals on the planet had their origin in a "vital spark" that set in motion life as we

A 1874 photo of Charles Darwin, long after he had become famous for his theory of natural selection.

know it. In his posthumously published book *Temple of Nature* (1803), Erasmus Darwin further speculated on a theory of evolution that included basic ideas on the unity of organic life and the importance of both sexual selection and the struggle for existence in the evolutionary process. Years later, in writing a biography about his grandfather, Charles Darwin acknowledged that his grandfather's ideas had influenced his thinking.

What were the *Beagle* voyages?

The HMS *Beagle* was a naval survey ship that left England in December 1831 to chart the coastal waters of Patagonia, Peru, and Chile—it carried onboard the not yet famous naturalist Charles Darwin. On the five-year voyage, Darwin's job as unpaid companion to the captain on board the *Beagle* allowed him time to satisfy his interests in natural history. On its way to Asia, the ship spent time in the Galapagos Islands off the coast of Ecuador; Darwin's observations of flora and fauna on those islands helped him to generate several ideas, especially his theory of natural selection. (For more about Darwin's writings, see "Further Reading.")

Who was Alfred Russel Wallace?

Alfred Russel Wallace (1823–1913) was an English naturalist whose work was presented with Charles Robert Darwin's (1809–1882) at the Linnaean Society of London in 1858. After extensive travels in the Amazon basin, Wallace independently came to the same conclusions as Darwin on the significance of natural selection in driving the diversification of species. Wallace also worked as a natural history specimen collector in Indonesia, and, like Darwin, also read the works of British cleric and scholar Thomas Malthus (1766–1834). During an attack of malaria in Indonesia, Wallace made the connection between the Malthusian concept of the struggle for existence and a mechanism for change within populations. From this, Wallace wrote the essay that was eventually presented with Darwin's work in 1858.

How did Thomas Malthus influence Charles Darwin?

Both Charles Darwin and Alfred Russel Wallace read the work of British cleric and scholar Thomas Malthus (1766–1834), who in 1798 had published *Essay on the Principle of Population*. In that book, Malthus argued that the reproductive rate of humans grows geometrically, far outpacing the available resources. This means that individuals must compete for a share of the resources in order to survive in what

Never heard of Alfred Russel Wallace? He was the English naturalist who independently came to the same conclusions as Darwin did about natural selection.

has become known as the "struggle for existence." Darwin and Wallace incorporated this idea as part of natural selection—that is, adaptations that made for a more successful competitor would be passed on to subsequent generations, leading to greater and greater efficiency.

What is the Darwin-Wallace theory?

The Darwin-Wallace theory can be summarized as follows: Species as a whole demonstrate descent with modification from common ancestors, and natural selection is the sum of the environmental forces that drive those modifications. The modifications or adaptations make the individuals in the population better suited to survival in their environment, more "fit" as it were.

Why is Wallace not as well known as Darwin?

While Darwin was well connected to the scientific establishment of the time, Wallace entered the scene somewhat later, so he was less well known. Although Darwin would become far more famous than Wallace in subsequent decades, Wallace became quite well known during his own time as a naturalist, writer, and lecturer—he was also honored with numerous awards for his work.

What is the significance of Charles Darwin's study of finches?

In his studies on the Galapagos Islands, Charles Robert Darwin observed patterns in animals and plants that suggested to him that species changed over time to produce new species. Darwin collected several species of finches; they were all similar, but each had developed beaks and bills specialized to catch food in a different way. Some species had heavy bills for cracking open tough seeds, while others had slender bills for catching insects; still another species used twigs to probe for insects in tree cavities. All the species resembled one species of South American finch. In fact, all the plants and animals of the Galapagos Islands were similar to those of the nearby coast of South America. Darwin felt that the simplest explanation for this similarity was that a few species of plants and animals from South America must have migrated to the Galapagos Islands, then changed as they adapted to the environment on their new home, eventually giving rise to many new species. These observations led to part of his theory on evolution—that species change over time in response to environmental challenges.

Why was the Darwin-Wallace theory so ridiculed?

The work of Darwin and Wallace generated controversy on at least two fronts. First, their theory directly countered Christian teaching on how species do not change and all organisms being special creations by God. Second, by presenting arguments for common descent, Darwin and Wallace showed that humans were related to other animals—apes in particular. This insulted those who felt that humans were unique and not part of the animal kingdom. However, it should be noted that as the work was read among scientists, it gained general acceptance in the late decades of the nineteenth century.

Who was "Darwin's bulldog"?

English biologist Thomas Huxley (1825–1895) was a staunch supporter of Charles Darwin's work; in fact, Huxley wrote a favorable review of Darwin's *On the Origin of Species* that appeared soon after its publication. When the firestorm of controversy began after the appearance of Darwin's book, Huxley was ready and able to defend Darwin, whose chronic public reticence about his theories was at that time exacerbated by illness. In 1860 at the British Association for the Advancement of Science, Huxley's defense of Darwin was so vigorous during a debate with English Bishop Samuel Wilberforce (1805–1873) that he earned the title "Darwin's bulldog."

What is the significance of Charles Darwin's *On the Origin of Species*?

Charles Darwin (1809–1882) first proposed a theory of evolution based on natural selection in his treatise *On the Origin of Species*—a publication that ushered in a new era in our thinking about the nature of man. In fact, it is said that the intellectual revolution this work caused, and the impact it had on man's concept of himself and the world, were greater than those by the works of English physicist Isaac Newton (1642–1727) and other individuals. The effect of the publication was immediate—the first edition sold out on the day of publication (November 24, 1859). *Origin* has been referred to as "the book that shook the world." Every modern discussion of man's future, the population explosion, the struggle for existence, the purpose of man and the universe, and man's place in nature rests on Darwin.

The work was a product of his analyses and interpretations of his findings from his voyages on the HMS Beagle. In Darwin's day, the prevailing explanation for organic diversity was the story of creation as told in the Bible's book of Genesis. Darwin's book was the first to present scientifically sound, well-organized evidence for the theory of evolution. The theory was based on natural selection, in which the best, or fittest, individuals survive more often than those who are less fit. If a difference exists in the genetic endowment among these individuals that correlates with fitness, the species will change over time and will eventually resemble more closely (as a group) the fittest individuals. It is a two-step process: the first consists of the production of variation and the second of the sorting of this variability by natural selection in which the favorable variations tend to be preserved.

What are the four postulates presented in Charles Darwin's *On the Origin of Species?*

The four postulates presented by Darwin in *On the Origin of Species by Means of Natural Selection, or the Preservation of Favoured Races in the Struggle for Life* (eventually shortened to *On the Origin of Species*) are as follows: 1) Individuals within species are variable; 2) Some of these variations are passed on to offspring; 3) In every generation, more offspring are produced than can survive; and 4) The survival and reproduction of individuals are not random; the individuals who survive and go on to reproduce the most are those with the most favorable variation, and they are naturally selected. It follows logically from these that the characteristics of the population will change with each subsequent generation until the population becomes distinctly different from the original; this process is known as evolution.

Who coined the phrase "survival of the fittest"?

Although frequently associated with Darwinism, this phrase was coined by Herbert Spencer (1820–1903), an English sociologist. It is the process by which organisms that are less well adapted to their environment tend to perish and better-adapted organisms tend to survive.

What is social Darwinism?

Social Darwinism is a social movement—one of a number of perversions of the Darwin-Wallace theory. These movements attempt to use evolutionary mechanisms as excuses for social change. Followers of social Darwinism believe that the "survival of the fittest" applies to socioeconomic environments as well as evolutionary ones. By this reasoning, the weak and the poor are "unfit" and should be allowed to die without societal intervention. This idea has nothing to do with Charles Darwin and Alfred Russel Wallace but was promoted by British philosopher Herbert Spencer (1820–1903) and is related to the works of Thomas Malthus, whose work did indeed inspire Darwin. Although social Darwinism has faded as a movement, it did help to spur the eugenics movement of Nazi Germany as well as a number of laws and policies in the United States in the twentieth century.

What is artificial selection?

Artificial selection is the selective breeding of organisms for a desired trait, such as breeding a rose plant to produce larger flowers or a chicken to lay more eggs. Darwin cited artificial selection as evidence that species are not immutable—that is, unable to be changed by selection.

What are the different types of natural selection?

Natural selection can cause a population to change in several ways. For example, natural selection can cause a trait to change in one direction only, such as when individuals within a population grow taller with each generation, and this is known as directional selection. Diversifying selection can cause the loss of individuals in the midrange of a trait. For example, if a certain prey species ranges in color from very dark to very light, those individuals in the midcolor range may not be able to hide from predators. The midcolor prey will then be selectively removed from the population, leaving the population with only two forms, the very dark and the very light. In stabilizing selection, those at either end of the range are removed more often, creating selection pressure for the midrange. Selection can also work on traits important to sexual reproduction; this is known as sexual selection.

Who was Ernst Haeckel?

Ernst Haeckel (1834–1919) was a physician who became a fervent evolutionist after reading Charles Darwin's *On the Origin of Species*, although he differed with Darwin over natural selection as the primary mode of evolution. Haeckel is best known for his

What does a peacock's tail have to do with sexual selection?

The elaborate tails of male peafowl indicate sexual selection—because that trait often represents successful breeding. The male feathers' large size and brilliant colors make it difficult for peacocks that spend most of their time on the ground to hide from predators (peahens have tails that are much smaller and not as colorful). But that is not the feathers' true "use"—the peacocks' tails are useful for attracting mates, and it's been demonstrated that the more eyespot feathers a male has in his tail, the more offspring he will father during mating season.

attempts to tie the stages of development to the stages of evolution ("ontogeny recapitulates phylogeny"); he thought that each stage of the developmental process was a depiction of an evolutionary ancestor. Haeckel is also credited with coining the terms "phylum," "phylogeny," and "ecology."

HIGHLIGHTS OF EVOLUTION

What is evolution?

Although it was originally defined in the nineteenth century as "descent with modification," evolution is currently described as the change in frequency of genetic traits (also known as the allelic frequency) within populations over time.

What were early ideas on evolution?

In the 1700s, "natural theology" (the explanation of life as the manifestation of the creator's plan) was a popular belief in Europe. This idea was the motive force behind the work of Swedish naturalist Carolus Linnaeus (Carl von Linné, 1707–1778), who was the first to classify all known living things by kingdom. Also popular prior to the work of Charles Darwin (1809–1882) were the theories of "special creation" (creationism), "blending inheritance" (that offspring were always the mixture of the traits of their two parents), and "acquired characteristics."

What is the "ladder of nature"?

Greek philosopher Aristotle (384–322 B.C.E.) tried to use logic to build a diagram of living things, but he realized that living things could not be so easily separated. Instead, he constructed a "ladder of nature" or "scale of perfection," which began with humans at one end and proceeded through animals and plants to minerals at the other end. Aristotle ranked these groups on a scale based on the four classical elements: fire, water, earth, and air.

What is Lamarckian evolution?

The French biologist Jean-Baptiste de Lamarck (1744–1829) is credited as the first person to propose a theory that attempts to explain how and why evolutionary change occurs in living organisms. The mechanism Lamarck proposed is known as "the inheritance of acquired characteristics," meaning that what individuals experience during their lifetime will be passed along to their offspring as genetic traits. This is sometimes referred to as the theory of "use and disuse." A classic example of this would be the giraffe's neck: Lamarckian evolution would predict that as giraffes stretch their necks to reach higher branches on trees, their necks grow longer. As a result, this increase in neck length will be transmitted to egg and sperm such that the offspring of giraffes whose necks have grown will also have long necks. While Lamarck's idea was analytically based on available data (giraffes have long necks and give birth to offspring with long necks as well), he did not know that, in general, environmental factors do not change genetic sequences in such a direct fashion.

Who disproved Lamarck's theory?

In the 1880s, the German biologist August Weismann (1834–1914) formulated the germ-plasm theory of inheritance. Weismann reasoned that reproductive cells (germ cells) were separate from the functional body cells (soma or somatic cells). Therefore, changes to the soma would not affect the germ-plasm and would not be passed on to the offspring. In order to prove that the disuse or loss of somatic structures would not affect the subsequent offspring, Weismann removed the tails of mice and then allowed them to breed. After twenty generations of this experimental protocol, he found that mice still grew tails of the same length as those who

A bronze statue of French biologist Jean-Baptiste de Lamarck can be seen at the Jardin des Plantes botanical garden in Paris, France.

had never been manipulated. This not only disproved Lamarck's theory of use and disuse, it also increased understanding of the new field of genetics.

Who was Comte de Buffon?

French naturalist Georges-Louis Leclerc, Comte de Buffon (1707–1788) was an early proponent of natural history, which is the study of plants and animals in their natural settings. He was also known for his work as a mathematician. Buffon was interested in the modes by which evolutionary change could occur. A prolific writer (his work *Natural History* comprised thirty-five volumes), he pondered the meaning of the term "species" and whether such groupings were immutable (unchanging) over time. In addition, he served as mentor to French naturalist Jean-Baptiste de Lamarck (1744–1829).

What was the Scopes (monkey) trial?

John T. Scopes (1900–1970), a high-school biology teacher, was brought to trial by the State of Tennessee in 1925 for teaching the theory of evolution. He challenged a law passed by the Tennessee legislature that made it unlawful to teach in any public school any theory that denied the divine creation of man. He was convicted and sentenced, but the decision was reversed later and the law repealed in 1967.

In the early twenty-first century, pressure against school boards still affects the teaching of evolution. Recent drives by anti-evolutionists either have tried to ban the teaching of evolution or have demanded "equal time" for "special creation" as described in the biblical book of Genesis. This has raised many questions about the separation of church and state, the teaching of controversial subjects in public schools, and the ability of scientists to communicate with the public. The gradual improvement of the fossil record, the result of comparative anatomy, and many other developments in biological science has contributed toward making evolutionary thinking more palatable.

What is gradualism?

The Darwin-Wallace theory of evolution is based on gradualism—the idea that speciation occurs by the gradual accumulation of new traits. This would allow one species to gradually evolve into a different-looking one over many, many generations, which is the scale of evolutionary time.

What is punctuated equilibrium?

Punctuated equilibrium is a model of macroevolution first detailed in 1972 by American biologist Niles Eldredge (1942–) and American paleontologist, evolutionary biologist, and historian of science Stephen Jay Gould (1941–2002). It can be considered either a rival or supplementary model to the more gradual-moving model of evolution posited by neo-Darwinism. The punctuated equilibrium model essentially asserts that most of geological history shows periods of little evolutionary change, followed by short (geologically speaking, a few million years) periods of rapid evolutionary change.

Gould and Eldredge's work has been helped by the discovery of the Hox genes that control embryonic development. Hox genes are found in all vertebrates and many other species as well; they control the placement of body parts in the developing embryo. Relatively minor mutations in these gene sequences could result in major body changes for species in a short period of time, thereby giving rise to new forms of organisms and therefore new species.

What scientific disciplines provide evidence for evolution?

Although information from any area of natural science is relevant to the study of evolution, several in particular directly support the work of Darwin and Wallace. Paleobiology, geology, and organic chemistry provide insight on how living organisms have evolved.

Scientists give skulls like these as strong evidence that humans have evolved over time in contrast to creationists, who believe that we have always existed in our current form.

Ecology, genetics, and molecular biology also demonstrate how living species are currently changing in response to their environments and therefore undergoing evolution.

What is scientific creationism?

Scientific creationism is an attempt to promote the teaching of creation theory in schools. By designating creationist theory "scientific," proponents hope to gain equal time with evolutionary theory in school curricula. Creationist theory proposes that species observed today are the result of "intelligent design" or special creation rather than the result of the effects of natural selection.

What is the value of fossils to the study of evolution?

Fossils are the preserved remains of once living organisms. The value of fossils comes not only from the information they give us about the structures of those animals, but the placement of common fossils in the geologic layers also gives researchers a method for dating other, lesser-known samples.

How do fossils form?

Fossils form rarely, since an organism is usually consumed totally or scattered by scavengers after death. If the structures remain intact, fossils can be preserved in amber (hardened tree sap), Siberian permafrost, dry caves, or rock. Rock fossils are the most common. In order to form a rock fossil, three things must happen: 1) the organism must be buried in sediment; 2) the hard structures must mineralize; and 3) the sediment surrounding the fossil must harden and become rock. Many rock fossils either erode before they can be discovered, or they remain in places inaccessible to scientists.

How are fossils dated?

Fossils are dated using two methods: Relative dating determines the age of the surrounding rock, giving an approximate age to the fossils therein based on their distance from the surface (older rocks are generally deeper from the surface). Other fossils within the rock can also be used to give an approximate date. The second method is absolute dating, based on the known rates of radioactive decay within rocks. By measuring the ratio between the radioactive forms of an element like uranium-238 to its nonradioactive, "decayed" form, scientists can determine when the rock formed.

What does species diversity have to do with evolution?

Species diversity is direct evidence that evolution has occurred. When species can be identified in which individuals share a number of significant traits while also having some unique adaptations, it is logical to assume that the common traits are the result of common origin while the unique ones demonstrate adaptive radiation. An example of the significance of species diversity would be Darwin's finches of the Galapagos Islands. While the species share a common body structure that is inherited from a common ancestor, each species also demonstrates variations in beak size and structure that are indicative of their adaptations to local environments and the type of food available.

What is coevolution?

Coevolution is a rare form of evolution. By definition, it requires that two species adapt to evolutionary changes occurring in each other. An example of this reciprocal adaptation would be the development of mutualistic relationships between plants and the insects that prey upon them. As the plant develops defenses (like the oils of species in the mustard family), the insects develop counter weapons (cabbage butterflies have metabolic adaptations that can safely break down these toxic compounds).

Why did sexual reproduction evolve?

The appearance and maintenance of traits like sexual reproduction will occur only if a net benefit exists to the fitness of individuals with those traits. Sexual reproduction comes with costs; for example, finding a mate can demand considerable time and energy. More importantly, individuals engaging in sexual reproduction only pass on half of their genes to each offspring, so their fitness is half that of asexually reproducing individuals. Scientists have examined the costs and benefits of sexual reproduction and determined that it is most likely to have evolved as a way to maintain genetic diversity. Experiments have shown that populations in erratic environments or those who are reproductively

What is industrial melanism?

Industrial melanism is the change in the coloration of species that occurs as a result of industrial pollution. Increased air pollution as a result of the Industrial Revolution in Great Britain during the eighteenth and nineteenth centuries led to an accumulation of soot on many structures, including tree trunks. As a result, organisms whose coloration allowed them to use the trees to hide from predators lost that advantage and were eaten more often by predators. A classic example of this was the peppered moth (*Biston betularia*), whose coloration is polymorphic. Prior to the Industrial Revolution, collection records indicate that the darker or melanistic form was almost unknown, but by 1895 it constituted about 98 percent of the moths collected. The two forms eventually reached a state of balanced polymorphism. Because the change in morphology could be directly linked to the change in industry, this process is described as industrial melanism.

isolated from the rest of their species are at an advantage when they reproduce sexually rather than asexually. By being able to mix and match alleles, individuals within these populations can maintain genetic diversity and phenotypic variation, expanding their tool kit as a hedge against an unpredictable future.

Is it possible to observe evolutionary change in other species?

Yes, the change within populations, which is driven by natural selection, is occurring around us constantly. However, whether we can actually observe those changes is a different matter. For example, elephants may be adapting to changing conditions in their environment, but since they tend to be extremely long-lived, it is unlikely that we will be able to follow enough generations to observe a trend. On the other hand, it is possible to notice such changes in populations of individuals with very short life spans, particularly when they are experimentally manipulated. For example, bacteria and guppies are just two species that have been observed evolving in response to changing environmental conditions.

EXTINCTION

What is extinction?

Extinction is the idea that some plants and animals that once inhabited the planet are no longer found in today's world. It has only been in the last 200 years that scientists have accepted the idea of extinction. In the seventeenth and eighteenth centuries, although scientists knew about fossils being the remains of plants and animals, the religious doctrines did not include the possibilities of organisms becoming extinct. They believed that the fossil animals and plants had not yet been discovered in such places as deep in the Amazon jungle or any other unexplored part of the globe.

What is a mass extinction?

A mass extinction—and many occurred over the over one billion years of life on Earth—occurs when a flora or fauna species suddenly (or gradually) dies out. This is usually seen in the fossil records. Mass extinctions are considered biological catastrophes because of the relative speed and range of their effects. The loss of so many species allows surviving populations to expand in numbers—they can change their habitat in new ways, adapting to new parts of the environment without facing competition from other species.

What mass extinctions have occurred throughout Earth's history?

Overall, about five major mass extinctions have occurred on Earth in the past around 450 million years. The following lists those extinctions (the dates are approximations):

- Around 438 million years ago, a mass extinction during the late Ordovician period killed off about 85 percent of the species on the planet
- Around 368 million years ago, during the late Devonian period, about 82 percent of the species on the planet were killed off
- Around 245 million years ago, between the Permian and Triassic periods, about 96 percent of all species were killed off—and is considered one of the most violent extinctions known
- Around 210 million years ago, between the Triassic and Jurassic periods, about 76 percent of the species on Earth were killed off
- About 65 million years ago, between the Cretaceous and Tertiary periods, about 76 percent of the species were killed off

What causes a mass extinction?

Scientists have discovered many reasons for mass extinctions, depending on when the extinctions occurred. Several theories for such extinction include the disease theory (that species died out because of some disease); overpopulation (that could spread disease and increase competition for food); impact theory (that a comet or asteroid struck the Earth, changing the climate); volcano theory (volcanic eruptions that changed the climate); changes in the Sun's output (again, changing the climate); and even a killer cloud theory (that the Earth encountered a cosmic cloud that increased species' exposure to radiation). A human theory also exists, but most mass extinctions occurred before what we call "humans" were even around on the planet.

What is one of the most famous mass extinctions and why?

One of the most famous mass extinctions occurred around sixty-five million years ago, between the Cretaceous and Tertiary periods: the demise of about 76 percent of species—

including the dinosaurs. Speculations over time as to the reason included a change in the climate due to volcanic eruptions or a difference in output from the Sun, making it difficult for the dinosaurs to adapt and survive.

The most "agreed-upon" solution for the dinosaur demise is the impact theory: objects from space collided with the Earth and changed the climate. Scientists know of impact craters on the Earth—evidence that comets and asteroids have struck our planet in the past. The one impact crater they are pointing to the most is found around the Yucatan peninsula in Mexico—the Chicxulub—an impact crater that dates to around the demise of the dinosaurs. Although it may only be partially responsible, scientists speculate that the impact would throw enough dust and debris high into the atmosphere to filter or block out the sunlight; the result would be a change in the worldwide climate. Vegetation would change, as would the normal habitats of the creatures. Eventually the food chain would collapse—as would the dinosaur species.

What were the dinosaurs?

The word dinosaur (meaning "fearful lizard") was first used by English paleontologist and comparative anatomist Richard Owen (1804–1892) to describe a group of large extinct reptiles. Today, they are considered an extremely diverse group of animals of the clade (a type of classification) Dinosauria that first appeared about 230 million years ago, dominated the land (no dinosaurs lived in water or in the air) for about 135 million years, and died out "suddenly" (in terms of geologic time) about sixty-five million years ago.

What scientists proposed the impact theory of dinosaur extinction?

Many scientists argue that a single disastrous event caused the extinction not only of the dinosaurs, but also of a large number of other species that coexisted with them. In 1980, the American physicist Luis Alvarez (1911–1988) and his geologist son Walter Alvarez (1940–) proposed that a large comet or meteoroid struck Earth sixty-five million years ago. They pointed out that the sediments at the boundary between the Cretaceous and Tertiary periods contain a high concentration of the element iridium—an element that is rare on Earth, but not in space. This iridium anomaly has since been discovered at more than fifty sites around the world.

In 1990, tiny glass fragments, which could have been caused by the extreme heat of an impact, were identified in Haiti. A 110-mile (177-kilometer) wide crater in the Yucatan Peninsula, long covered by sediments, has been dated to 64.98 million years ago, making it a leading candidate for the site of this impact. A hit by a large extraterrestrial object, perhaps as much as 6 miles (9.3 kilometers) wide, would have had a catastrophic effect upon the world's climate. Huge amounts of dust and debris would have been thrown into the atmosphere, reducing the amount of sunlight reaching the surface. Heat from the blast may also have caused large forest fires, which would have added smoke and ash to the air. Lack of sunlight would kill off plants and have a dominolike effect on other organisms in the food chain—including the dinosaurs.

How do biologists predict whether a species will become extinct?

In order to predict whether a species in a particular habitat will become extinct, biologists use a method called population viability analysis (PVA). Computer modeling generates PVAs using the species' life history data, genetic variability, and a population's response to environmental conditions—especially disturbances—to predict viability of a species.

How did the dodo bird become extinct?

A bird called the dodo—native to the Mascarene Islands in the Central Indian Ocean—became totally extinct around 1800. (They became extinct on the island of Mauritius soon after 1680, on Reunion about 1750, and on Rodriguez about 1800.) Thousands were slaughtered for meat, but pigs and monkeys destroyed the dodo eggs and were probably most responsible for the dodo's extinction.

SPECIES AND POPULATION

What is a species?

A species can be defined in several ways, and scientists will use different definitions depending on whether they are referring to a fossil (extinct) species or a living (extant) one. For example, an extant species can be defined as all the individuals of all the populations capable of interbreeding. A group of populations that are evolutionarily distinct from all other populations may also be defined as a species, even if they are incapable of interbreeding due to extinction.

What are a subspecies and a strain?

A subspecies is another way of describing a distinct population or variety. This term is used to describe the generation of hybrids that can occur when two different populations meet and interbreed. A strain or variety is a subcategory of a species. For example, Gregor Mendel's (1822–1884) work with garden peas involved various strains; one strain had purple flowers while another had white.

What do homology and analogy mean in terms of species?

Homology is the similarity in traits between two species that indicate their common ancestry. For example, the general characteristics of cheetahs, lions, tigers, and house cats are whiskers, retractable claws, tooth structure, and so forth. These similarities indicate that each of these traits was inherited from a feline ancestor. To an evolutionist, an analogous structure is one that looks similar or has the same purpose but is definitely not the result of common inheritance. For example, bats and birds both use wings to fly, and the wings have the same general shape (thin but broad in width). However, the structures were not inherited from the same ancestors. Bats were four-legged mammals before their front limbs became modified for flight, while birds are not descended from mammals at all. Scientists can determine whether a trait is homologous or analogous by comparing it in species thought to be of common origin and contrasting it to traits of unrelated species in similar habitats.

What is phylogeny?

Phylogeny is the evolutionary history of a group of species. This history is often displayed as a phylogenetic tree, in which individual species or groups of species are listed at the end of the branches. The points where the branches join indicate common ancestors.

What is speciation?

Speciation is the process by which new species are formed. This occurs when populations become separated from the rest of the species. At this point, the isolated group will respond independently to natural selection until the population becomes reproductively isolated. The group is then considered a new species. If a population becomes reproductively isolated, then individuals within the population will no longer exchange genetic material with the rest of the species. At that point environmental factors (for example, natural selection) will work on the genetic variation within that population until it has become a new species.

What is a population?

A population is a group composed of all members of the same species that live in a specific geographical area at a particular time. An example of a population might include all the gray squirrels that live in a certain urban park. The areas occupied by a population could include the small area (measured in square millimeters) occupied by bacteria in a rotting apple to the vast areas of ocean (square kilometers) that include the territory of migrating sperm whales. Population ecology is the branch of ecology that studies the structure and changes within a population. Studies of specific populations will indicate the dynamics of the population, in terms of active, ongoing growth; declining growth; or stability.

Who was the first person to study populations mathematically?

In 1798, British cleric and scholar Thomas Robert Malthus (1766–1834) attempted to inform people that the human population, like any other population, had the potential to increase exponentially. Malthusian ideas were not well received, as he predicted the rate of population growth would exceed the ability of the land to produce food. His work was later used by Charles Darwin to explain his theory of natural selection.

Can scientists predict how populations will grow?

Yes, using mathematical models scientists can predict growth when populations are growing at their maximum rates. This happens in two distinct patterns: logistic or exponential growth. In logistic population growth, the population grows in cycles, responding to limiting factors in the environment. An example would be a population of insects that is limited by the amount of available food. Exponential growth is growth at a constant rate of increase per unit of time and is used to model continuous population growth in an unlimited environment. An example of exponential growth would be the doubling rate of bacterial growth on the turkey left unrefrigerated after Thanksgiving dinner.

How can a population become reproductively isolated?

Reproductive isolation means that individuals of one population are unable to exchange gene sequences (eggs and sperm) with individuals from another. This means that natural selection will work on the isolated population independently from the rest of the species, therefore increasing the likelihood that those isolated organisms will become a separate species. Methods by which this can occur include geographic isolation, habitat isolation, and temporal isolation. In other words, two populations can become physically separated by a barrier like an ocean or mountain range; they can use different parts of the same habitat (birds that visit only the tops of trees as opposed to the lower branches); or they may be active at different times—for example, nocturnal and diurnal insects.

What is adaptive radiation?

As populations move into new environments and adapt to those local conditions, diversity increases. This splitting creates a divergence from the original population. When diagrammed on paper, the new populations appear to be radiating outward from the original like the spokes of a wheel.

What did Hugo de Vries discover about species populations?

Dutch botanist Hugo de Vries (1848–1935) discovered a way in which a population could become a separate species while still sharing the same environment with other members of the species. The process, known as sympatric speciation, occurs almost exclusively in plants rather than in animals and involves a series of rare genetic accidents that can occur during the formation of gametes (eggs and sperm). As a result, gametes are formed as polyploids—that is, they have extra copies of each chromosome and thus are unable to match their chromosomes to others of the same species. Since these poly-

---

What separates humans from other primates?

Morphologic and molecular studies suggest that our closest living relative species is the chimpanzee, although some of the evidence is conflicting. We do know that analysis of protein structure in both chimps and humans shows that approximately 98 percent of our gene sequences are functionally identical, meaning that if the gene sequences differ it is not enough to radically change the proteins produced from them. It is also estimated that our last common ancestor with the chimp would have lived at least five million years ago.

ploids are forced to mate only with other polyploids in the population, they are reproductively isolated and considered a new species.

What is cladogenesis and anagenesis?

Cladogenesis is the formation of a group of species that share a common ancestor. Cladogenesis can occur as a result of adaptive radiation, which is the divergence or splitting of one species into several. When a species gradually changes over time to the extent that it becomes a "new" species but does not give rise to additional species (no divergence), this is described as anagenesis.

What is balanced polymorphism?

When a trait exists in several forms within a population, it is said to be polymorphic. Polymorphisms that maintain a stable distribution within the population over generations are known as balanced polymorphisms. Balanced polymorphism can be maintained if heterozygotes (mixtures of two types) have a fitness advantage. When this occurs, both types of alleles are maintained in the population. Strangely enough, a classic example of this is sickle cell anemia and malaria: Individuals who are heterozygous (Hh) are resistant to malaria, dominant homozygotes (HH) are susceptible to malaria, and recessive homozygotes (hh) have sickle cell anemia. Because those who have both types of alleles and who live in malaria-prone regions are the most likely to survive long enough to produce children, both types are maintained in the population at a relatively stable rate.

What are Müllerian and Batesian mimicry?

In 1878, Fritz Müller (1821–1897), a German-born zoologist, described a phenomenon in which a group of species with the same adaptations against predation was also of similar appearance. This phenomenon is now called Müllerian mimicry; such mimics include wasps and bees, all of which have similar yellow-and-black-striped patterns that serve as a warning to potential predators. In 1861, Henry Walter Bates (1825–1892), a British naturalist, proposed that a nontoxic species can evolve (especially in color and color pattern) to look or act like a toxic or unpalatable species in order to avoid being eaten by a predator. The classic example is the viceroy butterfly, which resembles the unpalatable monarch butterfly—an adaptation called Batesian mimicry. However, it has more recently been proven that the viceroy is actually a Müllerian mimic, since it seems to be unpalatable to birds, too. Bates was also a col-

The viceroy butterfly was long considered an example of Batesian mimicry (its colors being similar to the toxic monarch butterfly). More recently, however, it appears to be a Müllerian mimic.

Why are dust mites examples of reverse evolution?

Scientists recently did a study of house dust mites and found that these members of the arachnids (related to spiders) demonstrated reverse evolution. Although Dollo's law states that evolution is not reversible, genetic studies have shown that house dust mites—little mites that thrive in our mattresses, rugs, and sheets, or any place where we shed dead skin cells and are known as one of the most diverse animals on Earth—evolved from parasites. These parasites, in turn, evolved from free-living organisms millions of years ago. Now dust mites are back to their roots—literally—and are once again free-living organisms.

league of Alfred Russel Wallace, and it was Bates who introduced Wallace to botany and field collecting of animals and plants (for more about Wallace, see above).

What are convergent and divergent evolution?

Convergent evolution occurs when diverse species develop similar adaptations in response to the same environmental pressure. For example, dolphins and sharks are descended from different ancestors, but as a result of sharing an aquatic environment, they have similar adaptations in body shape. When two species move away from the traits that they share with a common ancestor as they adapt to their own environments, the result is called divergent evolution. As an example, imagine the diversity among bird species. Ducks, hummingbirds, ostriches, and penguins are all descended from an ancestral bird species, yet they have all diverged as they adapted to their particular environments.

What are microevolution and macroevolution?

Microevolution is the change in allelic frequencies that occurs at the level of the population or species. When individuals with certain traits are more successful at reproduction, the ensuing generation will have more copies of that trait. Should the trend continue, eventually the traits will become so common in the population that the population profile will change. This is microevolution. Macroevolution is large-scale change that can generate entire new groups of related species, also known as a clade. One example would be the movement of plants onto land; all terrestrial plants are descended from that event, which occurred during the Devonian period about 400 million years ago.

What is the bottleneck effect?

The bottleneck effect is a term used to describe a population that has undergone some kind of temporary restriction that has severely reduced its genetic diversity. A bottlenecking event could be an epidemic or natural disaster like fire or flood. It is hypothesized that the lack of genetic diversity among African cheetahs is due to some bottlenecking event in the species' past.

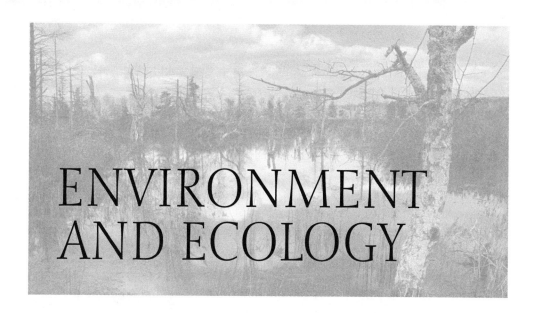

ENVIRONMENT AND ECOLOGY

THE EARTH'S ENVIRONMENT

What is the Gaia Hypothesis?

The Gaia Hypothesis (or Gaia Theory or Gaia Principle)—named for Gaia, the Earth goddess of ancient Greece—is a theory that all organisms coevolve and interact with their nonliving surroundings to create a complex system that helps maintain our planet. It was first proposed in 1974 by English environmentalist James Lovelock (1919–) and American biologist Lynn [Alexander] Margulis (1938–2011; she was also known for her theory on the origin of eukaryotic organelles). Precursors to the theory proposed that the world was a single living organism capable of self-maintenance and regulation—an idea that was eventually expanded upon by Lovelock. Many scientists have regarded the Gaia Hypothesis as a useful analogy, but a difficult theory to test scientifically; however, it has also helped to expand ideas in several fields, including biogeochemistry and geophysiology.

Why does the Earth have seasons?

Although many people believe that the Earth has seasons depending on our planet's distance from the Sun, that is not the case. The Earth's four seasons are due to one main factor: the Earth's inclination on its axis (23.5°) in relation to the plane of the Sun. For example, when the geographic North Pole is tilted toward the Sun, it is summer in the Northern Hemisphere and winter in the Southern Hemisphere; when the direct rays of the Sun are overhead at the Equator, and day and night are of equal length in the Northern and Southern Hemispheres, it is called the equinox—both of which occur in the respective spring (vernal equinox) and fall (autumnal equinox) of both hemispheres. And finally, when the geographic North Pole is pointed away from the Sun (at the 23.5° angle), it is winter in the Northern Hemisphere and summer in the Southern Hemisphere.

If the Earth is closest to the Sun in January, why is it so cold in the Northern Hemisphere at that time?

While the Northern Hemisphere is experiencing cold in January—and the Southern Hemisphere its warm summer—the Earth is closest to the Sun in its orbit. When the planet is farther from the Sun—and although it seems contrary to what we would think—the Northern Hemisphere is warmer. Scientists know that the average temperature increases about 4°F (2.3°C) when the Earth is farther from the Sun (the Northern Hemisphere summer), even though the sunlight is less intense from that distance. This is not only because of the tilt of our planet's axis (see above), but also because of a characteristic of our planet—the distribution of the continents and oceans. In particu-

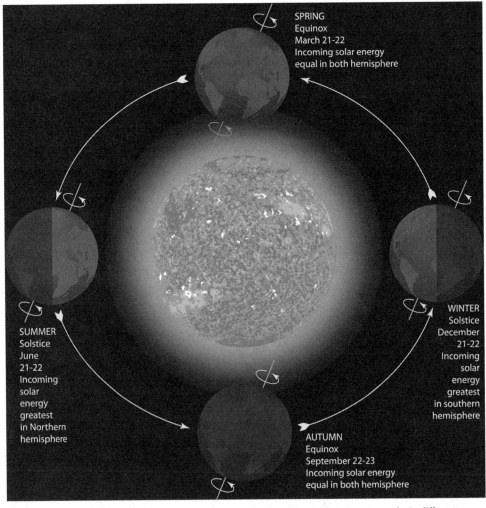

SPRING
Equinox
March 21-22
Incoming solar energy
equal in both hemisphere

WINTER
Solstice
December
21-22
Incoming
solar
energy
greatest
in southern
hemisphere

SUMMER
Solstice
June
21-22
Incoming
solar
energy
greatest
in Northern
hemisphere

AUTUMN
Equinox
September 22-23
Incoming solar energy
equal in both hemisphere

This diagram, which is obviously not to scale, shows how the tilt of the earth on its axis results in different seasons as our planet orbits the sun over the course of a year.

lar, the Northern Hemisphere contains more land, while the Southern Hemisphere has more oceans. Heating up the land is easier than heating the oceans; thus, when it's January, it's more difficult for the Sun to heat up the southern oceans—and we experience overall colder temperatures even though our star is closer.

Another factor when it comes to warmer Northern Hemisphere temperatures versus the Southern Hemisphere is that the duration of the summers in each hemisphere differs as the planet orbits the Sun. Following Johann Kepler's (1571–1630) second law of planetary motion, a planet moves more slowly when it is farther away from the Sun (called aphelion) than when it is close (perihelion). In fact, the Northern Hemisphere's summer is two to three days longer when the planet is farther from the Sun, giving the Sun more time to bake the continents.

What is a climate, and how is it characterized?

Climate refers to the long-term weather conditions of a region, based on long-term averaging of temperature. Climates often undergo cyclic changes over decades, centuries, and millennia, but it is difficult to predict future climate changes. A climate diagram summarizes seasonal variation in temperature, precipitation, length of wet/dry seasons, and portion of the year spent in specific temperature ranges. Weather and climate are important because they are the determining factors of biomes and ecosystems.

What is a microclimate?

When you notice that the temperature forecast in your local media is consistently warmer or colder than that which occurs in your neighborhood, you have identified a microclimate. Light, temperature, and moisture may all vary from one area to another because of changes in altitude, vegetation, or other factors. One good example may be in your own yard—in some areas, plants may grow better or faster if they are partially hidden from such extreme weather elements such as wind or snow and may experience more heat from the Sun because of their location, such as along a brick wall exposed to sunlight. In this case, that part of your yard would be called a microclimate.

What is an El Niño event?

Along the west coast of South America in the Equatorial Pacific Ocean, near the end of the calendar year, an unusually warm current of nutrient-poor tropical water replaces the cold, nutrient-rich surface water. It does not occur every year, but when it does—and because this condition frequently occurs around Christmas—local residents call it El Niño (Spanish for "child," referring to the Christ child). Scientists now know that the El Niño is normally accompanied by a change in atmospheric circulation called the Southern Oscillation. Together, the ENSO (El Niño-Southern Oscillation) phenomenon is one of the main sources of annual variability in weather and climate around the world.

The Walker circulation (also called the Walker cell)—in which air and water flows from west to east in the tropical Pacific Ocean due to differences in high and low surface pressures—also contributes to the formation of an El Niño. When the Walker circulation is normal, it is usually warm, and there is wet weather in the western Pacific and cool and dry weather in the eastern Pacific. But when the Walker cell weakens or reverses every few years, the winds weaken, and the warmer-than-usual Pacific waters flow to the east, creating an El Niño. In contrast, during a La Niña event (see below),the Walker circulation becomes stronger, the winds increase, and cooler waters rise because of upwelling to the east.

What is a La Niña event?

La Niña (or "little girl," and often called El Viejo, anti-El Niño, or simply "a cold event" or "a cold episode") is characterized by unusually cold ocean temperatures in the Equatorial Pacific Ocean, compared to El Niño that is characterized by unusually warm ocean temperatures (see above). El

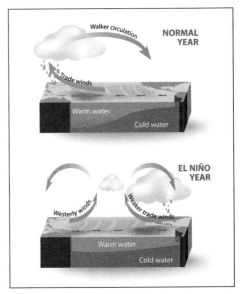

A high pressure system over the Pacific Ocean results in a variation of pressure that causes a Walker circulation (named after British statistician and physicist Sir Gilbert Walker [1868–1958]). When the Walker circulation weakens, an El Niño results. A stronger Walker circulation brings about a La Niña.

Niño and La Niña events tend to alternate about every three to seven years, but the time from one event to the next can vary from one to ten years.

What are some atmospheric and environmental effects of El Niño and La Niña?

Weather-wise, El Niño events can cause severe weather events, including such extremes as flooding in the U.S. Southeast and California, ice storms in the Northeast, and tornado outbreaks in Florida. During a La Niña year, the climate in various places around the world can change; for example, during a La Niña event, the average winter temperatures are warmer than normal in the Southeast United States and cooler than normal in the Northwest—just the opposite as what happens during El Niño years, in which temperatures in the winter are warmer than normal in the North Central states and cooler than normal in the Southeast and the Southwest.

The environment also suffers during a severe El Niño. Large numbers of fish and marine plants may die; the decomposing dead material depletes the water's oxygen supply, leading to the bacterial production of huge amounts of smelly hydrogen sulfide. A greatly reduced fish (especially anchovy) harvest affects the world's fishmeal supply, leading to higher prices for poultry and other animals that normally are fed fishmeal. Anchovies and sardines are also major food sources for marine mammals such as sea lions and seals; when the food source is in short supply, not only do many sea lions and seals starve, but also a large proportion of the infant animals die.

What is the hydrologic cycle?

The hydrologic cycle takes place in the hydrosphere, which is the region containing all the water in the atmosphere and the Earth's surface. Simply put, it involves five phases: condensation, infiltration, runoff, evaporation, and precipitation. Solar energy drives winds that evaporate water from the surface of the oceans. The water vapor cools and condenses as it rises and then falls to the ground as rain, snow, or some other form of precipitation—rain, snow, hail, etc., which then infiltrates into the ground (replenishing ground water) or runs off into streams, creeks, rivers, and oceans. From there, it evaporates again, thus completing the cycle.

What is the carbon cycle?

To survive, every organism must have access to carbon atoms, as carbon makes up about 49 percent of the dry weight of organisms. In general, the carbon cycle includes movement of carbon from the gaseous phase (carbon dioxide [CO_2] in the atmosphere) to solid phase (carbon-containing compounds in living organisms) and then back to the atmosphere via decomposers. For example, the atmosphere is the largest reservoir of carbon, containing 32 percent CO_2, and gets much of its "carbon cycling" from plants: The process of photosynthesis removes CO_2 from the atmosphere, and cell respiration returns CO_2 to the atmosphere.

How do protists fit into the global carbon cycle?

Protists—organisms that are found in nearly all aquatic and moist environments—are important to the global carbon cycle. In particular, just the sheer numbers of protists affect the carbon percents. For example, protists living in the open ocean, in algal beds and reefs, and in swamps and marshes—including algae, diatoms, and dinoflagellates—form part of the basis for major food webs in aquatic environments and are one of the greatest contributors to the global carbon cycle. (For more about protists, see the chapter "Bacteria, Viruses, and Protists.")

How are plants associated with the nitrogen cycle?

The primary way that plants obtain nitrogen compounds is through the nitrogen cycle—a series of reactions involving several different types of bacteria, including nitrogen-fixing bacteria and denitrifying bacteria. During nitrogen fixation, symbiotic bacteria that live in association with the roots of certain plants (such as legumes like peas) are able, through a series of enzyme reactions, to make nitrogen available for plants. Nitrogen is crucial to all organisms because it is an important element of proteins and nucleic acids. (For more about plants, see the chapter "Plant Structure, Function, and Use.")

What is ozone, and does it benefit life on Earth?

Ozone, a form of oxygen with three atoms instead of the normal two, is beneficial to all living organisms, shielding and filtering the Earth (and its organisms) from excessive, damaging, and life-threatening ultraviolet (UV) radiation generated by the Sun. But the "dark side" to ozone is that near ground level, ozone is a major pollutant that helps form photochemical smog and acid rain. In fact, ozone in this form is highly toxic—less than one part per million of this blue-tinged gas is poisonous to humans. For this reason, especially in big cities, the weather bureau often issues an ozone alert for people with upper-respiratory health problems (or for everyone if the concentration of ozone is high enough)—in particular, if an air mass that can hold the ozone in the lower atmosphere is stagnant.

What is an "ozone hole"?

In Earth's upper atmosphere (stratosphere), what is called the ozone layer contains about 90 percent of the planet's ozone. Since the 1980s, scientists measuring the ozone layer over the North (Arctic) and South (Antarctic) Poles have noted "holes," or areas that do not contain as much ozone. It is thought that the holes are caused mainly by chlorine from human-produced chemicals, and the chlorine levels are high in the poles' stratospheres. In particular, the three main ingredients for ozone destruction are chlorine from human-made chlorofluorocarbons (CFCs), very cold temperatures, and sunlight—all of which are found at both poles.

The fear has always been if either of the poles' holes became larger or more persistent—thus diminishing or depleting the planet's ozone layer—it could lead to increased

What are some differences between the Antarctic and Arctic ozone holes?

The Antarctic and Arctic ozone holes are slightly different (and these conditions can change depending on the conditions during a specific year). The first ozone "hole" over the Antarctic was discovered in 1985; in 2002, the hole split into two distinct holes, but by 2012, the hole had decreased to the second smallest in twenty years. Most scientists believe this is because of the warmer temperatures (the global average temperatures have gone up) in the Antarctic's lower stratosphere, the layer above the troposphere where humans live.

The Arctic ozone hole tends to be milder and shorter-lived than the one over the Antarctic, but the ozone levels are still often significantly higher than at the South Pole. The Arctic hole is found in a region of the so-called polar vortex—a place in which fast-blowing circular winds intensify in the fall, isolate an air mass within the vortex of the winds, and keep the area very cold. In most years, atmospheric waves push the vortex to lower latitudes in the late winter and break it up, unlike the Antarctic hole, which is much more stable and lasting until midspring. And "good news" may be in the far future for the ozone layer above the Arctic: according to a study in 2013, some scientists believe the ozone layer over the North Pole should recover to its original nonhole state by around 2100.

health problems for humans, such as skin cancer, cataracts, and weakened immune systems. In addition, an increase in ultraviolet radiation could reduce crop yield and disrupt aquatic ecosystems, especially affecting the marine food chain in the upper layers of the oceans.

What is the greenhouse effect in terms of the Earth's atmosphere?

The greenhouse effect is a warming of the Earth's atmosphere that results when gases in our atmosphere trap the Sun's heat. This effect was first described by Irish physicist John Tyndall (1820–1893) in 1861; it was given the greenhouse analogy much later, in 1896, by the Swedish chemist Svante Arrhenius (1859–1927). Simply put, greenhouse gases, such as methane, water vapor, and carbon dioxide, absorb some radiated heat (infrared radiation) from the ground that passes through the atmosphere toward space; the heat is then reradiated back to the Earth, keeping us warm. The greenhouse effect is not bad—it's what makes Earth habitable. Without the presence of the greenhouse gases, too much heat would escape, and Earth would be too cold to sustain life. It is only when an excess amount of greenhouse gases is present in the atmosphere that an increase in global temperatures occurs—and thus, as we've seen in the past decades—changes in the Earth's climate.

How has the greenhouse effect affected the Earth's atmosphere in the past century?

Various independent historical measurements conclude that the global average surface temperature has increased by about 3.3°F (0.5°C) over the past one hundred years. Whether the rise in the Earth's average surface temperature is due to human or natural increases in carbon dioxide output is a controversial topic, although more scientists are leaning toward human-driven causes. Cited for contributing to increased atmospheric carbon dioxide includes the burning of fossil fuels, volcanic activity, destruction of the rain forests, use of aerosols, and an increase in agricultural activity.

What is global climate change?

Global climate change is just as the phrase implies: the change in the global climate. The Earth has been through numerous global climate changes in the past; for example, the planet has had numerous periods of global temperate climates (for example, glaciations) followed by warming. These changes in the climate have always been attributed to natural causes: for example, a rise in temperatures caused by intense volcanic activity or changes in climate due to asteroids or comets striking the surface, throwing material into the atmosphere, changing the intensity of sunlight, thus altering the climate. Modern global climate change, according to what seems to be the majority of scientists, is due to human-generated greenhouse gases. Since the Industrial Revolution, the increase in greenhouse gases is thought to have caused an increase in global surface temperatures, raised worldwide sea levels, and changed the habitats of various organisms—not to mention caused extinctions—in its wake.

How could global climate change affect organisms in the future?

Many scientists believe that global climate change has already affected organisms on Earth—and will continue to do so in the near and far future. For example, a warmer world would cause the expansion of certain groups that spread in warmer climates, such as certain insects, algae, and nematodes; such changes could potentially change the landscape of infectious diseases. New species—from either mutations or those that thrive best in a warmer climate—could essentially introduce, or even reintroduce, harmful parasites into the world—ones that could affect not only humans, but other organisms, such as the beneficial insects that pollinate plants. New diseases—either from other organisms or more resistant bacteria and viruses—could also wreck havoc on organisms worldwide.

Why study global climate change?

The main reason for scientific study of global climate change is obvious: The impact on climate changes on organisms—in the oceans, on land, and in between—can (and will) be enormous. This includes changes in organism diversity and habitats; changes in vegetation (for example, areas once good for growing experiencing extended droughts); and other environmental factors that will create competition between organisms for food or shelter, or even becoming extinct. In addition, because the population is over seven billion (to date) and over 50 percent of humans live near the oceans, the rise in ocean waters will be nothing less than deleterious. So far, the biggest question eludes most scientists: How long will it take before we reach some of these changes—or are we already at those thresholds of climate change?

How fast is the Arctic sea ice disappearing due to global climate change?

One of the major indicators of global climate change is thought to be the rapid Arctic sea ice loss, or the opening of the sea ice above the Arctic Circle to the North Pole. Although no one knows when the area will be an ice-free Arctic summer, several theories exist. The first is based on observed sea ice trends—how the ice has decreased rapidly over the previous decade. Using these trends, it appears the area will be ice-free by 2020. Another theory is based on the assumption that in the future multiple, but random in time, large sea ice loss events will occur, as happened in 2007 and 2012. Using this idea, it is estimated that the area will be ice-free by 2030. And yet another theory is based on computer models, using not only sea ice date, but other factors of the climate, such as ocean, land, atmosphere (especially the amount of greenhouse gases), and ice conditions in the past. Most of these models estimate that the area would be ice-free by 2060. Whatever theory a person ascribes to, the reality is that almost all scientists who study such sea ice changes believe the Arctic will be ice-free within the next fifty years.

BIOMES

What is biogeography?

Biogeography is the study of the distribution, both current and past, of individual species in specific environments. One of the first biogeographers was Swedish naturalist Carolus Linnaeus (born Carl von Linné, 1707–1778), who studied the distribution of plants. Biogeography specifically addresses the questions of evolution, extinction, and dispersal of organisms in specific ecosystems. (For more about biogeography and species, see the chapter "Heredity, Natural Selection, and Evolution.")

What are the general characteristics of biomes?

A biome is a one of the world's prominent ecosystems, characterized by both vegetation and organisms particularly adapted to that environment. The following summarizes the major types of biomes:

Biome	Temperature	Precipitation	Vegetation	Animals
Arctic	-40–18°C (-40–64°F)	Dry season, wet season	Shrubs, tundra grasses, lichens, mosses	Birds, insects, mammals
Deciduous forest	Warm summers, cold winters	Low, distributed throughout year	Trees, shrubs, herbs, lichens, mosses	Mammals, birds, insects, reptiles
Desert	Hottest; great daily range	Driest, <10" (25 cm) rain per year	Trees, shrubs, succulents, forbs	Birds, small mammals, reptiles
Taiga	Cold winters, cool summers	Moderate	Evergreens, tamarack	Birds, mammals
Tropical	Hot short dry season	Wet season,	Trees, vines, Rainforest epiphytes, fungi	Small mammals, birds, insects
Tropical	Hot	Wet season, dry season	Tall grasses shrubs, trees, Savannah	Large mammals, birds, reptiles
Temperate	Warm summers, cold winters	Seasonal drought, occasional fires	Tall grasses grassland	Large mammals, birds, reptiles

What are bryophytes, and why are they important to ecosystems?

Bryophytes are a group of green land plants and include liverworts, hornworts, and mosses. Scientists know that bryophytes were one of the first plants to take root on land and are still important: They are also the second largest group of land plants after flow-

A rainforest is one type of biome, a prominent ecosystem to which animals and plants have adapted to certain climatic and geographical characteristics. Rainforests are areas of rich vegetation that receive between 98 and 180 inches (250 to 450 centimeters) of rain annually.

ering plants. Overall, they constitute a major part of the biodiversity in moist forests, wetlands, mountains, and even tundra ecosystems. Thus, all three types also play a significant part in the global carbon budget and carbon dioxide exchange, nutrient cycling, and water retention, to name just a few. They are also important environmental indicators and are even used as predictors and prognosticators of past—and now future—global climate changes. (For more about bryophytes, see the chapter "Plant Diversity.")

What is a wetland?

A wetland is an area that is covered by water for at least part of the year and has characteristic soils and water-tolerant plants. Examples of wetlands and their typical features include swamps (with tree species such as willow, cypress, and mangrove); marshes (with grasses such as cattails, reeds, and wild rice); bogs (with floating vegetation, including mosses and cranberries); and estuaries (specially adapted flora and fauna, such as crustaceans, grasses, and special types of fish that adapt, many of which adapt to the freshwater-saltwater environment).

What is an estuary?

Estuaries are wetlands in which freshwater streams and rivers meet the sea. Because of the mixing of freshwater and ocean waters, the salinity in an estuary is less than that of the open ocean, but greater than that of a typical river. Thus, organisms living in or near estuaries have special adaptations, for example, invertebrates such as clams, shrimps, and crabs, as well as fish such as striped bass, mullet, and menhaden. Unfortunately, es-

Wetlands, including bogs, swamps, marshes, and estuaries, are important for preventing flooding, purifying water, reducing erosion, and providing food and habitat for plants and wildlife.

tuaries are also popular locations for human developments and businesses. Contamination from shipping, household pollutants, and power plants (that is easily carried to the sea via the estuary) continually threaten the ecological health of many estuaries.

How do cattails help an estuary?

Scientists have long searched for a way to clean the pollutants in estuaries and other wetlands—and it may have been there all along in the form of the commonly seen cattail. These water-loving plants (*Typha latifolia*) are found growing in dense stands where shallow water or flooding is common, and unlike many other plants, they can live in water-saturated soils. The common cattail is found in most of North America, Europe, Asia, and Africa and provide an important habitat for wildlife—especially waterfowl and birds like the red-winged blackbirds. The plants have long been used as a food—roots, shoots, and even the pollen—for native peoples when food was scarce; as medicine, the roots being used to treat maladies and burns; and even as building materials. But more recently, scientists have been studying the plant to curb pollution: The cattails act as biological filters, removing silt, sand, and organic pollutants from runoff that flows into an estuary.

What is eutrophication?

Eutrophication is a process in which the supply of plant nutrients in a lake or pond is increased. Natural fertilizers, washed from the soil, result in an accelerated growth of   353 plants, producing overcrowding. As the plants die off, the dead and decaying vegetation depletes the lake oxygen supply, causing fish to die. The accumulated dead plant and animal material eventually changes a deep lake to a shallow one, then to a swamp, and finally it becomes dry land. While the process of eutrophication is a natural one, it is often accelerated enormously by human activities. Fertilizers from farms, sewage, industrial wastes, and some detergents all contribute to the problem.

What is limnology?

During his studies of Lake Geneva, Swiss physician François-Alphonse Forel (1841–1912) established many of the ideas behind the study of limnology and is called the father of limnology. In general, limnology is the study of freshwater inland ecosystems—especially lakes, ponds, and streams—and includes the chemistry, physics, geology, and biology of these bodies of water. These ecosystems are more fragile than ocean environments since they are subject to greater extremes in temperature (because they are much shallower).

What does it mean when a lake is brown or blue?

When a lake is brown, it usually indicates that eutrophication is occurring. This process refers to premature aging of a lake when a "megadose" of nutrients are added to the water, usually due to run-off that is either contaminated by agriculture or industry. Due to this rich supply of nutrients, blue-green algae begin to take over the green algae in the lake, disturbing the food webs and leading to an eventual loss of fish. When a lake is blue, this usually means that the lake has been damaged by acid precipitation. The gradual drop in pH is most often caused by exposure to acid rain, causing the disruption of the food webs and eventually killing most organisms. The end result is clear water, which is an indication of the lake's low productivity.

What is a red tide?

A red tide occurs when a population explosion occurs among toxic red dinoflagellates (members of the genera *Gymnodidium* and *Gonyaulax*, both protists that have an unusual cellular plate or armor). The population explosion, referred to as a "bloom," may tint the water orange, red, or brown and can be toxic to shellfish, birds, and humans who eat red tide-contaminated food.

What is a community?

The term used by ecologists to describe a group of populations of different species living in the same place at the same time is "community." This is a concise way of describing the organisms likely to be affected by any change to the local environment. For example, suppose you wanted to study not just the sparrows living in your backyard but also the insects they feed to their young and the plants that those insects eat; you would be studying a "community."

Where did a recent algae bloom affect manatee populations?

Each year off the west coast of Florida, a huge red algae bloom appears, closing beaches and causing chaos to wildlife of the area. The main reason is thought to be from phosphorus—algae thrive on the element—as the runoff flows through farms and lawns that use fertilizer, in which phosphorus is a main constituent. The year 2013 was no exception. The Caloosahatchee River, which runs through farmland and empties into the ocean at Fort Myers, is thought to have carried the phosphorus that fed the algae. The bloom was longer and more toxic, affecting humans, as the algae contain a nerve poison (brevetoxin) that is blown through the air when waves break; people can also become ill if they ingest local oysters and clams that absorb the brevetoxin. It also affects birds, dolphins, and other animals when they ingest the poison—as much of it clings to the ubiquitous sea grasses of the area. In fact, the toxin killed a record 241 (to date) manatees—animals that eat about 100 pounds (45.36 kilograms) of sea grass daily. Some scientists believe the algae were more virulent because a mild, semiwindless winter allowed the organisms to stay longer in 2012 and start earlier in 2013.

What is a niche?

A niche is similar to the job description of a population within a given species. However, instead of describing how many hours individuals will work, it describes features such as whether the species is active at night or during the day, what it eats, where it lives, and other aspects of day-to-day life. From one environment (or community) to another, the niche may vary depending on how much competition the species faces.

Can two populations occupy the same niche?

In 1932, Russian biologist Georgii Frantsevich Gause (1910–1988; he also is known for his research in antibiotics) proposed a new theory called the competitive exclusion principle: Two species that are in direct competition for the same resource cannot coexist if that resource is limited in some way. The work of Gause and others predicts that under such conditions, one population will drive the other to extinction in that local area.

What is the difference between a food chain and a food web?

A food chain refers to the transfer of energy from producers through herbivores through carnivores in a community. An example of a food chain would be little fish eating plankton and those little fish in turn being eaten by bigger fish. The term food web is broader, as it includes interconnected food chains within a specific ecosystem (perhaps the bigger fish eat the plankton as well!). A food web describes a nutritional portrait of the ecosystem. In 1927, English biologist Charles Elton (1868–1945) was one of the first scientists to diagram a food web—in his case, a description of feeding relationships on Bear Island in the Arctic.

What is a trophic level?

In 1942, American ecologist Raymond Lindeman (1915–1942) was one of the first ecologists to refer to the "trophic dynamics" of ecosystems. Today, scientists use the term trophic level to represent a step in the dynamics of energy flow through an ecosystem. The first trophic level is made up of the producers, or those within an ecosystem that harvest energy from an outside source (like the Sun or deep-sea thermal vents), and stabilize or "fix" it so that it remains in the system. The second level would be those who consume the producers, also known as the primary consumers; the next level would be the secondary consumers (those who consume the primary consumers), and so on. Because of the limited amount of energy available to each level, these trophic pyramids rarely rise above a third or fourth level of structure.

Large predators like this wolf were once thought to be keystone species that were essential to the health of an ecosystem because of their role in controlling populations. Now scientists know many less conspicuous species are vital, too.

What is a keystone species?

A keystone species is a species that is crucial or essential to the ecosystem's community structure. Originally, a keystone species was always thought to be the top predator, such as the gray wolf. Scientists have found that wolf population sizes influence populations of both their prey and other species in the environment. However, a more recent viewpoint recognizes that less conspicuous species are also very important, as all species are interconnected in a biological community. Other examples of keystone species include the sea star *Pisaster*, found along the coast of Washington state, that feeds on mussels and prevents the mussels from crowding out other species. Another example is the black-tailed prairie dog of the prairie ecosystem—not only is it a critical source of food for larger predators, its burrowing loosens the soil and its burrows act as home for other creatures.

What are producers and consumers?

"Producers" and "consumers" are terms used to describe the different roles played by species within ecosystems. Producers are those who "fix" energy—that is, they take energy from one source and convert it into a form (their biomass) that makes it accessible to others within the system (the consumers). Consumer levels are numbered according to their reliance on producers as a main source of energy. So primary consumers are those that rely heavily on producers, while secondary and tertiary (and even quaternary) consumers exploit other consumers as their preferred energy sources.

What is an ecological pyramid?

If the organisms in a food chain are arranged according to trophic levels, they form a pyramid, with a broad base representing the primary producers and usually only a few individuals in the highest part of the pyramid. Also known as a "pyramid of numbers," an ecological pyramid is a way of describing the distribution of energy, biomass, or individuals among the different levels of ecosystem structure.

Why do ecosystems need decomposers?

While energy flows through ecosystems in only one direction, entering at the producer level and exiting as heat and the transfer of energy (as biomass) to consumers, chemical compounds can be reused over and over again. In a well-functioning ecosystem, some organisms make their living (their niche) by breaking down structures and recycling the compounds. These organisms are known as decomposers. Without these organisms, the chemicals used to build a tree would remain locked in the tree biomass for eternity instead of being returned to the soil after the tree's death. From this soil will spring new growth, beginning the cycle once again.

What is meant by ecological succession?

With land-based communities, ecological succession is the process in which one community is slowly replaced by another, and that one by another, and so on. For example, a typical succession of plants in New York state can be: lichens and mosses; grasses; shrubs; coniferous woodlands; and finally, deciduous woodlands. Each type will remain relatively stable until the next succession stage, and the plants in each stage essentially "set the stage" for the next group. For example, the grasses and herbs stage can contribute organic matter to the soil and protect the soil from erosion; this adds depth to the soil, making it more conducive to the next stage, the shrubs, with their longer root systems.

What is a climax community?

Land-based (terrestrial) communities of organisms move through a series of stages from bare earth or rock to forests of mature trees in ecological succession. The last stage of succession is described as the "climax" because it is thought that, if left undisturbed, communities can remain in this stage forever. However, many scientists now believe that a climax community may be only one part of the continuous cycle of succession.

What is a survivorship curve?

A survivorship curve can indicate how long individuals survive in a population, or all members of the same species on one region. The curves come in three distinct types. In a Type I curve, the young have a high survival rate and typically live a long life. An example of this curve can be seen in the Daal sheep that live in Mt. Denali National Park in Alaska. Humans are also an example of a Type I curve. In a Type II curve, individuals have a relatively constant death rate throughout their life span. An example of this curve can be found in pop-

ulations of American robins. A Type III curve includes those species that have a large number of young—most of which die at a high rate, at an early age—but have a lower death rate
at later ages. An example of this survivorship curve can be found in lobsters and crabs.

What is a life history table?

A life history table, also referred to as a life table, is a table that shows both survival and
death rates in a specific population or organism. The life table is patterned after actu-
arial tables used by insurance companies.

What is adaptation?

As it sounds, adaptation refers to how well an organism adapts to its environment.
Adapted individuals survive and reproduce better than individuals without those adaptations. For example, an adaptation would be the long ears and limbs of rabbits living
in desertlike conditions. These adaptations allow the rabbits to radiate heat more efficiently over a large surface area, thus making it easier to survive in such a harsh climate.

What methods are used to estimate wildlife populations?

Since it is usually impossible (and often impractical) to count all individuals in a population, researchers use a variety of sampling techniques to estimate population densities.
One method is to count the individuals in a certain area. The larger the number and size
of sample plots, the more accurate the estimates. Population densities may also be estimated based on indirect indicators such as animal droppings or tracks, nests, or burrows.

What do MVP and PVA mean?

In this case, it does not mean "most valuable player," but in population studies, the MVP
is the minimum viable population size—the smallest number of individuals needed to
perpetuate a population, subpopulation, or species. PVAs (population viability analyses)
are especially helpful in predicting the MVP and is considered a process of identifying
the threats faced by a species, then evaluating the likelihood that it will persist for a
given time into the future or go extinct.

ENDANGERED PLANTS AND ANIMALS

What is biodiversity, and how is it measured?

Biodiversity or biological diversity refers to the breadth of species represented within
ecosystems, or even on Earth as a whole. Biodiversity may be defined at three levels:
Genetic diversity refers to the variety of genes found within a population or between
populations of the same species; species diversity, which may also be described as species
richness—how many different species are there within a habitat; and ecosystem diversity, which is an attempt to keep track of the loss of different types of habitats. This, in
turn, gives scientists a sense of what types of species are going extinct at any given time.

What is the difference between an endangered species and a threatened species?

An "endangered" species is one that is in danger of extinction throughout all or a significant portion of its range. A "threatened" species is one that is likely to become endangered in the foreseeable future.

What is the status of the African elephant?

The African elephant (*Loxodonta africana*) may actually be—according to genetic evidence—two species: the Savanna Elephant (*Loxodonta africana*) and the Forest Elephant (*Loxodonta cyclotis*), and maybe even a third, the West African elephant. This species is the largest terrestrial animal; its extent is not certain, as it has such an enormous range and wide variety of habitats it occupies. The African elephant is considered an endangered species, mainly because of habitat loss due to human development and expansion, and before conservation laws were enforced, illegal hunting for both meat and ivory (tusks). Between 1979 and 1989, Africa lost half of its elephants from poaching and illegal ivory trade, with the population decreasing from an estimated 1.3 million to 600,000. This led to the transfer of the African elephant from threatened to endangered status in October 1989 by CITES (the Convention on International Trade in Endangered Species). An ivory ban took effect on January 18, 1990, with Botswana, Namibia, and Zimbabwe agreeing to restrict the sale of ivory to a single, government-controlled center in each country. All countries have further pledged to allow independent monitoring of the sale, packing, and shipping process to ensure compliance with all conditions. In addition, all net revenues from the sale of ivory will be directed back into elephant conservation for use in monitoring, research, law enforcement, other management expenses, or community-based conservation programs within elephant range.

Have all the conservation measures worked to save the African elephant?

Yes, conservation efforts have worked in many ways. But the conservation laws seem to be working, thanks to such organizations such as the United Nations Environmental

When did the last passenger pigeon die?

Around 200 years ago, the passenger pigeon (*Ectopistes migratorius*) was the world's most abundant bird. Although the species was found only in eastern North America, it had a population of between three and five billion (25 percent of the North American land bird population). Overhunting caused a chain of events that reduced their numbers below minimum threshold, causing them to go extinct. In the 1890s, several states passed laws to protect the pigeon, but it was too late. The last known wild bird was shot in 1900; the last passenger pigeon in captivity, named Martha, died on September 1, 1914, in the Cincinnati Zoo.

Programme and the Convention on International Trade in Endangered Species (CITES). To date, and according to estimates, the African elephant populations seem to be growing, with one estimate showing an increase at an average annual rate of 4 percent per year. But that doesn't mean everyone pays attention. Organized criminal networks have been killing elephants in certain areas—in 2011, around 11,000 elephants were illegally killed by poachers. Authorities continue to try to control this slaughter before poaching outweighs the number of elephants born.

What are the most recent ways scientists categorize animals—from thriving to extinct?

Not all animals have just an "endangered" or "extinct" label. Over time, scientists have determined different levels of how an animal species is either thriving or becoming dangerously close to extinction. But extinction and how scientists categorize the status of a species happen in steps. According to the International Union for Conservation of Nature and Natural Resources, the following is a breakdown of some of these levels toward extinction: Not evaluated; Data deficient; Least Concern; Near threatened; Vulnerable; Endangered; Critically endangered; Extinct in the wild; and Extinct. For example, endangered animals in the United States include the Red-Cockaded woodpecker (*Picoides borealis*), a species in the southern United States; the Alabama Red-Belly turtle (*Pseudemys alabamensis*), a reptile that feeds almost entirely on aquatic plants; and, believed to be extinct in the 1950s but "rediscovered" in the 1970s, the 8-ounce (224-gram) Mount Graham Red Squirrel (*Tamiasciurus hudsonicus grahamensis*), found only in the area of Mount Graham in Arizona.

What has been the impact of zebra mussels on North American waterways?

Zebra mussels (*Dreissena polymorpha*) are black-and-white-striped, bivalve mollusks. They are hard-shelled species that adhere to hard surfaces with byssal threads. They were probably introduced to North America in 1985 or 1986 via discharge of a foreign ship's ballast water into Lake St. Clair. They have spread throughout the Great Lakes, the Mississippi River, and as far east as the Hudson River. High densities of zebra mussels have been found in the intakes, pipes, and heat exchangers of waterways throughout the world. They can clog the water intakes of power plants, industrial sites, and public drinking water systems; foul boat hulls and engine cooling water systems; and disrupt aquatic ecosystems. Water-processing facilities must be cleaned manually to rid the systems of the mussels. Zebra mussels are a threat to surface water resources because they reproduce quickly, have free-swimming larva and rapid growth, lack competitors for space or food, and have no predators.

How many species are threatened or endangered in the United States?

As of February 2014, the U.S. Fish and Wildlife Service lists 645 endangered of threatened animal species and 874 plant species in the United States.

How many species in the United States might have become extinct if the Endangered Species Act had not been passed in 1973?

Since the passage of the Endangered Species Act, it is estimated that fewer than a dozen species have become extinct in the United States—and some of them might

Coral reefs are like the rainforests of the oceans. The coral are a source of food and shelter for a wide range of aquatic life.

have actually already gone extinct by the time the Act passed. It is also estimated that, because of the ESA, over two hundred species in the United States have been spared extinction, including the spotted owl, American bald eagle, and black-footed ferret.

Why are reef-building corals economically important?

Coral reefs are among the most productive of all ecosystems. They are large formations of calcium carbonate (limestone) in tropical seas laid down by living organisms over thousands of years. Fish and other animals associated with reefs provide an important source of food for humans, and reefs serve as tourist attractions. Many terrestrial organisms also benefit from coral reefs, which form and maintain the foundation of thousands of islands. By providing a barrier against waves, reefs also protect shorelines against storms and erosion.

How is coral bleaching related to changes in the environment?

Although corals can capture prey, many tropical species are dependent on photosynthetic algae for nutrition—in other words, the algae that live within the cells lining the digestive cavity of the coral. The symbiotic relationship between coral and algae is mutually beneficial: The algae provide the coral with oxygen and carbon and nitrogen compounds while the coral supplies the algae with ammonia (waste product), from which the algae make nitrogenous compounds for both partners.

Coral bleaching is the stress-induced loss of colorful algae that live in coral cells. In coral bleaching, the algae lose their pigmentation or are expelled from coral cells. Without the algae, coral becomes malnourished and dies. The causes of coral bleaching are many, but it is believed the most adverse one involves environmental factors. Pollution, invasive bacteria, salinity changes, temperature changes (associated with global climate change), and high concentrations of ultraviolet radiation (associated with the destruction of the ozone layer) all seem to contribute to coral bleaching. And if the healthy coral reef becomes diseased or dying, it slowly erodes away. When this happens, marine life disappears, floodwaters are not absorbed, the waves pass over the reef and break directly onshore, and most importantly for areas prone to such events as hurricanes, the coastal structures bear the brunt of any storm surge.

CONSERVATION

Who is considered the founder of modern conservation?

American naturalist John Muir (1838–1914) is the father of conservation and the founder of the Sierra Club. He fought for the preservation of the Sierra Nevada in California and the creation of Yosemite National Park. He directed most of the Sierra Club's conservation efforts and was a lobbyist for the Antiquities Act, which prohibited the removal or destruction of structures of historic significance from federal lands. Another prominent influence was George Perkins Marsh (1801–1882), a Vermont lawyer and scholar. His book *Man and Nature* emphasized the mistakes of past civilizations that resulted in destruction of natural resources. As the conservation movement swept through the country in the last three decades of the nineteenth century, a number of prominent citizens joined the efforts to conserve natural resources and to preserve wilderness areas. Writer John Burroughs (1837–1921), forester Gifford Pinchot (1865–1946), botanist Charles Sprague Sargent (1841–1927), and editor Robert Underwood Johnson (1857–1937) were early advocates of conservation.

Who started Earth Day?

The first Earth Day, April 22, 1970, was coordinated by environmental activist Denis Hayes (1944–) at the request of Gaylord Nelson (1916–2005), U.S. senator from Wisconsin. Nelson is sometimes called the father of Earth Day. His main objective was to organize a nationwide public demonstration so large it would get the attention of politicians and force the environmental issue into the political dialogue of the nation. Important official actions that began soon after the celebration of the first Earth Day were the establishment of the Environmental Protection Agency (EPA); the creation of the President's Council on Environmental Quality; and the passage of the Clean Air Act, establishing national air quality standards. In 1995, Nelson received the Presidential Medal of Freedom for his contributions to the environmental protection movement. Earth Day continues to be celebrated each spring.

Why is Henry David Thoreau associated with the environment?

Henry David Thoreau (1817–1862) was a writer and naturalist from New England. His most familiar work, *Walden,* describes the time he spent in a cabin near Walden Pond in Massachusetts. He is also known for being one of the first to write and lecture on the topic of forest succession. His work, along with that of John Muir (1838–1914) and others, has served to inspire those others to understand the natural world and provide for its conservation.

What was the United States's first national park?

On March 1, 1872, an act of Congress signed by President Ulysses S. Grant (1822–1885) established Yellowstone National Park as the first national park, followed by Mackinac National Park in 1875 and Sequoia and Yosemite in 1890. The action inspired a worldwide national park movement and since that time, fifty-nine official national parks in twenty-seven states—and a multitude of monuments and historic centers that boost the numbers to about 388—have been added to the list.

What are the five largest national parks in the United States?

The five largest national parks in the United States are Alaska's Wrangell-St. Elias, Gates of the Arctic, Denali, and Katmai National Parks, plus the largest in the "lower 48 states"—Death Valley, California.

What is the importance of the rain forest?

Half of all medicines prescribed worldwide are originally derived from wild products, and the United States National Cancer Institute has identified more than 2,000 tropical rain forest plants with the potential to fight cancer. Rubber, timber, gums, resins and waxes, pesticides, lubricants, nuts and fruits, flavorings and dyestuffs, steroids, latexes, essential and edible oils, and bamboo are among the products that would be drastically affected by the depletion of the tropical forests. In addition, rain forests greatly influence patterns of rain deposition in tropical areas; smaller rain forests mean less rain. Large groups of plants, like those found in rain forests, also help control levels of carbon dioxide in the atmosphere.

In what way can forest fires be good for the environment?

Wildfires are critical to maintaining the integrity of forest and grassland ecosystems. Forest and grass fires, usually started by lightning, act as an ecologically renewing force by creating necessary conditions for plant germination and continued healthy growth. The primary goal of fire management is to simulate the actual and natural aspects of fire cycles. Fire management also attempts to prevent large catastrophic wildfires from occurring by removing accumulated debris from forests. Seen throughout the American West every summer, these extremely intense fires are caused primarily by decades of fire suppression, which has allowed heavy fuels—accumulated debris—to build up. Ironically, by attempting to prevent natural fires, humans have only increased their prevalence. Only in the past decade have firefighters and scientists realized that nature actually does know best when it comes to maintaining their forests through fire and allowed nature to take its course.

When was the EPA created, and what does it do?

In 1970 President Richard M. Nixon (1913–1994) signed an executive order that created the Environmental Protection Agency (EPA) as an independent agency of the U.S.

government. The creation of a federal agency by executive order rather than by an act of the legislative branch is somewhat uncommon. The EPA was established in response to public concern about unhealthy air, polluted rivers and groundwater, unsafe drinking water, endangered species, and hazardous waste disposal. Responsibilities of the EPA include environmental research, monitoring, and enforcement of legislation regulating environmental activities.

What is the Superfund Act?

In 1980, the United States Congress passed the Comprehensive Environmental Response, Compensation, and Liability Act, commonly known as the Superfund program. This law (along with amendments in 1986 and 1990) established a $16.3-billion Superfund financed jointly by federal and state governments and by special taxes on chemical and petrochemical industries (which provide 86 percent of the funding). The purpose of the Superfund is to identify and clean up abandoned hazardous waste dump sites and leaking underground tanks that threaten human health and the environment. To keep taxpayers from footing most of the bill, cleanups are based on the "polluter-pays principle." The EPA is charged with locating dangerous dump sites, finding the potentially liable culprits, ordering them to pay for the entire cleanup, and suing them if they don't. When the EPA can find no responsible party, it draws money out of the Superfund for cleanup.

ENVIRONMENTAL CHALLENGES

What is the Pollutant Standard Index?

The U.S. Environmental Protection Agency and the South Coast Air Quality Management District of El Monte, California, devised the Pollutant Standard Index to monitor concentrations of pollutants in the air and inform the public concerning related health effects. The scale, which measures the amount of pollution in parts per million, has been in use nationwide since 1978. The following lists the PS Index and the cautionary status:

PS Index	Health effects	Cautionary status
0	Good	
50	Moderate	
100	Unhealthful	
200	Very unhealthful	Alert: elderly or ill should stay indoors and reduce physical activity.
300	Hazardous	Warning: general population should stay indoors and reduce physical activity.
400	Extremely hazardous	Emergency: all people should remain indoors with windows shut and no physical exertion.
500	Toxic	Significant harm; same as above.

What is the Toxic Release Inventory?

Toxic Release Inventory (TRI) is a government-mandated, publicly available compilation of information on the release of over 650 individual toxic chemicals and toxic chemical categories by manufacturing facilities in the United States. The law requires manufacturers to state the amounts of chemicals they release directly to air, land, or water, or state that they transfer to offsite facilities that treat or dispose of wastes. The U.S. Environmental Protection Agency compiles these reports into an annual inventory and makes the information available in a computerized database.

What is DDT?

Dichlorodiphenyl-trichloro-ethene (DDT) was synthesized as early as 1874 by Austrian chemist Othmar Zeidler (1859–1911); it was the Swiss chemist Paul Müller (1899–1965) who recognized its insecticidal properties in 1939. He was awarded the 1948 Nobel Prize in Physiology or Medicine for his development of DDT. Unlike the arsenic-based compounds then in use, DDT was effective in killing insects and seemed not to harm plants and animals. In the following twenty years, it proved to be effective in controlling disease-carrying insects (mosquitoes that carry malaria and yellow fever and lice that carry typhus) and in killing many plant crop destroyers. Publication of Rachel Carson's book *Silent Spring* in 1962 alerted scientists to the detrimental effects of DDT. Increasingly DDT-resistant insect species and the accumulative hazardous effects of DDT on plant and animal life cycles led to its disuse in many countries during the 1970s. In fact, DDT and PCBs have been added to the list of chemicals known as estrogenic compounds—that is, synthetic substances in the environment that cause the mammalian body to respond as if to estrogen.

How are hazardous waste materials classified?

The four types of hazardous waste materials are:

- Corrosive materials can wear away or destroy a substance. Most acids are corrosive and can destroy metal, burn skin, and give off vapors that burn the eyes.
- Ignitable materials can burst into flames easily. These materials pose a fire hazard and can irritate the skin, eyes, and lungs. Gasoline, paint, and furniture polish are ignitable.
- Reactive materials can explode or create poisonous gas when combined with other chemicals. Combining chlorine bleach and ammonia, for example, creates a poisonous gas.
- Toxic materials or substances can poison humans and other life. They can cause illness or death if swallowed or absorbed through the skin. Pesticides and household cleaning agents are toxic.

What are PCBs and CfCs?

A group of chemicals with the same general chemical structure and physical properties as DDT are known as polychlorinated biphenyls or PCBs. Because of their physical properties (nonflammability, chemical stability, high boiling point, and electrical insulating properties), PCBs can be used in a variety of applications. Formerly, many products contained these compounds—from electrical circuitry to the dyes and pigments used in paint to carbonless copy paper—all were manufactured with PCBs. Before production ceased in 1977, the United States produced about 1.5 billion pounds (6.8 billion kilograms) of PCBs.

Chlorofluorocarbons (CfCs) are commonly used as aerosol sprays, refrigerants, solvents, and foam-blowing agents. They are in and of themselves nontoxic and nonflammable molecules containing chlorine, fluorine, and carbon. However, they are thought to have a deleterious effect on ozone concentrations in the atmosphere (for more about the ozone and ozone depletion, see this chapter).

Who was Rachel Carson?

American marine biologist and conservationist Rachel Carson (1907–1964) was one of the first to describe to the general public the consequences of chemical contamination in the environment. In her book *Silent Spring*, published in 1962, Carson exposed the dangers of hydrocarbons, particularly DDT, to the reproduction of species that prey upon the insects for whom the pesticide was intended.

What is a "green" building?

Green building involves an integrated approach to design and construction that makes less waste and emphasizes energy conservation and efficiency. Some additional approaches to green building involve solar water-heating systems, "smart-house" systems that save on utility use, and more efficient heating and cooling systems.

What is zero population growth?

Zero population growth, or ZPG, is the estimation of the birth rate necessary to maintain the size of the human population at its current level. As of now, the rate is esti-

What is meant by ecoterrorism?

Ecoterrorism is the term used to describe actions taken by individuals or organizations to prevent what may loosely be termed as "environmental change." This change may be seen as the clear-cutting of forests for wood products or land for housing or the use of genetically modified plants or animals for human consumption. Ecoterrorists are those who are willing to take violent and potentially harmful action in order to prevent these types of changes. Beginning in the 1980s, industrial sabotage such as tree-spiking (the process of inserting metal spikes into trees so that they cannot be cut down by chainsaws) was used to prevent logging—and tree-spiking can seriously injure the loggers who are cutting down such trees.

mated as 2.1, which means that each set of existing parents would need to have (on average) slightly more than two children during their lifetime. The extra 0.1 allows for infant mortality.

What was the distribution of radioactive fallout after the 1986 Chernobyl nuclear accident?

On April 25 to 26, 1986, the world's worst nuclear power accident occurred at Chernobyl in the former USSR (now Ukraine). While scientists were testing one of the four reactors at the Chernobyl nuclear power plant, located 80 miles (129 kilometers) north of Kiev, an unusual chain reaction occurred in the reactor. This subsequently led to explosions and a fireball that blew the heavy steel and concrete lid of the reactor. Radioactive fallout containing the isotope cesium-137 and nuclear contamination covered an enormous area, including Byelorussia, Latvia, Lithuania, the central portion of what was then the Soviet Union, the Scandinavian countries, the Ukraine, Poland, Austria, Czechoslovakia, Germany, Switzerland, northern Italy, eastern France, Romania, Bulgaria, Greece, Yugoslavia, the Netherlands, and the United Kingdom. The fallout, extremely uneven because of the shifting wind patterns, extended 1,200 to 1,300 miles (1,930–2,090 kilometers) from the point of the accident. The accident led to the release of roughly 5 percent, or 7 tons, of the reactor fuel containing fifty to one hundred million curies. Estimates of the effects of this fallout range from 28,000 to 100,000 deaths

The power plant at Chernobly, shown here in 2012, still sits abandoned long after the 1986 nuclear disaster that killed tens of thousands Soviet Union citizens.

from cancer and genetic defects within the subsequent fifty years. In particular, livestock in high rainfall areas received lethal dosages of radiation.

What is acid rain?

The term acid rain was coined by the British chemist Robert Angus Smith (1817–1884), who in 1872 published *Air and Rain: The Beginnings of a Chemical Climatology*. Since then acid rain has become an increasingly used term for rain, snow, sleet, or other precipitation that has been polluted by acids such as sulfuric and nitric acids. When gasoline, coal, or oil are burned, their waste products of sulfur dioxide and nitrogen dioxide combine in complex chemical reactions with water vapor in clouds to form acids. The United States alone discharges 40 million metric tons of sulfur and nitrogen oxides into the atmosphere. This, combined with natural emissions of sulfur and nitrogen compounds, has resulted in severe ecological damage. Hundreds of lakes in North America (especially in northeastern Canada and the United States) and in Scandinavia are so acidic that they cannot support fish life. Crops, forests, and building materials such as marble, limestone, sandstone, and bronze have been affected as well, but the extent is not as well documented as it is with fish life. However, in Europe, where many trees are stunted or have been killed, a new word—Waldsterben ("forest death")—has been coined to describe this phenomenon.

What is indoor air pollution, and how is it caused?

Indoor air pollution, also known as "tight building syndrome," results from conditions in modern, highly energy-efficient buildings, which have reduced outside air exchange or have inadequate ventilation along with chemical contamination and microbial contamination. Indoor air pollution can produce various symptoms, such as headaches, nausea, and eye, nose, and throat irritation. In addition, houses are affected by indoor air pollution emanating from consumer and building products and from tobacco smoke. Below are some pollutants found in a typical household:

Pollutant	Sources	Effects
Asbestos	Old or damaged insulation, fireproofing, or acoustical tile	Many years later, chest and abdominal cancers and lung diseases
Biological pollutants	Bacteria, mold and mildew, viruses, animal dander and mites, cockroaches, and pollen	Eye, nose, and throat irritation; shortness of breath; dizziness; lethargy; fever; digestive problems; asthma; influenza and other infectious diseases
Carbon monoxide	Unvented kerosene and gas heaters; leaking chimneys and furnaces; wood stoves and fireplaces; gas stoves; automobile exhaust from garages; tobacco smoke	At low levels, fatigue; at higher levels, impaired vision and coordination; headaches; dizziness; confusion; nausea. Fatal at very high concentrations

Pollutant	Sources	Effects
Formaldehyde	Plywood, wall paneling, particle board, fiber-board; foam insulation; fire and tobacco smoke; textiles and glues	Eye, nose, and throat irritations; wheezing and coughing; fatigue; skin rash; severe allergic reactions; may cause cancer
Lead	Automobile exhaust; sanding or burning of lead paint; soldering	Impaired mental and physical development in children; decreased coordination and mental abilities; kidneys, nervous system and red blood cell damage
Mercury	Some latex paints	Vapors can cause kidney damage; long-term exposure can cause brain damage
Nitrogen dioxide	Kerosene heaters and unvented gas stoves and heaters; tobacco smoke	Eye, nose, and throat irritation; may impair lung function and increase respiratory infections in young children
Organic gases	Paints, paint strippers, solvents, and wood preservatives; aerosol sprays; cleansers and disinfectants; moth repellents; air fresheners; stored fuels; hobby supplies	Eye, nose and throat irritation; headaches; loss of coordination; nausea; damage to liver, kidney, and nervous system; some organics cause cancer in animals and are suspected of causing cancer in humans
Pesticides	Products used to kill household pests and products used on lawns or gardens that drift or are tracked inside the house	Irritation to eye, nose, and throat; damage to nervous systems and kidneys; cancer
Radon	Earth and rock beneath the home; well water, building materials	No immediate symptoms; estimated to cause about 10 percent of lung cancer deaths; smokers at higher risk

What is the NIMBY syndrome?

NIMBY is the acronym for "Not In My Back Yard" (another acronym, although not as popular, is NIMFY or "Not In My Front Yard.") It refers to major community resistance to construction of new incinerators, landfills, prisons, roads, electric windmill generators, mining operations, hydrofracking, and so forth, especially near residential areas. Most people do not want what goes with such construction mainly because of the pollution (air, water, and noise) that is often associated with such activities.

What is nuclear waste?

Nuclear waste consists either of fission products formed from atom splitting of uranium, cesium, strontium, or krypton or from transuranic elements formed when uranium atoms absorb free neutrons. Wastes from transuranic elements are less radioactive than fission products; however, these elements remain radioactive far longer than fission products. Transuranic wastes include irradiated fuel (spent fuel) in the form of twelve-foot- (four-meter-) long rods, high-level radioactive waste in the form of liquid or sludge, and low-level waste (nontransuranic or legally high-level) in the form of reactor hardware, piping, toxic resins, water from fuel pools, and other items that have become contaminated with radioactivity.

How is nuclear waste stored?

Most spent nuclear fuel in the United States is safely stored in specially designed pools at individual reactor sites around the country. If pool capacity is reached, licensees may move toward use of above-ground dry storage casks. The three low-level radioactive waste disposal sites are Barnwell, South Carolina; Hanford, Washington; and Envirocare, Utah. Each site accepts low-level radioactive waste from specific regions of the country, but only Envirocare uses above-ground storage.

Most high-level nuclear waste has been stored in double-walled stainless-steel tanks surrounded by 3 feet (1 meter) of concrete. The current best storage method, developed by the French in 1978, is to incorporate the waste into a special molten glass mixture, then enclose it in a steel container and bury it in a special pit. The Nuclear Waste Policy Act of 1982, as amended in 1987, specified that high-level radioactive waste would be disposed of underground in a deep geologic repository. Yucca Mountain, Nevada, was chosen as the single site to be developed for disposal of high-level radioactive waste. On July 23, 2002, President George W. Bush signed House Joint Resolution 87, allowing the Department of Energy to establish a repository in Yucca Mountain to safely store nuclear waste. However, some scientists still expressed concerns about the estimates of how long it would take for rainwater and snow to infiltrate the mountain and corrode the containers.

How much garbage does the average American generate?

According to the Environmental Protection Agency, in 2010 (to date, the latest figures), Americans generated about 250 million tons of trash and recycled and composted over 85 million tons of this material, equivalent to a 34.1 percent recycling rate. On average, they recycled and composted 1.51 pounds out of our individual waste generation of 4.43 pounds per person per day. In general, the total amount of waste is distributed as follows: Paper and paperboard (38 percent); yard waste (12.1 percent); food wastes (10.9 percent); plas-

The symbol for recycled or recyclable products is now very familiar to almost everyone. Numbers are added to this symbol to indicate different types of plastics.

tics (10.5 percent); metals (7.8 percent); glass (5.5 percent); wood (5.3 percent); textiles (3.9 percent); other waste (3.2 percent); rubber and leather (2.7 percent).

What do the numbers inside the recycling symbol on plastic containers mean?

The Society of the Plastics Industry developed a voluntary coding system for plastic containers to assist recyclers in sorting plastic containers. The symbol is designed to be imprinted on the bottom of the plastic containers. The numerical code appears inside a three-sided triangular arrow. A guide to what the numbers mean is listed below. The most commonly recycled plastics are polyethylene terephthalate (PET) and high-density polyethylene (HDPE). The following lists the numbers, codes, and examples:

Code	Material	Examples
1	Polyethylene terephthalate (PET/PETE)	2-liter soft drink bottle
2	High-density polyethylene (HDPE)	Milk and water jugs
3	Vinyl	(PVC) Plastic pipes, shampoo bottles
4	Low-density polyethylene (LDPE)	Produce bags, food storage containers
5	Polypropylene (PP)	Squeeze bottles, straws
6	Polystyrene (PS)	Food packaging

When offered a choice between plastic or paper bags for your groceries, which should you choose?

The answer is neither. Both are environmentally harmful, and the question of which is more damaging has no clear-cut answer. On one hand, plastic bags degrade slowly in landfills and can harm wildlife if swallowed; in addition, producing them uses vast amounts of water and pollutes the environment. On the other hand, producing the brown paper bags used in most supermarkets uses trees and also pollutes the air and water. Overall, white or clear polyethylene bags require less energy for manufacture and cause less damage to the environment than do paper bags not made from recycled paper. Instead of having to choose between paper and plastic bags, you can bring your own reusable canvas or string containers to the store, or you can save and reuse any paper or plastic bags you receive (if the store allows it—some places have health codes that do not allow you to reuse paper or plastic bags).

What is the Kyoto Protocol?

The Kyoto Protocol was an international summit held in Kyoto, Japan, in December 1997. Its goal was for governments around the world to reach an agreement regarding emissions of carbon dioxide and other greenhouse gases. The Kyoto Protocol called for industrialized nations to reduce national emissions over the period 2008–2012 to 5 percent below the 1990 levels; they also made a second commitment period to lower emissions between 2013 and 2020, but as of this writing, it has not been legally verified. The protocol covers these greenhouse gases: carbon dioxide, methane, and nitrous oxide, with other chemicals such as hydrofluorocarbons, perfluorocarbons, and sulfur hexa fluoride added in subsequent years. Of course, without most people signing on, the Kyoto Protocol is essentially one-sided, with some of the major polluters in the world not adhering to the lowering of greenhouse gases. Unfortunately, by 2010, the countries signing on to the second commitment period emitted a mere 13.4 percent of the total troubling greenhouse gases each year.

What is a bioinvader?

A bioinvader is an exotic organism usually introduced into an ecosystem accidentally. These bioinvaders are either non-native plants or animals that often overwhelm the native species. For example, the kudzu vine, first introduced in the 1930s to control erosion, quickly spread in the southeastern United States and now grows uncontrollably. (For more about vines, see the chapter "Plant Structure, Function, and Use.") Other bioinvader species include zebra mussels (that have taken over aquatic habitats in the Great Lakes), purple loosestrife (an invasive wild plant found in northern United States and Canada), the Asian long-horned beetle (first reported in New York and that has spread into the Midwest), and the emerald ash borer (one of many invasive beetles—this one currently spreading in the northeast and destroying ash trees).

BIOLOGY IN THE LABORATORY

HISTORICAL INTEREST IN BIOTECHNOLOGY

What is biotechnology?

Biotechnology is the use of a living organism to produce a specific product. It includes any technology associated with the manipulation of living systems for industrial purposes. In its broadest sense, biotechnology includes the fields of chemical, pharmaceutical, and environmental technology as well as engineering and agriculture.

What were some major biotechnological achievements of the mid-twentieth and beginning of the twenty-first centuries?

Numerous (too many to mention here) advancements in biotechnology have been made in the mid-twentieth and beginning of the twenty-first centuries. The following lists only a few of those achievements:

Year	Achievement
1968	American biochemist Stanley Cohen (1922–) uses plasmids to transfer antibiotic resistance to bacterial cells.
1970	American biochemist Herb Boyer (1936–) discovers that certain bacteria can "restrict" some bacteriophages by producing enzymes (restriction enzymes).
1972	American biochemist Paul Berg (1926–) splices together DNA from the SV 40 virus and *E. coli*, making recombinant DNA; shares 1980 Nobel Prize with American molecular biologist Walter Gilbert (1932–) and British biochemist Fred Sanger (1918–2013; he has won the Nobel twice, also in 1958).
1974	American biochemist Stanley Cohen (1922–), then research technician Annie Chang, and American biochemist Herb Boyer (1936–) splice frog DNA into *E. coli*, producing the first recombinant organism.

Year	Achievement
1975	DNA sequencing developed by American molecular biologist Walter Gilbert (1932–), American molecular geneticist Allan Maxam (1952–), and British biochemist Fred Sanger (1918–2013).
1978	Human insulin cloned in *E. coli* by a biotech company called Genentech.
1986	America biochemist Kary Banks Mullis (1944–) develops the polymerase chain reaction (PCR), in which DNA polymerase can copy a DNA segment many times in a short period of time.
1989	Human Genome Project (HGP) begins; it is complete by 2003.
1990	Researchers at National Institutes of Health (NIH) use gene therapy to treat a human patient.
1994	Introduction of the first transgenic food, the Flavr Savr tomato; it is engineered for a longer shelf life.
1996	Dolly the Finn Dorset lamb—the first mammal—is cloned by English embryologist Ian Wilmut (1944–).
1997	First human artificial chromosome is developed.
2000	Completion of the first working draft (90 percent complete) of the Human Genome Project.
2003	Glofish—a fish that fluoresces and is the first genetically modified pet—are marketed and sold in the U.S.
2004	By now, the rat, mouse, and human genomes are the first mammals to be sequenced—and all have roughly the same number of genes—between 25,000 and 30,000.
2005	The International Rice Genome Sequencing Project publishes its "Map-based sequence of the rice genome," covering 95 percent of the genome of the world's most important staple crop.
2007	Human artificial chromosomes were created and patented, and companies appear to use this new technology.
2008	A study suggests that some RNAi drugs work by activating the immune system rather than by silencing genes.
2010	The first synthetic bacterial cell is created.
2010	The Neanderthal Genome Project points to genetic evidence that interbreeding did likely take place between Neanderthals and "modern" humans—and that a small but significant portion of this Neanderthal mix is present in modern non-African populations.
2011	The U.S. Court of Appeals for the District of Columbia lifts a lower-court injunction—thus allowing further research on the controversial use of embryonic stem cells.
2013	Companies develop more "smart drugs"; for example, drug design based on understanding on how genes and proteins work—unlike the past, when many drugs were based on random hit-and-miss experiments with organic molecules.

Who was the first individual to find the gene for breast cancer?

American human geneticist Mary Claire King (1946–) determined that in 5 to 10 percent of those women with breast cancer, the cancer is the result of a mutation of a gene on chromosome 17, the BRCA1 (Breast Cancer 1). The BRCA1 gene is a tumor suppressor gene and is also linked to ovarian cancer. Subsequently, other researchers were able to clone the gene and pinpoint its exact location on chromosome 17.

A LOOK AT A GENETICS LAB

What is a gene library?

A gene library is a collection of cloned DNA, usually from a specific organism. Just as a conventional library stores information in books and computer files, a gene library stores genetic information either for an entire genome, a single chromosome, or specific genes in a cell. For example, one can find the gene library of a specific disease such as cystic fibrosis, the chromosome where most cystic fibrosis mutations occur, or the entire genome of those individuals affected by the disease. To create a gene library, scientists extract the DNA of a specific organism and use restriction endonucleases to cut it. The scientists then insert the resulting fragments into vectors and make multiple copies of the fragments. The number of clones needed for a genomic library depends on the size of the genome and the size of the DNA fragments. Specific clones in the library are located using a DNA probe. (For more about DNA and genes, see the chapter "DNA, RNA, Chromosomes, and Genes.")

How are genes physically found in a specific genome?

Finding one gene out of the possible tens of thousands of genes in the human—or any organism's—genome is a difficult task, but the process is made easier if the protein product of the gene is known. For example, if a researcher is looking to find the gene for mouse hemoglobin, he or she would isolate the hemoglobin from mouse blood and determine the amino acid sequence. The amino acid sequence could then be used as a template to generate the nucleotide sequence. Working backward again, a complementary DNA probe to the sequence would be used to identify DNA molecules with the same sequence from the entire mouse genomic library. However, if the protein product is not known, the task is more difficult; for example, the difficulty of finding the susceptibility gene for late-onset Alzheimer's disease in humans.

What is a gene probe?

A gene probe is a specific segment of single-strand DNA that is complementary to a desired gene. For example, if the gene of interest contains the sequence AATGGCACA, then the probe will contain the complementary sequence TTACCGTGT. When added to the appropriate solution, the probe will match and then bind to the gene of interest. To help locate the probe during the procedure, scientists usually label it with a radioisotope or a fluorescent dye so that it can be seen and identified.

What is a gene gun?

The gene gun, developed by Cornell University plant scientists in the early 1980s, is a method of direct gene transfer used in plant biotechnology. In order to transfer genes into plants, gold or tungsten microspheres (1 micrometer in diameter) are coated with DNA from a specific gene. The microspheres are then accelerated toward target cells (contained in a petri dish) at high speed. Once inside the target cells, the DNA on the outside of the

375

microsphere is released and can be incorporated into the plant's genome. This method is also known as "microprojectile bombardment" or biolistics. The survival rate of the bombarded cells varies with the rate of penetration. For example, if the particle penetration rate reaches twenty-one per cell, approximately 80 percent of bombarded cells may die.

What is terminator gene technology?

This is a method of biotechnology in which crops that are bioengineered for a specific desirable trait (such as drought resistance) would contain a lethal gene that would cause any seeds produced by the plant to be nonviable (not capable of developing another plant). The lethal gene could be activated by spraying with a solution sold by the same company that originally marketed the bioengineered plant. Thus, the plants would still provide seeds with nutritional value, but these seeds could not be used to produce new plants. This technique would allow companies to control the product's genes so that they would not spread into the general plant population, and seeds would have to be purchased by the grower for each season.

What is gene therapy?

Gene therapy involves replacement of an "abnormal" disease-causing gene with a "normal" gene. The normal gene is delivered to target cells using a vector, which is usually a virus that has been genetically engineered to carry human DNA. The virus genome is altered to

Who was the first person to receive gene therapy?

In 1990, Ashanti DeSilva was a four-year-old with a rare, life-threatening immune disorder known as ADA deficiency, which made her vulnerable to even the mildest infections (it is also often referred to as severe combined immune deficiency, or SCID—in which the immune system lacks the ability to produce an important enzyme called adenosine deaminase). Doctors then removed her white blood cells and replaced them with genetically altered white blood cells—thus essentially altering the structure of her DNA. Although the treatment proved to strengthen her immune system, the treated cells failed to produce additional healthy cells. In order to maintain normal levels of adenosine deaminase, DeSilva—who is relatively healthy at this writing—still has to get periodic gene therapy to maintain the necessary levels of the enzyme in her blood. But because she also receives doses of the enzyme itself (called PEG-ADA), it is unknown whether or not the gene therapy would have actually worked if it was the only therapy administered.

Such genetic therapies still have a long way to go. And although scientists have made thousands of gene therapy attempts since DeSilva's time, many treatments have failed to correct the disease being treated—and some have also, unfortunately, caused other diseases.

remove disease-causing genes and insert therapeutic genes. Target cells are infected with the virus. The virus then integrates its genetically altered material containing the human gene into the target cell, where it should produce a functional protein product.

Can genetic engineering be used to save endangered species?

As endangered species disappear from natural habitats and are only found in zoos, researchers are looking for ways to conserve these species. Using cryopreservation, the Zoological Society of San Diego has created a "frozen zoo" that stores viable cell lines from more than 3,200 individual mammals, birds, and reptiles, representing 355 species and subspecies. Researchers maintain that efforts to preserve species in their natural habitats should still be continued, but by preserving and studying animal DNA, scientists can learn genetic aspects crucial to the species' survival.

CLONING

What is cloning?

A clone is a group of cells derived from the original cell by fission (one cell dividing into two cells) or by mitosis (cell nucleus division with each chromosome splitting into two). Cloning perpetuates an existing organism's genetic makeup—in simpler terms, cloning produces genetically identical copies of a biological entity. Gardeners have been making clones of plants for centuries by taking cuttings of plants to make genetically identical copies. Such simple cloning starts with taking a cutting of a plant that best satisfies such criteria as reproductive success, beauty, or some other standard. Since all of the plant's cells contain the genetic information that will allow the entire plant to be reconstructed, in most cases, the cutting can be taken from any part of the plant (although some plants do better with cuttings from a stem, others from leaves, and some even the time of the year the cutting is made). The cutting is then added to a culture medium having nutritious chemicals (such as a fertilized soil) and sometimes a growth hormone (not all plants need a root-growing hormone). The cells in the cutting eventually divide, doubling in size every six weeks until the mass of cells produces small white globular points called embryoids. These embryoids develop roots, or shoots, and begin to look like tiny plants. Transplanted into rich soils and compost, these plants grow into exact copies of the parent plant, with the process taking only a few months to over a year for the cloned plant to mature. This process is called tissue culture and has been used to make clones of asparagus, pineapples, strawberries, bananas, carnations, ferns, and others. Besides making highly productive copies of the best plant available, this method often controls viral diseases that are passed through normal seed generations.

Can human beings be cloned?

In theory, yes—but in reality, human cloning has many technical obstacles, as well as moral, ethical, philosophical, religious, and economic issues to be resolved before a human being can be cloned. At the present time, most scientists would agree that cloning a human being is unsafe—too many variables are involved in such an endeavor—and many continue to say it is ethically wrong. But that doesn't mean some researchers will not stop trying to push the issue: For example, in 2004, South Korean researchers claimed to have cloned a human embryo in a test tube, but no proof was found, and the researchers retracted their claim in 2006.

What was the first animal to be successfully cloned?

In 1970, the British developmental biologist John B. Gurdon (1933–) cloned a frog. He transplanted the nucleus of an intestinal cell from a tadpole into a frog's egg that had had its nucleus removed. The egg developed into an adult frog that had the tadpole's genome in all of its cells and was therefore a clone of the tadpole.

What was the first mammal to be successfully cloned?

The first mammal cloned from mature (somatic) cells was Dolly, a ewe born in July 1996 in Scotland—it only took 276 attempts! English embryologist Ian Wilmut (1944–) led the team of biologists that removed a nucleus from a mammary (udder) cell of a six-year-old adult ewe and transplanted it into an enucleated egg extracted from a second ewe. Electrical pulses were administered to fuse the nucleus with its new host. When the egg began to divide and develop into an embryo, it was transplanted into a surrogate mother ewe. Dolly was the genetic twin of the ewe that donated the mammary cell nucleus. On April 13, 1998, Dolly gave birth to Bonnie—the product of a normal mating between Dolly and a Welsh mountain ram. This event demonstrated that Dolly was a healthy, fertile sheep, able to produce healthy offspring.

Do cloned animals look identical to the original and other offspring?

No, contrary to popular science fiction stories, cloned animals do not look like the original donor of the mature cells or their identical clones. As with most people—even human twins—the environment plays an important role in how an organism looks. For example, in 2001, the first cloned cat was born, called Cc, or Carbon Copy or CopyCat—from a genetic donor named Rainbow, but put into a "surrogate mother" that was a female tabby (Cc was the only one of eighty-seven embryos in the experiment that was successful). The resulting cat looks very different from Rainbow or her surrogate mother. This difference is due to the fact that the color and pattern of cat coats are not attributed exclusively to genes, and in this case, the cloning process also "changed" the cat's color and pattern. In 2006, Cc became the first cloned cat that had ever given birth to kittens.

Have any other animals been successfully cloned?

Dolly is considered to be the first mammal cloned from a mature cell taken from an adult animal, but in 1979, scientists produced the first genetically identical mice by splitting mouse embryos in a test tube. They then implanted the embryos into the wombs of adult female mice with success. After that, the same procedure was used to produce genetically identical cows, sheep, and chickens by transferring the embryos into the wombs of their respective adult female animals.

After the success of Dolly, most animal cloning has used the mature cells from adult animals such as a skin cell or udder cells. Two years after Dolly, Japanese scientists cloned eight calves from a single cow, with only half of them surviving. And since then, other mammals have been cloned in the same way, including cats, deer, dogs, horses, mules, ox, rabbits, and rats.

DNA IN THE LAB

What is genetic engineering?

In general, genetic engineering—also popularly known as molecular cloning or gene cloning—is the artificial recombination of nucleic acid molecules in a test tube; their insertion into a virus, bacteria, or other system; and the subsequent incorporation of the molecules into a host organism where they are able to propagate. The construction of such molecules has also been termed gene manipulation because it usually involves unique genetic combinations using biochemical means.

Genetic engineering techniques include cell fusion and the use of recombinant DNA or gene splicing; the following describes these two techniques:

Cell fusion—In cell fusion the tough outer membranes of sperm and egg cells are removed by enzymes, then the fragile cells are mixed and combined with the aid of chemicals or viruses. The result may be the creation of a new life form from two species (called a chimera; see below).

Recombinant DNA techniques—These techniques (also called gene-splicing) transfer a specific genetic activity from one organism to the next through special DNA and enzymes. Simply put, the recombinant DNA process begins with the isolation and fragmentation of suitable DNA strands. After these fragments are combined with vectors, they are carried into bacterial cells, where the DNA fragments are "spliced" on to plasmid DNA that has been opened up. These hybrid plasmids are then mixed with host cells to form transformed cells. Since only some of the transformed cells will exhibit the desired characteristic or gene activity, the transformed cells are separated and grown individually in cultures. This methodology has been successful in producing large quantities of hormones (such as insulin) for the biotechnology industry. And although it is more difficult to transform animal and plant cells in this way,

What is a chimera?

The chimera from Greek mythology is a fire-breathing monster with a lion's head, a goat's body, and a serpent's tail. The chimera of biotechnology is an animal formed from two different species or strains—that is, a mixture of cells from two very early embryos. Most chimeras used in research are made from different mouse strains—and they cannot reproduce.

this technique is often used to generate plants that are more resistant to diseases and to make animals grow larger.

What is recombinant DNA?

Recombinant DNA is hybrid DNA that has been created from more than one source. For example, the splicing of human DNA into bacterial DNA so that a human gene product is produced by a bacterial cell results in recombinant DNA.

What are some uses of genetic engineering techniques?

When it comes to humans, the uses of genetic engineering have been very controversial. This is because genetic engineering interferes with the processes of heredity and can alter the genetic structure of our own species—thus, the concern over the ethical ramifications of such power, as well as the possible health and ecological consequences of the creation of these bacterial forms. That being said, genetic engineering of other organisms for various fields are many and continue to grow. The following lists only some of the applications of genetic engineering in various fields:

Agriculture—Crops having larger yields, disease- and drought-resistance; bacterial sprays to prevent crop damage from freezing temperatures; and livestock improvement through changes in animal traits.

Industry—Use of bacteria to convert old newspaper and wood chips into sugar; oil- and toxin-absorbing bacteria for oil spill or toxic waste cleanups; and yeasts to accelerate wine fermentation.

Medicine—Alteration of human genes to eliminate disease (experimental stage); faster and more economical production of vital human substances to alleviate deficiency and disease symptoms (but not to cure them); substances include insulin, interferon (cancer therapy), vitamins, human growth hormone ADA, antibodies, vaccines, and antibiotics.

Research—Modification of gene structure in medical research, especially cancer research.

Food processing—Use of rennin (an enzyme) in cheese ageing.

What is a polymerase chain reaction and its connection to DNA?

Polymerase chain reaction—also known by its acronym PCR—is a laboratory technique that amplifies or copies any piece of DNA very quickly without using cells; DNA amplification is a method in which a small piece of DNA is copied thousands of times using PCR. DNA amplification is used in cloning, to detect small amounts of DNA in a sample, and to distinguish different DNA samples, as in DNA fingerprinting (see below).

Simply put, the process is as follows: DNA is incubated in a test tube with a special kind of DNA polymerase, a supply of nucleotides, and short pieces of synthetic, single-strand DNA. With a special machine, PCR can make billions of copies of a particular segment of DNA in a few hours (each cycle of the PCR procedure takes only about five minutes). At the end of the cycle, the DNA segment—even one with hundreds of base pairs—will be doubled. PCR is much faster than the days it took to clone a piece of DNA by making a recombinant plasmid and letting it replicate within bacteria. PCR was developed in 1983 by the American biochemist Kary Mullis (1944–) at Cetus Corporation, a California biotechnology firm; in 1993 Mullis, along with British-born Canadian chemist Michael Smith (1932–2000), won the Nobel Prize in Chemistry for development of PCR.

How is PCR used for DNA profiling?

The PCR (polymerase chain reaction) technique is one of the methods used for DNA profiling. The success of the method depends on identifying where the DNA of two individuals varies the most and how this variation can be used to discriminate between two different DNA samples. In fact, PCR-based DNA fingerprinting is a rapid, inexpensive method and requires only a very small amount of DNA—as little as fifty white blood cells.

What was the first commercial use of genetic engineering?

Commercial recombinant DNA technology was first used to produce human insulin in bacteria. In 1982, genetically engineered insulin was approved by the FDA for use by diabetics. Insulin is normally produced by the pancreas, and for more than fifty years the pancreas of slaughtered animals such as swine or sheep was used as an insulin source. To provide a reliable source of human insulin, researchers harvested the insulin gene from cellular DNA; they made a copy of DNA carrying this insulin gene and spliced it into a bacterium. When the bacterium was cultured, the microbe split from one cell into two cells, and both cells got a copy of the insulin gene. Those two microbes grew, then divided into four, those four into eight, the eight into sixteen, and so forth. With each cell division, the two new cells each had a copy of the gene for human insulin. Because the cells had a copy of the genetic "recipe card" for insulin, they could make the insulin protein. In fact, in this case, using genetic engineering to produce insulin was both cheaper and safer for patients, as some patients were allergic to insulin from other animals.

What is the basis for DNA fingerprinting?

DNA fingerprinting, also known as DNA typing or DNA profiling, is based on the unique genetic differences that exist between individuals. Most DNA sequences are identical, but out of one hundred base pairs (of DNA), two people will generally differ by one base pair. Since human DNA contains three billion base pairs, one individual's DNA will differ from another's by three million base pairs.

When was DNA fingerprinting developed?

British geneticist Sir Alec Jeffreys (1950–) developed DNA fingerprinting in the early 1980s, when he was studying inherited genetic variations between people. He was one of the first scientists to describe small DNA changes, referred to as single nucleotide polymorphisms (SNPs). From SNPs, he began to look at tandem repeat DNA sequences in which a short sequence of DNA was consecutively repeated many times.

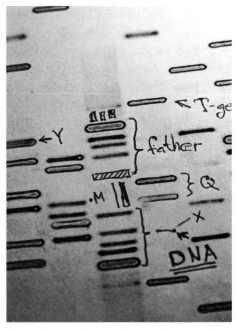

Using only a small sample from a subject, scientists can use DNA fingerprinting to quickly identify a species, a person, or a person's parents. It is a very useful tool in forensics.

What types of samples can be used for DNA fingerprinting?

Any body fluid or tissue that contains DNA can be used for DNA fingerprinting, including hair follicles, skin, earwax, bone, urine, feces, semen, or blood. In criminal cases, DNA evidence may also be gathered from dandruff; from saliva on cigarette butts, chewing gum, or envelopes; and from skin cells on eyeglasses.

What was one of the largest forensic DNA investigations in U.S. history?

The identification of the remains of the victims from the September 11, 2001 terrorist attacks in New York City has comprised the largest and most difficult DNA identification project to date. After 1.6 million tons of debris was removed from the site of the attacks on the World Trade Center, only 239 intact bodies (out of 2,795) were found, along with about 20,000 pieces of human remains. In order to match DNA profiles to the bodies, personal items such as razor blades, combs, and toothbrushes were collected from the victims' homes. When possible, cheek swabs were taken from the victims' family members for comparison with remains. The identification process was still ongoing as of this writing, with about 1,121 victims still not identified.

What is the accuracy of DNA fingerprinting?

Sometimes procedural sources of error (such as contamination from another sample or improper sample preparation) occur, but in many cases, DNA fingerprinting has become more reliable. It would be nice to have more twins (DNA from identical twins will have identical DNA profiles), but of course, that is not practical when determining the connections of people for heredity interpretation or crime cases. Overall, DNA testing is said to be 99.99 percent accurate for most of the tests, thanks to modern technology and the rules for DNA data analysis. In most cases, tests for such things as a crime scene are carried out at least two times from two independent analytical facilities. If a discrepancy occurs between the two tests, then fresh samples are often collected and the entire process is carried out again.

What are applications of DNA fingerprinting?

DNA fingerprinting is used to determine paternity, in forensic crime analysis, in population genetics to analyze variation within populations or ethnic groups, in conservation biology to study the genetic variability of endangered species, to test for the presence of specific pathogens in food sources, to detect genetically modified organisms either within plants or food products, in evolutionary biology to compare DNA extracts from fossils to modern-day counterparts, and in the identification of victims of a disaster.

Can an innocent person be convicted based on DNA analysis?

Current methods of DNA analysis are very sensitive, as only a few cells are needed for DNA extraction. However, it is possible for an innocent person's DNA to be found at any crime scene, either from accidental contamination or direct contamination by a third party. Also, a partial DNA profile from a crime scene could match that of an innocent person whose DNA is already in a DNA data bank. In addition, close relatives of suspected criminals could also be partial matches to a DNA profile.

What is CODIS?

CODIS is the FBI Laboratory's Combined DNA Index System, which allows federal, state, and local police agencies to compare DNA profiles electronically. CODIS uses two indexes: 1) the Forensic Index, which contains DNA profiles from crime scenes; and 2)

Can DNA be extracted from a mummy?

Yes, DNA can be (and has been) extracted from a mummy. However, the problem with extraction of ancient DNA has to do with contamination from modern DNA. In order to minimize contamination, researchers usually try to get DNA from inside teeth or bone. Ancient DNA is being used to study the genealogy of the pharaohs of Egypt.

the Offender Index, which contains DNA profiles of individuals convicted of sex offenses and other violent crimes.

What is a gene chip?

A gene chip is part of the process of microarray profiling; it is also known as a biochip or a DNA chip. It is about the size of a postage stamp and is based on a glass wafer, holding as many as 400,000 tiny cells. Each tiny cell can hold DNA from a different human gene and can perform thousands of biological reactions in a few seconds. These chips can be used by pharmaceutical companies to discover what genes are involved in various disease processes. They can also be used to type single nucleotide polymorphisms (SNPs), which are base pair differences that are found approximately every 500 to 1,000 base pairs in DNA. More than three million SNPs are in the human genome. They are very important in DNA typing because they represent about 98 percent of all DNA polymorphisms.

What is pharmacogenomics?

Pharmacogenomics is the use of DNA technology to develop new drugs to treat individual patients. For example, the interaction of a drug with a specific protein can be studied and then compared to a cell in which a genetic mutation has inactivated that protein. Its potential is to tailor drug therapy to an individual's genome—a tailoring that could reduce adverse drug reactions and increase the efficacy of drug treatment.

INSIDE OTHER BIOTECH LABS

What is bioprospecting?

Bioprospecting involves the search for possible new plant or microbial strains, particularly from the world's largest rain forests and coral reefs. These organisms are then used to develop new phytopharmaceuticals. Who owns the resources of these countries is a controversial subject: the countries in which the resources reside or the company that turns them into valuable products.

What is a bioreactor?

A bioreactor is a large vessel in which a biological reaction or transformation occurs. Bioreactors are used in bioprocessing technology to carry out large-scale mammalian cell culture and microbial fermentation. For example, penicillin (an antibiotic) is made from the fungus *Penicillium chrysogenum* in a bioreactor; the vitamin riboflavin is made from the bacteria *Eremothecium ashbyii*; and the hormone insulin is made from the bacteria recombinant *Escherichia*.

What is cell culture?

Cell culture is the cultivation of cells (outside the body) from a multicellular organism. This technique is very important to biotechnology processes because most research pro-

Bioreactors are machines in which bacteria can be processed to make useful drugs such as penicillin and insulin.

grams depend on the ability to grow cells outside the parent animal. Cells grown in culture usually require very special conditions (for example, specific pH, temperature, nutrients, and growth factors). Cells can be grown in a variety of containers, ranging from a simple petri dish to large-scale cultures in roller bottles (bottles that are rolled gently to keep culture medium flowing over the cells).

What is a control group?

A control group is the experimental group tested without changing the variable. For example, to determine the effect of temperature on seed germination, one group of seeds may be heated to a certain temperature. The researcher will then compare the percent of seeds in this group that germinate and the time it takes them to germinate to another group of seeds (the control group) that have not been heated. All other variables, such as light and water, will remain the same for each group.

What is a double-blind study?

In a double-blind study, neither the subjects of the experiment nor the people administering the experiment know the critical aspects of the experiment. This method is used to guard against both experimenter bias and placebo effects.

What is FISH?

FISH (an acronym that stands for fluorescent in situ hybridization) is a method in which a clone gene is "painted" with a fluorescent dye and mapped to a chromosome. In FISH,   385 cells are arrested in the metaphase stage of mitosis and placed on a slide where they burst open, spreading chromosomes over the surface. A fluorescent-labeled DNA "piece" of interest is placed on the slide and incubated long enough for hybridization to occur. The slide is then viewed under a fluorescence microscope that focuses ultraviolet light on the chromosomes, and the researcher will view hybridized regions of the chromosome that fluoresce. This can generate a physical map matching clones and gene markers to specific parts of a chromosome.

What is bioremediation?

Bioremediation is the use of organisms to remove toxic materials from the environment. Bacteria, protists, and fungi are good at degrading complex molecules into waste products that are generally safe and recyclable. Sewage treatment plants perform bioremediation in a limited way. An example of bioremediation is the massive cleanup in Alaska following the Exxon Valdez oil spill in 1989. The superficial layer of oil was removed by suction and filtration, but the oil-soaked beach was cleaned by bacteria that could use oil as an energy source.

What is tissue engineering?

Tissue engineering is used to create semisynthetic tissues that are used to replace or support the function of defective or injured body parts. It is a broad field, encompassing cell biology, biomaterial engineering, microscopic engineering, robotics, and bioreactors, in which tissues are grown and nurtured. Tissue engineering can improve on current medical therapies by designing replacements that mimic natural tissue function. In fact, commercially produced skin is already in use for treating patients with burns and diabetic ulcers.

What is xenotransplantation?

Xenotransplantation is the transplantation of tissue or organs from one species to another. The development of this technique has led to the breeding of animals specifically for use as human organ donors. Because humans would reject a nonhuman organ as foreign, transgenic animals (for example, pigs) are genetically altered with human DNA with the hope of suppressing any eventual rejection. One of the major risks of xenotransplantation is the risk of transplanting animal viruses along with the transfer cells or organs. Since the patient is already immunosuppressed, the patient could die from the viral infection, or the virus could be spread to the general population. But that doesn't mean xenotransplantaion has not been successful for humans: For example, inert heart valves from pigs have been used in human heart valve-replacement operations, and scientists continue to test animals to treat human disease, such as using fetal pig cells, to treat strokes, epilepsy, and spinal cord injuries.

Why have there been so many men's health studies—but few for women?

No, it is not your imagination, ladies; yes, there is, indeed, a discrepancy between male and female scientific studies—and those health studies favor men (even when it comes to using male animals for studies versus female animals on experiments that have nothing to do with gender). For example, in clinical trials, much of the reason for ignoring females seems to have to do with hormones (males don't have a menstrual cycle, so their hormones don't fluctuate as much; thus, the results are more "stable") and pregnancy (in 1977, the U.S. Food and Drug Administration banned women who had the potential to become pregnant from participating in early stage clinical trials)—ideals that have lingered. But in the past few decades, gender research has changed, albeit slowly. More women's studies have emerged, such as the Nurse's Health Study that started in 1976 and, over the years, has included over 200,000 nurse-participants; there is also the Women's Health Initiative, started in 1991, with over 150,000 participants, which studies postmenopausal women.

The reason for more emphasis on female studies (besides the fact that they represent over half the world's population) has to do with necessity: These recent studies have shown that males and females *do* have some significant health differences. For example, it's been found that women only need half as much of the influenza vaccine for the same level of protection as men, but they are often given the same dose as men. It has also been shown that daily aspirin helps reduce a first heart attack in men by a third, but not so much for women (yet it does help with reducing strokes in women); but if you are a healthy woman over sixty-five years of age, a daily aspirin will help reduce strokes *and* heart attacks similar to men who take the drug (but it also can cause gastrointestinal problems for both genders). And in the controversial world of stem cell research, researchers have discovered that cells from women can regenerate skeletal muscle tissue better than cells extracted from men, but most of the research has been done on male cells.

How does an *in vivo* study differ from an *in vitro* study?

An *in vivo* study uses living biological organisms and specimens to obtain results. In contrast, an *in vitro* biological study is carried out in isolation from a living organism, such as in a petri dish or test tube.

SEEING SMALL

What distinguishes the different types of microscopes?

Microscopes have played a central role in the development of cell biology, allowing scientists to observe cells and cell structures that are not visible to the human eye. The two basic types of microscopes are light microscopes and electron microscopes, the major differences being the source of illumination and the construction of the lenses. Light microscopes utilize visible light as the source of illumination and a series of glass lenses; electron microscopes utilize a beam of electrons emitted by a heated tungsten filament as the source of illumination, and the lens system consists of a series of electromagnets.

Advances using optical techniques also led to the development of specialized light microscopes, including fluorescence microscopy, phase-contrast microscopy, and differential interference-contrast microscopy. In fluorescence microscopy, a fluorescent dye is introduced to specific molecules. Both phase-contrast microscopy and differential interference-contrast microscopy use techniques that enhance and amplify slight changes in the phase of transmitted light as it passes through a structure that has a different refractive index than the surrounding medium.

Who invented the compound microscope?

The principle of the compound microscope, in which two or more lenses are arranged to form an enlarged image of an object, occurred independently, at about the same time, to more than one person. Certainly many opticians were active in the construction of telescopes at the end of the sixteenth century, especially in Holland, so it is likely that the idea of the microscope may have occurred to several of them independently. In all probability the date may be placed within the period 1590–1609, and the credit should go to three spectacle makers in Holland. Dutch spectacle makers Zacharias Janssen (or Jansen; c. 1580–c. 1638), German-Dutch lensmaker Hans Lippershey (1570–1619), and although disputed, possibly Zacharias Janssen's father Hans Janssen have all been cited at various times as deserving chief credit. English scientist and inventor Robert Hooke (1635–1703) was the first to make the best use of a compound microscope, and his book *Micrographia*, published in 1665, contains some of the most beautiful drawings of microscopic observations ever made.

What is the difference between magnification and resolution?

Magnification—making smaller objects seem larger—is the measure of how much an object is enlarged. Resolution is the minimum distance that two points can be separated and still be seen as two distinct points.

How do dissecting microscopes differ from compound microscopes?

Compared to compound microscopes, dissecting microscopes—also called stereoscopic microscopes—provide a much larger working distance between the lens and stage in order to dissect and manipulate specimens. The light source on a dissecting microscope is above the specimen since the specimen is often too thick to allow light to be transmitted from a light source below the specimen. Dissecting microscopes are always binocular, which provides a three-dimensional image.

Who invented the electron microscope?

The theoretical and practical limits to the use of the optical microscope were determined by the wavelength of light. But when the oscilloscope was developed, scientists realized that cathode-ray beams could be used to resolve much finer detail because their wavelength was so much shorter than that of light. The electron microscope revolutionized biological research, and for the first time scientists could see the molecules of cell structures, proteins, and viruses.

The electron microscope we know today was developed in numerous steps, with many companies and scientists inventing better and more precise instruments. In 1928, German physicist Ernst August Friedrich Ruska (1906–1988) and German electrical engineer Max Knoll (1897–1969), using magnetic fields to "focus" electrons in a cathode-ray beam, produced a crude instrument that gave a magnification of 17; by 1932, they had developed an electron microscope having a magnification of 400. By 1938, Canadian inventor and physicist James Hillier (1915–2007) and others had advanced this magnification to 7,000—the first practical electron microscope; he also helped push to make the instrument standard equipment in hospitals, universities, and laboratories around the world. More improvements came later; for example, Russian-American inventor and scientist Vladimir Zworykin (1889–1982; he was also a pioneer in television technology) helped, in 1939, to develop an electron microscope with a magnification of 100,000 times, whereas today's electron microscope can display a power magnification of up to 2,000,000.

How does a transmission electron microscope differ from a scanning electron microscope?

The electrons used to visualize the specimens in transmission electron microscopes are transmitted by the material. The scanning electron microscope beams the electrons onto the surface of the specimen from a fine probe that rapidly passes back and forth. Electrons reflected back from the surface of the specimen, along with other electrons emitted by the specimen itself, are amplified and transmitted to a television screen for viewing.

What is scanning tunneling microscopy?

Scanning tunneling microscopy (STM), also called a scanning probe microscopy, was developed in the 1980s to explore the surface structure of specimens at the atomic level. This technique uses electronic methods to move a metallic tip (a conducting material such as platinum-iridium), composed ideally of a single atom, across the surface of a specimen. As the tip is moved across the surface of the specimen, electrical voltage is applied to the surface. If the tip is close enough to the surface and the surface is electrically conductive, electrons will begin to leak or "tunnel" across the gap between the probe and the sample. The tip of the probe is automatically moved up and down to maintain a constant rate of electron tunneling across the gap as the probe scans the sample. The movement is presented on a video screen. Successive scans then build up an image of the surface at atomic resolution.

What are the common types of slide preparations for investigation with a microscope?

Commonly prepared slide preparations are wholemounts, smears, squashes, and sections. Smears, squashes, and sections are techniques used to make specimens thinner or smaller; wholemounts are often used to examine an entire organism or specific organ structure in some detail. Smears are mostly prepared for bacteriological and blood specimens, squashes are prepared to study chromosomes, and sections are prepared to examine tissues and cells.

What are the basic steps when preparing a specimen for examination?

The three basic steps to prepare a specimen are fixation (preservation), staining, and mounting. Preservation prevents destruction or decay of the specimen as well as inhibiting microbiological growth. Different stains and dyes attach to different parts of a cell, such as the nucleus. Specimens are mounted in a medium that is also a preservative and often covered with a cover slip. "Fixing" a biological specimen retains a reasonably good semblance of the object as it appeared when it was alive. It allows the scientists to observe details of the external and internal anatomy of the specimen.

What are simple stains?

Simple stains highlight an entire microorganism so that cellular shapes and basic structures are visible. Simple stains commonly used include methylene blue, carbolfuchsin,

Who first developed the modern technique for the isolation of cell parts?

American-Canadian anatomist Robert R. Bensley (1867–1956) and American anatomist Normand Louis Hoerr (1902–1958) disrupted the liver cells in a guinea pig and isolated mitochondria in 1934. Between 1938 and 1946, Albert Claude (1899–1983) continued the work of Bensley and Hoerr and isolated two fractions—a heavier fraction consisting of mitochondria and another fraction of lighter submicroscopic granules, which he called microsomes. Further developments led to the development of centrifugal techniques of cell fractionation commonly used now. The development of this procedure was one of the earliest examples of differential centrifugation. It initiated the era of modern experimental cell biology.

crystal violet, and safranin. A stain is applied to a fixed smear for a certain amount of time and then washed off, and the slide is dried and examined.

What is centrifugation?

Centrifugation is the separation of immiscible liquids or solids from liquids by applying centrifugal force. Since the centrifugal force can be very great, it speeds the process of separating these liquids instead of relying on gravity. Biologists primarily use centrifugation to isolate and determine the biological properties and functions of subcellular organelles and large molecules. They study the effects of centrifugal forces on cells, developing embryos, and protozoa, thus allowing scientists to determine certain properties about cells, including surface tension, relative viscosity of the cytoplasm, and the spatial and functional interrelationship of cell organelles when redistributed in intact cells. (For more about cells, see the chapter "Cellular Basics.")

What is chromatography?

The Russian-Italian biochemist Mikhail Semyonovich Tsvet (or Tswett) (1872–1919) coined the term "chromatography" and published the first paper on the method in 1903. The term comes from the Greek words *chroma*, meaning "color," and *graphein*, meaning "writing or drawing." Chromatography has many applications in biology. For example, it is used to separate and identify amino acids, carbohydrates, fatty acids, and other natural substances. Environmental testing laboratories use chromatography to identify trace quantities of contaminants such as PCBs in waste oil and pesticides such as DDT in ground water. It is also used to test drinking water and test air quality. Pharmaceutical companies use chromatography to prepare quantities of extremely pure materials.

What is gas chromatography?

Gas chromatography, specifically gas-liquid chromatography, involves a sample being vaporized and injected onto the head of the chromatographic column. The sample is transported through the column by the flow of an inert, gaseous mobile phase. The column itself contains a liquid stationary phase that is absorbed onto the surface of an inert solid. The carrier gas must be chemically inert. Commonly used gases include nitrogen, helium, argon, and carbon dioxide. Gas-liquid chromatography is the most widely used chromatographic technique for environmental analyses. Analysis of organic compounds is possible for a variety of matrices such as water, soil, soil gas, and ambient air. It is often used to test hazardous waste sites for determining personal protective equipment (PPE) levels and emergency response testing.

What is spectroscopy?

Spectroscopy includes a range of techniques to study the composition, structure, and bonding of elements and compounds. The different methods of spectroscopy use different wavelengths of the electromagnetic spectrum to study atoms, molecules, ions, and the bonding between them. The types of spectroscopy are: Nuclear magnetic resonance spectroscopy (uses radio waves); infrared spectroscopy (uses infrared radiation); atomic absorption spectroscopy, atomic emission spectroscopy, and ultraviolet spectroscopy (use visible and UV radiation); and X-ray spectroscopy (uses X-rays).

What is the electromagnetic spectrum?

The electromagnetic spectrum is the range of wavelengths. It ranges from gamma rays with very short wavelengths (10^{-5} nanometers) and high energy to radio waves with longer wavelengths (10^3 meters) and less energy. Visible light, seen as color (in other words, all the colors of the rainbow), occurs between 380 and 750 nanometers; ultraviolet light has a shorter wavelength, while infrared light has a greater wavelength.

What are X-rays?

X-rays are electromagnetic radiation with short wavelengths (10^{-3} nanometers) and a great amount of energy. They were discovered in 1898 by William Conrad Roentgen (1845–1923). X-rays are frequently used in medicine because they are able to pass through opaque, dense structures such as bone and form an image on a photographic plate.

What is X-ray crystallography?

X-ray crystallography, also called X-ray diffraction, is used to determine crystal structures by interpreting the diffraction patterns formed when X-rays are scattered by the electrons of atoms in crystalline solids. X-rays are sent through a crystal to reveal the pattern in which the molecules and atoms contained within the crystal are arranged.

What important scientific discoveries were made using X-ray diffraction?

In 1951, the protein a-helix was discovered by American chemist and biochemist Linus Pauling (1901–1994) using X-ray diffraction; it was used to reveal the double-helix structure of DNA in 1953 by New Zealand-born English physicist and molecular biologist Maurice Wilkins (1916–2004), English molecular biologist Francis Crick (1916–2004), British biophysicist and X-ray crystallographer Rosalind Franklin (1920–1958), and American molecular biologist and geneticist James Watson (1928–). British chemist Dorothy Mary Crowfoot Hodgkin (1901–1994) also used the technique to determine the structure of vitamin B_{12} in 1956; she is also credited with the development of protein crystallography.

What is nuclear magnetic resonance (NMR) and magnetic resonance imaging (MRI)?

Nuclear magnetic resonance (NMR) is a process in which the nuclei of certain atoms absorb energy from an external magnetic field. Scientists use NMR spectroscopy to identify unknown compounds, check for impurities, and study the shapes of molecules. They use the knowledge that different atoms will absorb electromagnetic energy at slightly different frequencies.

Magnetic resonance imaging (MRI), sometimes called nuclear magnetic resonance imaging (NMRI), is a noninvasive diagnostic technique. It is useful in detecting small tumors, blocked blood vessels, or damaged vertebral disks. Because it does not involve the use of radiation, it can often be used where X-rays are dangerous. Large magnets beam energy through the body, causing hydrogen atoms in the body to resonate. This produces energy in the form of tiny electrical signals. A computer detects these signals, which vary in different parts of the body and according to whether an organ is healthy or not. The variation enables a picture to be produced on a screen and interpreted by a medical specialist.

Why is using MRI often "better" than using X-rays?

What distinguishes MRI from computerized X-ray scanners is that most X-ray studies cannot distinguish between a living body and a cadaver, while MRI "sees" the difference between life and death in great detail. More specifically, it can discriminate between healthy and diseased tissues with more sensitivity than conventional radiographic instruments like X-rays or CAT scans. CAT (computerized axial tomography) scanners have been around since 1973 and are actually glorified X-ray machines. They offer three-dimensional viewing but are limited because the object imaged must remain still. The main advantages of MRI are that it not only gives superior images of soft tissues (like organs), it can also measure dynamic physiological changes in a noninvasive manner (without penetrating the body in any way). A disadvantage of MRI is that it cannot be used for every patient. For example, patients with implants, pacemakers, or cerebral

What is nanotechnology?

The term "nanotechnology" was coined in 1974 by Japanese scientist Norio Taniguchi (1912–1999) at the University of Tokyo. It includes a number of technologies that deal with the miniaturization of existing technology down to the scale of a nanometer (one-billionth of a meter) in size, about the size of molecules and atoms—1/40,000th the width of a human hair.

The potential of nanotechnology is enormous; for example, it includes microcomputers capable of storing trillions of bytes of information in a space smaller than a dime; portable fluids containing nanobots that are programmed to destroy cancer cells or deliver medicines; the ability to sense and adapt to environmental stimuli such as heat, light, sound, surface texture, and chemicals; and to perform complex calculations faster and more efficiently—singularly or *en masse*. A push is even in place to develop nanobots that will be able to move, communicate, and work together, assemble things on a molecular level—and to even possibly repair or replicate themselves.

aneurysm clips made of metal cannot be examined using MRI because the machine's magnet could potentially move these objects within the body, causing damage.

What is ultrasound?

Ultrasound is another type of 3-D computerized imaging. Using brief pulses of ultra-high frequency acoustic waves (lasting 0.01 second), it can produce a sonar map of the imaged object. The technique is similar to the echolocation used by bats, whales, and dolphins.

What is lypholization?

Lypholization is a freeze-drying technique for preservation and storage of bacteria and other microorganisms. Bacteria can be stored for extended periods of time (three to five years) as frozen cell suspensions or as freeze-dried (lypholized) cultures. This technique is achieved by placing bacteria in a nutrient broth containing 15 to 25 percent glycerol and freezing at temperatures of -94°F (-70°C) or lower. The glycerol reduces ice crystal formation that would cause subsequent cell damage and disrupt biological structures.

BIOTECH LABS AND FOOD

How long have humans been "genetically modifying" organisms?

Today, we talk about how GMOs (Genetically Modified Organisms) are genetically modified organisms created using new techniques of recombinant DNA technology. But the term is misleading because almost all domesticated animals and crop plants have been genetically modified over thousands of years by human selection and cross-breeding.

What do GMO, GE, and GM stand for in terms of food?

These terms all have a different definition, but they are close in terms of meaning. GMOs have had certain changes in their DNA through genetically engineered techniques. GE means Genetically Engineered and often refers to crops. For example, GE seeds were introduced commercially in 1996 and are now found throughout certain areas of the world, especially places that grow corn, soybeans, and cotton; the GE crops are engineered to withstand the direct application of chemical herbicides and/or GE crops that are engineered to produce toxins that kill certain insect pests. GM stands for Genetic Modification and usually refers to food, as in GM foods, which are produced from GMOs.

How widespread is the use of GMO crops all over the world?

As of 2002, more than 120 million acres (40 million hectares) of fertile farmland were planted with GMO crops. The acreage is confined to four countries: United States (containing 68 percent of the total acreage), Argentina (22 percent), Canada (6 percent), and China (3 percent). In 2012, GMO crops grew on about 420 million acres of land in

The modern farm industry in some countries like the United States have produced Genetically Modified Organisms, such as these vegetables, that they hope will resist disease and survive better in bad weather. However, there are concerns that GMOs might not be healthy for us.

twenty-eight countries worldwide, a record high according to the International Service for the Acquisition of Agri-biotech Applications, an industry trade group. Thus it is estimated that the area of land devoted to genetically modified crops has ballooned by one hundred times since farmers first started growing the crops commercially in 1996. Over the past seventeen years, millions of farmers in twenty-eight countries have planted and replanted GMO crop seeds on a cumulative 3.7 billion acres of land—an area 50 percent larger than the total land mass of the United States, the group adds. But in about thirty other countries around the world, including Australia, Japan, and all of the countries in the European Union, the production of GMOs has significant restrictions or outright bans—mostly because they have not been proven to be safe.

How widespread is the use of GMO crops in the United States?

According to the United States Department of Agriculture (USDA), in 2009, 93 percent of soy, 93 percent of cotton, and 86 percent of corn grown in the U.S. were based on GMOs. It is estimated that over 90 percent of canola grown is genetically modified; sugar beets, squash, and Hawaiian papaya also have GM varieties. Thus it is estimated that GMOs are now present in more than 80 percent of packaged products in the average United States or Canadian grocery store. This means the ingredients made from them—including ingredients you read on a label in a processed food product that may have

395

What was the Flavr Savr tomato?

The Flavr Savr tomato, also called also known as CGN-89564-2, was developed in response to consumer complaints that tomatoes were either too rotten to eat when they arrived at the store or too green. Growers had found that they could treat green tomatoes in the warehouse with ethylene, a gas that causes the tomato skin to turn red—but the tomato itself stayed hard. In the late 1980s, researchers at the biotech company Calgene discovered that the enzyme polygalactouronase (PG) controlled rotting in tomatoes. The scientists reversed the DNA sequence of PG; the effect was that tomatoes turned red on the vine, yet the skin of the tomatoes remained tough enough to withstand the mechanical pickers. However, before the Flavr Savr tomato was introduced to the market in the mid-1990s, Calgene disclosed to the public how the tomato was bioengineered—thus causing a public protest that led to the worldwide movement against genetically modified organisms (GMOs).

corn syrup, sugar, or vegetable oil—are also genetically modified. At this time, none of these GM products are required to be labeled as genetically engineered.

What is Starlink corn and Bt?

Starlink was a bioengineered corn variety that was genetically modified to include a gene from the bacterium *Bacillus thuringiensis* (Bt), which produces a protein (called an endotoxin) that kills some types of insects. Bt endotoxin has been registered as a biopesticide in the United States since 1961, and the Bt endotoxin has been used by organic farmers for biological pest control. The endotoxins only become activated in the guts of susceptible insects. Because of the significant losses to corn crops caused by the European corn borer, scientists targeted the corn plant itself as a candidate for insertion of the Bt gene.

What is "Frankenfood"?

"Frankenfood" is a term invented by environmental and health activist groups to denote any food that has been genetically modified (GM) or that contains genetically modified organisms (GMO). Opposition to GM food is based on concerns that the gene pool of "natural" plants could be altered permanently if exposed to pollen from genetically altered plants. A fear also exists that people and animals that consume GM food might have allergic reactions to altered protein or could develop health problems later.

What is biopreservation?

Biopreservation refers to the preservation and enhanced safety of food using biological materials. An example of this is nisin, a bacterial protein that can act as a broad-spectrum antibiotic. Nisin cannot be synthesized chemically, so the nisin-producing bacteria *Lactobacillus* must be used to generate the protein.

What is a biopesticide?

A biopesticide is a chemical derived from an organism that interferes with the metabolism of another species. An example is the Bt toxin *Bacillus thuringiensis*, which interferes with the absorption of food in insects but does not harm mammals (for more information on Bt toxin, see above).

What are biosensors?

A biosensor is a unique combination of biological substances (for example, a microbe, cell, enzyme, or antibody) linked to a detector. It can be used to measure very low concentrations of a particular substance. An example of a biosensor currently on the market is the insulin pump, which maintains correct blood glucose concentrations for diabetics.

What is bovine growth hormone, and why is it so controversial?

One of the earliest applications of biotechnology was the genetic engineering of a growth hormone produced naturally in the bovine pituitary. The recombinant Bovine Growth Hormone (rBGH), or Bovine Somatotropin (rBST), a genetically engineered hormone manufactured by Monsanto, was reported to increase milk production in lactating cows. Using biotechnology, scientists bioengineered the gene that controls bovine growth hormone production into *E. coli* bacteria, grew the bacteria in fermentation chambers, and thus produced large quantities of hormone. The bioengineered hormone, when injected into lactating cows, resulted in an increase of up to 20 percent in national milk production. Using bovine GH, farmers were able to stabilize milk production in their herds, avoiding fluctuations in production levels.

But bovine growth hormone is controversial for those who drink cow's milk, especially infants. The hormone actually makes the cow produce more milk than it naturally would, thus reportedly making the cow more susceptible to disease. This is attached to a concern that the amount of antibiotics given to the cow—if it does become diseased— would be passed along to the humans who consume the milk. Still another claim is that the hormone stimulates another hormone called Insulin-Like Growth Factor-1, or IGF-1, which promotes cell division and is associated, in this instance, as a possible impetus for cancer growth.

BIOLOGY AND YOU

BEING HUMAN

How did humans evolve?

Evolution of the Homo lineage of modern humans (*Homo sapiens sapiens*) is a highly debated subject, mainly because so few fossil remains exist that it is difficult to come to a definite conclusion. But suggestions abound: For example, several researchers propose that modern humans originated from a hunter of nearly 5 feet (1.5 meters) tall, *Homo habilis*, who may have evolved from an australopithecine ancestor. Near the beginning of the Pleistocene about two million years ago, *Homo habilis* is thought to have evolved into *Homo erectus* (Java Man), who used fire and possessed culture. Middle Pleistocene populations of *Homo erectus* are said to show steady evolution toward the anatomy of *Homo sapiens* (Neanderthals, Cro-Magnons, and modern humans), 120,000 to 40,000 years ago and were known to build huts and make clothing. Of course, all this is still speculation, and many scientists would disagree with each lineage described above.

What is a race?

The term "race" was originally used to describe subspecies. However, as genetic analysis of humans has shown, greater genetic variation within geographic subpopulations (races) exists than among the human population as a whole. In other words, so much genetic overlap exists between groups formerly designated as races that the term is meaningless and biologically indefensible.

Are humans evolving?

This question depends on your view of "evolution": if you consider that minute changes have occurred in the human body over the past centuries, then yes, humans are evolving. For example, research suggests that the human brain has been shrinking over the past 5,000 years. Another study showed that wisdom teeth are beginning to vanish from the human mouth—with about 35 percent of people already born without such teeth. Another sign is thought to be the blending of populations, or gene flow, in which racial differences disappear. Still another study suggests that humans of the future—a trend we are starting now—will no longer require big muscles to perform feats of strength, thanks to machines and technology—meaning we may be weaker and smaller in the future. But this may also mean, say some researchers, that we may be weaker in terms of our immune systems, too—becoming more susceptible to pathogens such as viruses and bacteria.

YOU AND YOUR CELLS

What is a stem cell, and why is "stem cell research" so controversial?

Stem cells are undifferentiated cells—meaning that they do not have a specific function—that are capable, under certain conditions, of producing cells that can become a specific type of tissue. Stem cells present in adult humans are found in bone marrow and other tissues (such as fat). However, most research interest is focused on stem cells present in fetal tissue, which, in a laboratory setting, can reproductively divide indefinitely and be stimulated into becoming a variety of different cell types. The potential benefits of stem cells have made the research of these cells an exciting research topic. In fact, stem cells could be used to grow new hearts that could be transplanted without fear of rejection or to renew the function of injured structures like the spinal cord. They could also be used as cell models for drug testing, thereby increasing the speed for finding cures. But the main controversy stems from the actual "harvesting" of the most useful stem cells—from fetal tissues from human embryos—especially the question of what point in cellular growth constitutes a "human."

Other than fetal cells, what are human sources of stem cells?

Many types of cells have been found that can be used as stem cell sources, but according to some researchers, these cells may not be as useful as embryonic stem cells. The following lists the possibilities:

Cell Source	Possible Use
Brain	Neurodegenerative diseases; spinal cord injury
Hair and skin	Burn healing
Breast (from cosmetic surgery)	Breast duct regeneration
Fatty tissue (liposuction leftovers)	Cartilage, bone, fat
Bone marrow	Almost any tissue; embryonic healing capacity
Pancreas	Diabetes treatment
Heart	Healing following myocardial infarction
Baby teeth (with associated tissue)	Can be used similarly to bone marrow

Can cells conduct electricity?

Yes, in fact, all living cells have membranes that allow them to maintain a difference in concentration of atoms located on the inside and outside of the cell, with some of these atoms being ions with a charge. This ability to maintain an imbalance of ions is called cell membrane potential and is similar to the electrical potential of a battery.

What is the most common blood cell in the human body?

Red blood cells, also known as erythrocytes, are the most common blood cells. A milliliter (ml) of blood contains approximately five billion red blood cells. The average per-

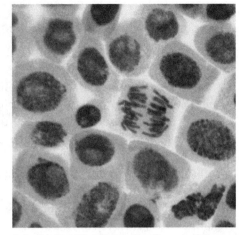

Scientists hope that stem cells will one day help cure genetic and other injuries in which cell regeneration is crucial, such as spinal injuries.

son has a total of twenty-five trillion red blood cells in their bloodstream! Red blood cells are very small; it would take a string of 2,000 red blood cells to circle a pencil.

How do skin cells synthesize vitamin D?

Vitamin D is crucial to normal bone growth and development. When UV light shines on a lipid present in skin cells, the compound is transformed into vitamin D. People native to equatorial and low-latitude regions of the earth have dark skin pigmentation as a protection against strong, nearly constant exposure to UV radiation. Most people native to countries at higher latitudes—where UV radiation is weaker and less constant—have lighter skin, allowing them to maximize their vitamin D synthesis.

How are drugs detoxified by cells?

Drugs are detoxified by the smooth endoplasmic reticulum of liver cells (for more about endoplasmic reticulum, see the chapter "Cellular Basics"). Detoxification usually involves changing the molecular structure of a toxin, a modification that increases the toxin's solubility, allowing it to be safely carried away by the blood and excreted via urine. Cells are able to increase their detoxification efforts when drug levels increase. Investigations conducted with rats that have been injected with a sedative known as phenobarbital have shown a striking increase in the amount of smooth endoplasmic reticulum of the liver cells.

What is a malignant cell?

Cells that reproduce quickly and expand beyond the tissue or organ where they originated are described as malignant or cancerous. Malignant tumors are more difficult to eradicate because they may grow in organs far removed from their origin. For example, cancers that originate in the lung can quickly spread (or metastasize) to the brain and other organs through the circulatory system.

401

Why do some people lack vitamin D?

During the shorter days of winter, the vitamin D synthesis that occurs in people who live in higher latitudes is limited to small areas of skin exposed to sunlight. Increased melanin pigmentation, present in people native to lower latitudes, reduces the production of vitamin D. Susceptibility to vitamin D deficiency is increased in these populations by the traditional clothing of many cultural groups native to low latitudes, which attempts to cover the body completely to protect the skin from overexposure to UV radiation. Most clothing effectively absorbs irradiation produced by ultraviolet B rays. The dose of ultraviolet light required to stimulate skin synthesis of vitamin D is about six times higher in African-Americans than in people of European descent. The presence of darker pigmentation and/or veiling may significantly impair sun-derived vitamin D production, even in sunny regions like Australia.

What are beta blockers?

Beta blockers—also called beta-adrenergic blocking agents—are medications that reduce blood pressure by blocking the effects of a hormone called epinephrine, also known as adrenaline. When you take beta blockers, the heart beats more slowly and with less force, thus reducing your blood pressure; they also help blood vessels open up to improve blood flow. An example of a beta blocker is propranolol, which is used to treat high blood pressure and angina, and works by protecting the heart against sudden surges of stress hormones like adrenaline.

How do statin drugs affect cell function?

Statins are a group of drugs that work to lower cholesterol levels, particularly the "bad cholesterol", or the low-density lipoprotein known as LDL. The drugs work in two ways: they can block an enzyme that is needed for cholesterol production, and they can increase LDL membrane receptors in the liver. Cholesterol can only get into cells by binding to specific receptors that remove the LDL from blood. The extra receptors that statins create help decrease the cholesterol levels. As Americans are more aware that high cholesterol is a major risk factor for heart disease, statins have become increasingly popular.

YOU AND YOUR BODY

What are certain organisms' normal body temperatures?

Normal body temperature is the acceptable temperature for an animal—whether human or not. The following chart lists normal body temperatures for a variety of organisms—

both ectotherms (can raise and maintain a steady body temperature) and endotherms (can raise but not maintain a steady body temperature)—including humans:

Animal	Normal Temperature °F	Normal Temperature °C
Human (endotherm)	98.6	37
Cat (endotherm)	101.5	38.5
Dog (endotherm)	102	38.9
Cow (endotherm)	101	38.3
Mare (endotherm)	100	37.8
Pig (endotherm)	102.5	39.2
Goat (endotherm)	102.3	39.1
Rabbit (endotherm)	103.1	39.5
Sheep (endotherm)	102.3	39.1
Pigeon (endotherm)	106.6	41
Lizard (ectotherm)	87.8–95	31–35
Salmon (ectotherm)	41–62.6	5–17
Rainbow trout (ectotherm)	53.6–64.4	12–18
Rattlesnake(ectotherm)	59–98.6	15–37
Grasshopper(ectotherm)	101.5–108	38.6–42.2

What animal has almost humanlike fingerprints?

While it is known that primates—not including humans—have a type of "fingerprint" (they are more like ridges for a better grip), one animal has fingerprints that are almost humanlike: the cuddly koala bear. These creatures from Australia are sometimes diffi-

cult to distinguish from a human's, as they are similar in size, shape, and pattern. No other animals have such a distinction, but they do have other "prints"—such as animals like pigs and dogs that have hairless snouts—and unique nose prints.

What is tooth decay?

Tooth decay is actually an infectious disease caused by bacteria called *Streptococcus mutans*—tiny critters that are able to ferment carbohydrates (the sugars and starches we eat) to form acids—which can lead to the demineralization of the tooth surface and eventual decay if it is left untreated. When you have a toothache from tooth decay, it means that the tooth's pulp (inner structure that contains the nerves and blood vessels) has become involved—

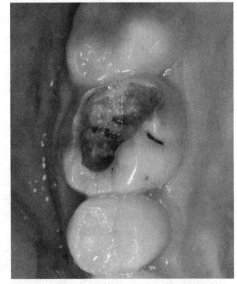

Not keeping one's teeth clean can encourage the growth of harmful bacteria that digest food particles and excrete acids that dissolve teeth.

403

and it's also why hot and cold foods or drinks can cause tooth pain and make your life miserable. In many cases, a root canal (removal of the pulp) is required to prevent the infection from spreading into the bone that holds the tooth in place. If the infection is not caught in time, it can lead to root decay, too, especially in older people whose gums normally recede as they get older. If not caught, this could also destroy the bone, causing the tooth to loosen or even abscess, and the affected tooth may need to be extracted.

Who discovered how muscles work?

British biologist Hugh Huxley (1924–2013) and English physiologist and biophysicist Andrew Huxley (1917–2012; the scientists were unrelated) researched theories regarding muscle contraction. Hugh Huxley was initially a nuclear physicist who entered the field of biology at the end of World War II. He used both X-ray diffraction and electron microscopy to study muscle contraction. Andrew Huxley obtained data similar to Hugh's, indicating that the contractile proteins thought to be present in muscles are not contractile at all, but rather slide past each other to shorten a muscle. This theory is called the "sliding filament theory of muscle contraction."

How do muscles work?

Muscle cells—whether the skeletal muscles in the arms or legs, the smooth muscles that line the digestive tract and other organs, or the cardiac muscle cells in the heart—work by contracting. Skeletal muscle cells are comprised of thousands of contracting units known as sarcomeres. The proteins actin (thin filament) and myosin (thick filament) are the major components of the sarcomere. These units perform work by moving structures closer together through space; in the skeletal muscles, they pull parts of the body through space relative to each other (for example, when you walk or swing your arms).

To visualize how a sarcomere works, try the following: Interlace the fingers of your two hands with the palms facing toward you (represents actin, myosin); push the fingers together so that the overall length from one thumb to the other is decreased (sarcomere length decreases). And from there, realize that any object attached to either thumb would be pulled through space as the fingers move together (sliding filament theory).

How do muscles respond to activities like weight lifting?

Weight lifting will cause muscles to grow—the size of the muscle cells increases, but the body does not actually grow more muscle cells. Rather, weight lifting causes the body to grow more of the thick (myosin) and thin (actin) proteins that aid muscle contraction. This process makes the muscle not only bigger, but stronger as well. Some muscles gain strength faster than others. In general, large muscles, like those present in your chest and back, grow faster than smaller ones, like those in your arms and shoulders. Most people can increase their strength between 7 and 40 percent after ten weeks of training each muscle group at least twice a week.

What sources do muscles use for energy?

Muscles (actually the muscles' cells) use a variety of energy sources to power their contractions. For quick energy, the cells use their stores of ATP and creatine phosphate, which is another phosphate-containing compound (for more about ATP, see the chapter "Basics of Biology"). These stored molecules are usually depleted within the first twenty seconds of activity. The cells then switch to other sources, most notably glycogen, a carbohydrate that is made of glucose molecules strung together in long-branching chains. The following lists the sources and where energy is stored in the human body:

Source	Storage Site
Carbohydrates	Glycogen (approx. 500 grams in average human) in liver and skeletal muscles.
Lipids	Adipose tissue. Healthy adult males have 12–18 percent body fat; healthy adult females have 12–25 percent body fat.
Protein	Throughout the body; last choice as energy source.

How much water does the human body contain?

The average human body is between 50 and 65 percent water; a newborn baby's body can be up to 78 percent water (by age one, it drops to about 65 percent). Adult males on the average have a higher percent of body water than adult females—around 60 percent—because males have less body fat (in fact, body fat contains 10 percent water, and bone has 22 percent water). Various organs of the body also contain various percents of water. For example, on the average, the adult human brain is about 70 percent water; the lungs are about 90 percent water; and our blood is about 83 percent water.

How does your stomach survive that organ's natural acid?

The inner layer of the human stomach produces hydrochloric acid with a pH of about 2.0—in other words, the acid is so strong, if a piece of wood were to drop into your stom-

What is body mass index, or BMI?

Body mass index, or BMI, is a way of measuring your body's mass—a statistical measurement that gives an estimate of a healthy body weight based on the height of a person. Overall, it gives you and your doctor a good idea of how you stand weight-wise. In general, for an adult female (males have a bit higher BMI numbers), if you have a BMI less than 18.5, you are considered underweight; 18.5 to 24.9 means normal weight; 25 to 29.9 means overweight; and over 30 is considered obese. To determine your BMI, take your weight in pounds and your height in total inches. Then multiply your weight times 703, and divide that number by your height squared. For example, if you are 5 feet 4 inches tall (or 64 inches tall) and weigh 133, the calculation would be as follows: $133 \times 703 = 93,449$; 64 inches squared $= 4,096$; divide $93,449/4,096 = 22.83$—a BMI in the normal range.

ach, it would break down very rapidly. Thus, special epithelial cells that line your stomach produce a bicarbonate-rich solution that is alkaline (basic), with a higher pH that counteracts the acid produced by the stomach. Because of this, cells are protected from the digestive enzymes at work in the stomach, but if stomach acid manages to reach the tissue below the protective mucus layer, gastric ulcers can result. (Another type of stomach ulcer, caused by the bacteria *Helicobacter pylori*, can impair the stomach's defenses.)

What prevents urine from leaking out of the bladder and into the body?

The cells that form your bladder are held together by tight junctions, which are connections between cells that hold them together so closely that urine can't slip through to reach the rest of the body. These connections, formed by protein strands that bind the cell membranes, also play an important role in keeping food in the digestive tract until it has been completely processed.

How do sperm work?

Cells called spermatogonium become sperm through the process of meiosis. A mature sperm has only half the DNA required for a functional cell, so it can't survive on its own. However, they do have flagella and mitochondria that power their journey through the reproductive tract in search of an egg. Human sperm typically die within forty-eight hours if they have not fertilized an egg. (For more about meiosis, see the chapter "Cellular Basics.")

Are a female human's unfertilized eggs considered functional cells?

Amazingly, a female human egg is about 2,000 times the size of a sperm; it also has all the organelles and proteins required of a living cell but only half of the necessary DNA. Thus, because of this, the egg cannot be considered a functional cell. Without fertilization, human egg cells survive about twenty-four hours after ovulation. Prior to that, the egg is maintained by the support structures known as follicular cells, which transfer nutrients to the egg.

What is parthenogenesis?

In a process called parthenogenesis in animals, an organism's egg cell can develop into an embryo without being fertilized (it occurs more often in plants). This type of fertilization most often occurs in invertebrates, such as flatworms, aphids, and nematodes; in a few vertebrates, such as whiptail lizards and some fish (reportedly once, a hammerhead shark); and even rarely in some birds, such as chickens and turkeys. (For more about animal reproduction, see the chapter "Physiology: Animal Function and Reproduction.")

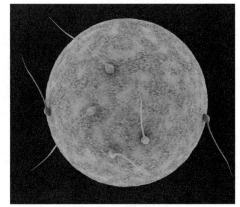

Sperm cells work to fertilize a human egg. Only one of them (usually), manages to enter the egg, combining the DNA of the male and female parents.

What is cancer?

Cancer is caused by the unrestrained growth of cells. Cells that do not "follow the rules" of normal cell cycling may eventually become cancerous. This means that the cells reproduce more often than normal, creating tumors. Usually this happens over an extended period of time and begins with changes at the molecular level. More than one hundred distinct types of cancer exist, each of which behaves in a specific fashion and responds to treatment differently.

How do cancerous tumors form?

When the reproductive rate of cells exceeds their death rate, the tissue becomes enlarged, forming a tumor. Although these cells are initially identical to the others in the tissue, they gradually take on characteristics of malignancy. The cancer cells reproduce rapidly and tend to be abnormally large or small. Malignant tumors grow very quickly and invade other tissues. Cancer types are named for the location of the tissue that gives rise to the tumor and the organs involved. Genetics, viruses, or even environmental exposure to substances like those in cigarette smoke may cause tumor formation. However, not all tumors are malignant; tumors that grow within a well-defined capsule are benign and unlikely to be life-threatening.

Are some forms of cancer related to infectious diseases?

Yes, although most cancer cells arise from within the body, certain types of cancer may actually be related to infectious diseases—mostly by viruses, but also bacteria and parasites may have an effect. Currently, it is estimated that infectious diseases may play a role in as much as 18 percent of human cancers worldwide. For example, infection by hepatitis B and hepatitis C can cause a chronic viral infection that leads to liver cancer—the most recent estimates are about 1 in 200 people for hepatitis B and 1 in 45 people for hepatitis C annually.

How do cancer cells feed themselves?

In the 1960s, American medical scientist Judah Folkman (1933–2008) realized that malignant tumors could not grow without nourishment, which is delivered by the blood. Rapidly growing tumors actually cause the formation of new blood vessels in a process known as angiogenesis. Folkman's hypothesis was that by identifying the substances used to cause angiogenesis, drugs could be formulated to prevent new vessel formation, thus starving the tumors. This work has led to the identification of at least two substances that inhibit angiogenesis: endostatin and angiostatin. These drugs are still being studied as of this writing and hold promise as new therapies to combat aggressive tumors.

How do most anticancer treatments target cancer cells?

Anticancer drugs attempt to slow down or stop the ongoing cell division that occurs in cancerous tissues. Treatment protocols include radiation, heat exposure, freezing, surgery, and/or drug therapy. The purpose of most anticancer drugs is to target the overproduction of malignant cells. Different types of cancers arise from different types of tissue. Since every tissue is made up of cells specialized for a certain function, it is not surprising that different forms of cancers will have different responses to the same drug. For example, a drug that targets the overproduction of liver cells (which are adapted to filtering and monitoring the blood supply) might have little effect on nerve cells that specialize in carrying messages.

How does carbon monoxide affect humans?

Carbon monoxide is a highly poisonous gas. Because of its molecular similarity to oxygen, hemoglobin can bind to carbon monoxide instead of oxygen, which subsequently disrupts hemoglobin's efficiency as an oxygen carrier. Carbon monoxide actually has a much greater affinity (about 300 times more!) for hemoglobin than oxygen. When carbon monoxide replaces oxygen, this causes cell respiration to stop, leading to death. The particular danger of carbon monoxide poisoning lies in the fact that a person exposed to high levels of this toxin cannot be saved by being transported to an environment free of the poison and rich with oxygen. Since the hemoglobin remains blocked, artificial respiration with overpressurized, pure oxygen must first be performed to return the hemoglobin to its original function and the body to normal cell respiration.

How does caffeine affect humans?

Caffeine is probably the most common drug ingested by people worldwide. Caffeine affects cells by stimulating lipid metabolism and slowing the use of glycogen as an energy source. As a whole, the body responds to caffeine by extending endurance, allowing you to stay awake for longer periods of time or perform extra activities. Adverse effects of excess caffeine intake include stomach upset, headaches, irritability, and diarrhea. The following lists caffeine sources and the average dose of caffeine in each:

Source	Average Dose (milligrams)
Coffee (12 oz) brewed	300
Coffee (12 oz) decaffeinated	7
Tea (12 oz)	100
Tea (12 oz) iced	70
Soft drinks (12 oz)	30–46
Jolting soft drinks (8–20 oz)	50–208
Dark chocolate (1 oz)	20
Milk chocolate (1 oz)	6
Cold remedies	0–30
Pain relievers	0 (aspirin)–130 (Excedrin)
Diet pills	200–280

How does alcohol affect humans?

Alcohol causes varying effects on different cells. In general, alcohol increases tissue sensitivity to injury and prevents postinjury recovery. Alcohol stimulates brain cells by disrupting calcium channels within the cell membranes. It is thought that alcohol affects the fluidity of the membrane phospholipids. Alcohol also causes mitochondrial damage, depressed platelet function, decreased synthesis and transportation of proteins from the liver, and the activation of pancreatic enzymes that may subsequently damage the lining of the lung.

Are fevers dangerous?

In most cases, no—but we need to qualify this: A fever is usually an indication of a bacterial or viral infection. While debate about whether fevers actually speed up the body's inflammatory response to infection is ongoing, no clinical evidence is apparent that reducing fevers by taking medication that quells the fever actually slows the healing process. But for most doctors,

Fevers can be a bit frightening, especially when it is your child who is sick, but medical professionals generally agree that a fever indicates your body is fighting an infection and actually helps a sick person get better.

fevers are considered an important part of the body's defense against infection, especially because most viruses and bacteria that cause disease thrive best at our body's normal temperature—98.6°F. Thus, a fever means the body is fighting off the pathogens and is actually helping you. A fever for an adult, for example, is usually considered to be above 99 to 99.5°F (37.2–37.5°C)—but all higher temperature fevers should be checked out by a physician.

YOU AND YOUR GENES

What are some human characteristics controlled by one gene?

Some human traits are controlled by just one gene. They include the hitchhiker's thumb (a recessive trait); tongue rolling (a dominant trait that is often debated as to whether it's controlled by genetics); thumb crossing right over left (a recessive trait); a widow's peak hairline (a dominant trait); attached earlobes (a recessive trait, although this is also highly debated by some scientists).

What is lactose intolerance, and is it a genetic disorder?

Lactose intolerance—a person's inability to break down milk products into sugars—is due to reduced activity of the enzyme lactase. Lactase changes the lactose of mammalian milk into the monosaccharides—glucose and galactose. But if you don't have enough lactase in your system, it can often lead to mild or severe symptoms from lactose intolerance, which range from mild gastrointestinal discomfort to vomiting and diarrhea. Some genetic connections exist: In particular, congenital lactase deficiency, especially in infants, is a rare, inherited recessive disorder that results in no lactase production and is caused by mutations in what is called the LCT gene. This is in contrast to childhood and adult-onset lactase deficiencies—also inherited and caused by variations in the element that regulates what is called the MCM6 gene.

Are mental disorders inherited?

It is thought that most mental disorders have a genetic component. A study on a large family in 1993 identified a link between a particular type of mental retardation, which included frequent aggressive and violent outbursts, with a region on the X chromosome in some of the males. This is one of the few cases in which a direct correlation between a single gene defect and a particular type of mental disorder is apparent. But in 2013, National Institutes of Health-funded researchers discovered that people with disorders traditionally thought to be distinct—mainly autism, ADHD (Attention Deficit Hyperactivity Disorder), bipolar disorder, major depression, and schizophrenia—were more likely to have a possible genetic variation at the same four chromosomal sites. More studies will need to be done to verify these genetic possibilities.

Do genes affect your mood?

Mood disorders are conditions that go beyond an occasional "bad day" into the realm of severe emotional disturbance. Mood disorders include depression (the most common mood disorder), bipolar disorder, and schizophrenia. By examining family medical histories and adoption studies, researchers have concluded that bipolar illness and genetics are linked. However, the presence of certain gene sequences is not a mandate, as only about 60 percent of monozygotic ("identical") twins with the bipolar sequence actually become ill. It is obvious that environmental and societal factors also play roles in determining one's mood.

Is alcoholism inherited?

It has been estimated that about one out of six adult people in the United States is classified as alcoholic; of that number, the ratio of males to females is approximately 4 to 1—a proportion that may be attributed to both environment and genetics. Research involving male relatives, both biological and adoptive, indicate that alcoholism has many sources. In other words, while sons and brothers of male alcoholics are likely to be alcoholic as well (25 percent and 50 percent, respectively), genetic inheritance is not solely responsible for all of the cases of alcoholism.

What is obesity?

According to the Centers for Disease Control and Prevention, obesity results when human body fat accumulates over time, mainly due to what is called a chronic energy imbalance—in other words, that calories consumed exceed calories expended in such activities as exercise. It is now known that, as of this writing, obesity is a major health hazard worldwide; in fact, in the United States alone, it is estimated that around 27 percent of adults are obese. Obesity is also costly, as it is associated with workplace absenteeism as a result of such obesity-related diseases as diabetes, hypertension, heart disease, and some cancers.

Is obesity inherited?

To date, researchers studying obesity genetics have identified more than thirty candidate genes on twelve chromosomes associated with body mass index—how much weight we carry around (for more about body mass index, see this chapter). For example, in 2007, the first "fat mass and obesity-associated" gene (or FTO) was found on chromosome 16; it's estimated that people who have this gene variant carry a 20 to 30 percent higher risk of obesity. Another obesity-associated gene is located on chromosome 18. But as many researchers mention, even though these genes are found on a person, they only account for a small part of the gene-related susceptibility to obesity. According to Harvard University's School of Public Health, recent research shows that genetic factors identified so far in obesity make only a small contribution to a person's obesity risk—and that our genes are "not our destiny." In other words, many people who have the so-called "obesity genes" do not necessarily become obese, or even overweight—and many times can counteract such potential overweight problems because of their genes with exercise and healthy eating habits.

What is meant by genetic discrimination?

Genetic discrimination refers to the use of genetic information by health insurers and employers to determine eligibility, set premiums (such as for health insurance), or hire

---

Who are some famous figures in history with genetic disorders?

The list of people who suffered from genetic disorders throughout history is lengthy—and fascinating. Many of the disorders actually give you a good sense of why a person acted or looked a certain way. For example, President Abraham Lincoln (1804–1865) was tall and lanky, a symptom of Marfan syndrome, a genetic disorder that affects 1 in 5,000 people; it most often affects the body's connective tissues. Another genetically affected person was King Charles II of Spain. The inbreeding of the early European royal families took its toll on many of the offspring, making it possible for genetic defects to pair up. This was the case with King Charles of the House of Hapsburg, who had what is called the "Hapsburg Jaw"—an exaggerated protruding jaw from inbreeding (called prognathism) that caused him to have problems not only with chewing, but with talking.

and fire people. The U.S. Senate passed the Genetic Information Nondiscrimination Act of 2003 on October 14, 2003, establishing basic legal protections that will prohibit discrimination in health insurance and employment on the basis of predictive genetic information. It was signed into law in 2008 and is supposed to fully protect the privacy of a person's genetic information (although it does not cover life insurance and long-term care insurance).

Is IQ genetically controlled?

This is actually two questions. First, can intelligence be measured quantitatively? And second, does a correlation exist between intelligence and certain genotypes? The answer to both questions is a definite no. As our definition of intelligence has evolved, it has become more difficult to assign a single number to the trait of intelligence. Therefore, it is almost impossible to demonstrate what gene sequences would correlate to high IQ values. And while our understanding of the genome is broadening, so too is our definition of intelligence. It becomes less and less likely that we will be able to find one or two genes that determine all of the facets of intelligence, considering the importance of environment in determining "smartness." Thus, our specific IQ is most likely created by many factors other than genetics.

YOU, BACTERIA, AND VIRUSES

Do good bacteria exist?

Yes, plenty of good bacteria are in and on the human body. For example, one type of bacteria helps you digest your food in the digestive tract and even helps you synthesize certain vitamins; another type barricades your skin from disease-causing bacteria; there are even bacteria on your tongue that help you protect the inside of your mouth. In fact, it's estimated that one hundred trillion good bacteria live in or on the human body at one time. And these small creatures are prolific—it's thought that we are actually only 10 percent human because every human cell that is necessary for our body comes with about ten "resident" microbes. These are called commensals, or generally harmless "freeloaders," on and in our bodies; the mutualists, or doing us a favor as long as we don't eliminate them; and in small numbers, pathogens that can harm us.

What is the Human Microbiome Project?

In an effort comparable to the Human Genome Project (for more about the HGP, see the chapter "Biology in the Laboratory"), a five-year federal endeavor was started called the Human Microbiome Project, trying to understand more about humans' good bacteria. Two hundred scientists at eighty institutions sequenced the genetic material of these bacteria from about 250 healthy individuals to discover even more about the good bacteria we take for granted in our bodies. For example, one offshoot of the project is the

What bacteria are "good" for infants?

Good bacteria are even present in—and good for—infants. In fact, recent studies show that a young baby may not develop a healthy mix of "good" intestinal bacteria if they are delivered by Caesarean section or don't drink breastmilk. The scientists believe certain microbes in the gut are linked to a healthy digestive tract and also help to stimulate the immune system. These microbes apparently signal the immune cells to not overreact to certain bacteria and react to others. And if the bacteria are not present, the immune system seems to overreact to something benign, say, dust or food—developing certain medical problems, such as asthma and allergies to foods.

American Gut project in which citizen-scientists help by comparing the microbes in a person's gut with people in other places around the world. In particular, the participants need to share information about factors that can affect the gut's microbiome—from your age and what you eat or drink to where you live or have lived before—so scientists can understand if the creatures we hold in our guts are dependent on such things as climate, your genetics, or maybe even something else.

What is West Nile Virus?

West Nile Virus (an arbovirus) is most commonly transmitted by infected mosquitoes; the insects become infected with the virus when they feed on infected birds, eventually spreading the virus to humans and other animals. It was first detected in North America in 1999 and has since spread across the continental United States and Canada. West Nile Virus has an incubation period of about two to fourteen days, but most people—around 70 to 80 percent—do not develop symptoms. Others who are infected can develop a fever, along with a headache, body aches, joint pains, vomiting, diarrhea, or a rash. Most people recover completely, but a feeling of fatigue and weakness can last for weeks or months. And finally, less than 1 percent of people develop a serious neurologic illness, such as encephalitis (inflammation of the brain) or meningitis (inflammation of the lining of the brain and spinal cord). As of this writing, no treatment or vaccine exists for the virus.

What is Lyme disease?

Lyme disease is caused by bacteria and is transmitted to humans through the bite of infected ticks: By the western black-legged ticks (*Borrelia burgdorferi*) in the Western, Northeastern, mid-Atlantic, and in North-Central United States, the disease is spread by deer ticks, or *Ixodes scapularis*. The smaller, immature tick nymphs usually spread the disease because they are so small and difficult to see. In most cases, the tick must be attached for thirty-six to forty-eight hours or more before the Lyme disease bacterium can be transmitted. If infected, a person can experience flulike symptoms, such as fever, headache, and fatigue; for about 80 percent of the cases, a "bull's-eye" skin rash (called erythema migrans) can be seen around the bite area. The typical treatment is a few weeks of antibiotics, but if it is not treated, the infection can eventually affect joints, the heart, and the nervous system.

When was the term "antibiotic" first used?

The term "antibiotic" is from the Greek meaning "against life." In 1889, French mycologist Jean Paul Vuillemin (1861–1932), a pupil of Louis Pasteur's, used the term to describe the substance pyocyanin, which he had isolated several years earlier. Pyocyanin inhibited the growth of bacteria in test tubes but was too lethal to be used in disease therapy. Now we know antibiotics as chemical products or derivatives of certain organisms that inhibit the growth of other organisms, or even a chemical substance produced by one organism that is destructive to another.

How do antibiotics destroy an infection?

Antibiotics function by weakening the cell wall or interfering with the protein synthesis or RNA synthesis of the bacterial cell. For example, penicillin weakens the cell wall to the point that the internal pressure causes the cell to swell and eventually burst. Certain antibiotics are more effective against gram-negative bacteria, while others are more effective against gram-positive bacteria.

What pathogens are antibiotic resistant?

Pathogens that are antibiotic or antimicrobial resistant produce diseases and infections that cannot be treated with standard antibiotics. These microbes have changed and mutated in ways that greatly reduce or eliminate the effectiveness of antibiotic drugs in curing or preventing infections. For example, methicillin-resistant *Staphylococcus aureus* (also called MRSA or golden staph) that can cause urinary tract infections and bacterial pneumonia and multidrug-resistant *Mycobacterium* tuberculosis (called MDR-TB), are bacteria that are not affected by antibiotics. Most doctors agree that avoiding unnecessary prescription of antibiotics can reduce these resistant bacteria.

Is antibiotic resistance an evolutionary trend?

Antibiotic resistance is the loss of susceptibility in bacteria to drugs like penicillin and erythromycin. The reason this is interesting from an evolutionary perspective is that it demonstrates evolution occuring in real time—that is, within a period easily observable by humans. The variation in response to antibiotics within a population of bacteria is similar to that described by Darwin and Wallace. When some bacterial cells survive after an incomplete course of antibiotic treatment, they form the basis for a new drug-resistant strain. What is troubling, however, is that different types of bacteria can actually share the genes that make them drug resistant so that this ability is becoming more prevalent among different strains of disease-causing organisms.

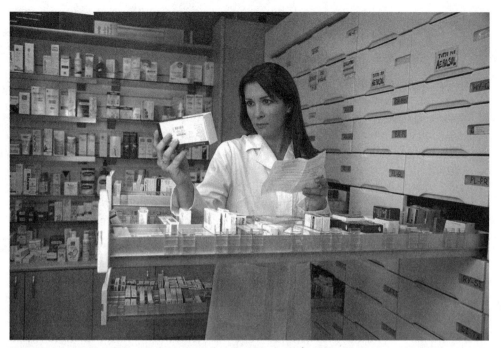

Medical professionals are becoming concerned that the overprescription of antibiotics is resulting in drug-resistant bacteria that could be much more difficult to treat in patients.

What factors have contributed to an increase in the number of resistant bacteria?

Bacteria mutate in order to adapt to new conditions. A mutation that enables a microbe to survive in the presence of an antibiotic quickly spreads throughout a microbial population. Since bacteria replicate very rapidly, a mutation can swiftly become prevalent. The overuse of antibiotics promotes the emergence of resistant bacteria. Antibiotics may sometimes be prescribed for viral infections, but they are not effective against such infections. Furthermore, patients often fail to follow the directions for taking antibiotics precisely. A prescribed dose of antibiotics should be taken until the course is completed. Although an individual can feel better shortly after starting a treatment, not completing the full course of antibiotics often destroys only the most vulnerable bacteria. Relatively resistant bacteria are able to survive and prosper in a human's body. Because antibiotic-resistant strains of bacteria do not respond to standard treatments, illnesses are able to last for longer periods of time and can result in death. The proliferation of resistant bacteria has made it more difficult to establish effective treatments.

How do antibiotics differ from antibacterials?

Antibiotics and antibacterials both interfere with the growth and reproduction of bacteria. Antibiotics are medications for humans and animals. Antibacterials, found in soaps, detergents, health and skincare products, and household cleaners, are used to disinfect surfaces and eliminate potentially harmful bacteria. Nonresidue-producing antibacterials—such as alcohols—quickly destroy bacteria, rapidly disappear from the surface they were applied to due to evaporation or chemical breakdown, and leave no active residue behind. Residue-producing antibacterials—such as triclocarban—have prolonged effects because they leave long-acting residues on the surfaces to which they are applied.

YOU AND FOOD

What is cholesterol?

Cholesterol belongs to the subclass of lipids known as steroids. Steroids have a unique chemical structure. They are built from four carbon-laden ring structures that are fused together. The human body uses cholesterol to maintain the strength and flexibility of cell membranes. Cholesterol is also the molecule from which steroid hormones and bile acids are built.

Do humans really need cholesterol?

Although it is thought that humans don't need cholesterol in their diet—the body makes all it needs—it is an important component in the cell membranes of animals. Cholesterol, a lipid-based molecule (steroid), actually has two functions: 1) to help stabilize the membrane, and 2) to maintain membrane flexibility as temperature changes. Normally the human body is capable of producing all the cholesterol it needs. Dietary ingestion of excess saturated fats and cholesterol is currently thought to be the source of the plaque that builds up in arteries and can cause heart attacks and strokes. Dietary cholesterol can be found in all animal sources, including shellfish. Plant phytosterols have the same function as cholesterol found in animals, but they do not affect the human body in the same way.

What are the "good" and "bad" cholesterols in humans?

According to our doctors, "good" and "bad" elements are in our bodies that can affect us—especially our circulatory system, or in other words, our heart. In particular, when too much LDL (or low-density lipoprotein, called the "bad" cholesterol) circulates in your blood, it can create a buildup of what is called plaque on the inner walls of the arteries that feed your heart and brain. This can, in turn, cause your arteries to narrow and become less flexible (in excess, such plaque buildup is called atherosclerosis). Another type of cholesterol is HDL (or high-density lipoprotein, called the "good" cholesterol), and it is thought to protect the body against heart attacks by carrying the cholesterol away from the arteries and to the liver, where it is passed from the body—and may even slow plaque buildup. Yet another player in your body's cholesterol is Lp(a)—a genetic variation of LDL cholesterol. In particular, a high level of Lp(a) may be a high risk factor for the premature formation of fatty deposits in your arteries.

What are the best cholesterol levels for an adult human?

Many factors affect our body—one of them has to do with our cholesterol levels, especially in terms of our heart. The following, from the American Heart Association, lists what is currently thought to be the best cholesterol levels for an adult:

Total Cholesterol	Category
Less than 200 mg/dL	Desirable level that puts you at lower risk for coronary heart disease. A cholesterol level of 200 mg/dL or higher raises your risk.
200 to 239 mg/dL	Borderline high
240 mg/dL and above	High blood cholesterol. A person with this level has more than twice the risk of coronary heart disease as someone whose cholesterol is below 200 mg/dL.

Source: American Heart Association.

How are carbohydrates classified?

Carbohydrates are classified in several ways. Monosaccharides (single-unit sugars) are grouped by the number of carbon molecules they contain: triose has three, pentose has five, and hexose has six. Carbohydrates are also classified by their overall length (monosaccharide, disaccharide, polysaccharide) or function. Examples of functional definitions are storage polysaccharides (glycogen and starch), which store energy, and structural polysaccharides (cellulose and chitin), which provide support for organisms without a bony skeleton.

What are vitamins and minerals?

A vitamin is an organic, nonprotein substance that is required by an organism for normal metabolic function but cannot be synthesized by that organism. In other words, vitamins are crucial molecules that must be acquired from outside sources. While most vitamins are present in food, vitamin D, for example, is produced as a precursor in our

417

skin and converted to the active form by sunlight. Minerals, such as calcium and iron, are inorganic substances that also enhance cell metabolism. Vitamins may be fat- or water-soluble. Recommended amounts of vitamins are to ensure normal enzymatic function, and excessive intake can be toxic.

Vitamin	Major Sources	Major Function
A	Animal products; plants contain only vitamin A building blocks	Aids normal cell division and development; particularly helpful in the maintenance of visual health
B-complex	Fruits and vegetables (folate); meat (thiamine, niacin, vitamin B_6, and B_{12}); milk (riboflavin, B_{12})	Energy metabolism; promotes harvesting energy from food
C	Fruits and vegetables, particularly citrus, strawberries, spinach, and broccoli	Collagen synthesis; antioxidant benefits; promotes resistance to infection
D	Egg yolks; liver; fatty fish; sunlight	Supports bone growth; maintenance of muscular structure and digestive function
E	Vegetable oils; spinach; avocado; shrimp; cashews	Antioxidant
K	Leafy, green vegetables; cabbage	Blood clotting

Why are some fats "good" and others "bad?"?

This question can be answered in a number of ways. One could argue that no fats are "bad," as fats are excellent sources of energy and help to maintain the health of the body. From this perspective, fat is only bad if one eats too much of it. Another way to answer would be to point out that several fats are considered essential (the omega-6 and omega-3 fatty acids)—in other words, they are substances that our bodies require for maintenance but that we cannot manufacture. These are considered to be "good" fats, and comparatively, the fats we don't need to ingest are often dubbed as "bad." Finally, artificial fats should be mentioned. While these may have been created to maintain the flavor and texture of food while reducing the caloric content, they may be difficult to metabolize and therefore "bad" for us in the long run.

The spinach and strawberries in this salad are high in vitamin C, which has antioxidant benefits and helps the body resist infection.

What is the difference between a fat and a lipid?

Lipids are bioorganic molecules that are hydrophobic. In other words, they do not mix with or dissolve in water. Among lipids is a category known as "fats." Each fat molecule is comprised of a glycerol (alcohol) molecule and at least one fatty acid (hydrocarbon chain with an acid group attached).

What is the difference between saturated and unsaturated fat?

Fat is a type of lipid molecule constructed by glycerol and three fatty acids. The molecular structure of the fatty acids determines whether the fat is saturated or unsaturated. Fats with hydrogen atoms but without double bonds are "saturated." Unsaturated fatty acids have double bonds and therefore have fewer hydrogen atoms.

How are fats used in the human body?

The following lists how fats are used in the human body:

Function	Specific structures
Cell structure	Cell membranes use phospholipids and cholesterol for membrane structure and stability
Energy storage	Triglycerides stored in adipose tissue
Insulation	Protects body from changes in outer temperatures; promotes signal conduction in the nervous system (similar to electrical wire insulation)
Messengers	Steroid hormones; HDL and LDL carry cholesterol through bloodstream

What is a trans fat?

The term "trans" fat refers to the arrangement of hydrogen atoms around the carbon backbone of the fatty acid. A trans-fatty acid is a molecule that has a carbon backbone with hydrogen atoms attached in a manner that is not normally found in nature. Most naturally occurring fatty acids have their hydrogen arranged in the "cis" form. In trans fats, some of these hydrogens are attached on opposite sides of the fatty acid molecule in what is known as a "trans" (as opposed to "cis") formation.

Can one type of fat be changed into another type?

The process of hydrogenation can convert an unsaturated fatty acid into a hydrogenated fatty acid. This process, which is achieved by adding extra hydrogen atoms to unsaturated fat, has become both a bane and a blessing. Hydrogenation is the process that allows unsaturated vegetable oils to be turned into margarine. This method prevents oxidation, and thus rancidity, and has allowed for the development of foods with less animal and saturated fats. However, the consumption of hydrogenated fatty acids may be linked to increased risk of heart disease because the fats cause a change in the structure of targeted unsaturated fatty acids. The consumption of trans fats has been shown to slightly increase the levels of bad cholesterol (LDL) in the blood.

What is a calorie?

Two kinds of calories exist, actually. Ask a chemist and you will learn that a calorie (with a lowercase "c") is the amount of energy (heat) required to raise 1 gram (1 millileter) of water by 1°C. A nutritionist, on the other hand, would describe a "big C" or kilocalorie (kcal) as the amount of energy required to raise 1 kilogram (1 liter) of water by 1°C. The kcal is the unit used to describe the energy value in food. For example, if a chocolate chip cookie is completely incinerated, the amount of heat energy released would be enough to raise the temperature of 1 liter of water by approximately 572°F (300°C)! However, as this system adheres to the laws of thermodynamics, the conversion is not totally efficient (only about 25 percent of energy actually performs cellular work).

Can humans make all the lipids they need?

No, we cannot make all the lipids we need. In fact, we need two types of fatty acids that we can't make: linoleic acid, an omega-6 fatty acid, and linolenic acid, which is also known as an omega-3 fatty acid. These acids are used to maintain cell membranes and to build hormonelike messengers known as eicosanoids, or the twenty carbon fatty acids such as prostaglandins. Good sources of omega-3 and omega-6 lipids are nuts, grains, and both fats and oils from vegetables or fish.

Why are carbohydrates a major part of the human diet?

Our cells have evolved in a manner that has made carbohydrates the human body's primary energy source; in fact, the entire metabolic system begins with glucose. While most of the cells in our bodies can adapt, at least temporarily, to using lipids and proteins as energy sources, our brain cells require glucose. This means that when blood glucose levels fall too low, the brain cells shut down and we faint. By fainting, the flow of blood and glucose to the brain increases.

What is so good about fish oil?

Fish oils, extracted from species such as mackerel, salmon, anchovy, sardines, and tuna, are excellent sources of the essential omega-3 fatty acids—however, they are not the only possible source. Nuts and seeds also provide omega-3 fatty acids. Nutritionists recommend that these fatty acids be acquired through a healthy and diverse diet rather than by taking nutrient supplements.

What is a better source of energy—fat or sugar?

Fats contain 9 kilocalories per gram, while carbohydrates average about 4 kilocalories

Some people take fish oil tablets like these to supplement their need for omega-3 fatty acids commonly found in fish, although they are also found in nuts and seeds.

per gram. Fats are useful for storing energy, but sugars are easily broken down during metabolism. Nutritionists and diet gurus are debating which source of energy is better. The absolute answer lies in individual situations. For example, babies require fat in their diets for both energy and to build a healthy nervous system. Middle-aged adults quite often take in too much fat for their metabolic requirements. As one ages, average energy needs decrease by 5 percent per decade.

What is dietary fiber?

Dietary fiber is a type of carbohydrate that cannot be broken down by digestive enzymes. Because of this, the fiber passes through the digestive tract more quickly, aiding in elimination. The term "dietary fiber" includes the cellulose found in plant cell walls and the chitin that makes up the support tissues of fungi (mushrooms), crustaceans, and insects.

Why does eating chocolate make some people feel good—or possibly in love?

Chocolate contains over 300 known chemicals, some of which can alter mood, such as caffeine. Chocolate also contains a small amount of the chemical phenylethylamine (PEA), a stimulant to the nervous system that makes people feel more alert and gives a sense of overall well-being.

Of course, it's interesting to note that chocolate has another hidden factor—that connected to love. Neurochemicals are released in the physical bodies of two people who are attracted to each other, so one can say that a chemical reaction does indeed occur between two people. The neurochemicals involved are phenylethylamine (also found in chocolate), which speeds up the flow of information between nerve cells; dopamine, which makes us feel good; and norepinephrine, which makes the heart race.

How are fungi related to soy sauce?

Aspergillus tamari and other deuteromycetes are used to produce soy sauce by slowly fermenting boiled soybeans. Soy sauce provides foods with more than its special flavor; the soybeans and fungi give soy sauce amino acids that are vital to human life. Fungi have been used in many cultures to improve the nutrient quality of the diet. (For more about fermentation, see the chapter "Basics of Biology.")

What cheeses are associated with fungi?

The unique flavor of cheeses such as Roquefort, Camembert, and Brie is produced by members of the genus *Penicillium*. Roquefort is often referred to as "the king of cheeses"; it is one of the oldest and best known cheeses in the world. This "blue cheese" has been enjoyed since Roman times and was a favorite of Charlemagne, king of the Franks and emperor of the Holy Roman Empire (742–814). Roquefort is made from sheep's milk that has been exposed to the mold *Penicillium roqueforti* and aged for three months or more in the limestone caverns of Mount Combalou, near the village of Roquefort in southwestern France. This is the only place true Roquefort can be aged. It has a creamy, rich texture and a flavor that is simultaneously pungent, piquant, and salty. It has a creamy white interior marked by blue veins; the cheese is held together with a snowy white rind. True Roquefort is authenticated by the presence of a red sheep on the emblem present on the cheese's wrapper. *Penicillium camemberti* gives Camembert and Brie cheeses their special qualities. Napoleon is said to have christened Camembert cheese with its name; supposedly the name comes from the Norman village where a farmer's wife first served it to Napoleon. This cheese is formed of cow's milk and has a white, downy rind and a smooth, creamy interior. When perfectly ripe and served at room temperature, the cheese should ooze thickly. Although Brie is made in many places, Brie from the region of the same name east of Paris is considered one of the world's finest cheeses by connoisseurs. Similar to Camembert, it has a white, surface-ripened rind and a smooth, buttery interior.

YOU AND THE "OTHER" ANIMALS

Can animals suffer from psychological disorders?

Although animals are used as models for research into disorders like anxiety, depression, and even schizophrenia, these conditions have been induced in them either through surgery or behavioral treatments. Scientists are currently developing strains of mice and rats that are genetically predisposed toward these diseases in hopes of gaining a better understanding of symptoms and developing more effective treatments. As for animals in the wild, the answer lies in the still unresolved issues of whether animals experience emotions and whether they have consciousness or the awareness of self.

Why do cats swish their tails?

A cat's tail is a means of communication and can represent a range of moods. Every cat owner is familiar with the slow sweep of a cat's tail from side to side. A happy cat will greet you with a tail raised high, while a slight movement indicates pleasure and/or anticipation. As the cat becomes more hunterlike, the flick of the tip becomes first a twitch and then a thrashing. At this point the cat is highly engaged and likely to pounce.

Why will my dog play fetch but not my cat?

Domestication occurred differently for cats and dogs, with dogs helping humans ear-

The game of fetch is an expression of a canine's cooperative hunting relationship with his or her owner.

> ## Who was Mr. Ed, and do "talking animals" really exist?
>
> **M**r. Ed was "the talking horse" and a star of a television show in the 1960s. When he appeared to talk, the horse was actually responding to cues from his trainer. Movement of a small rope running from his halter through his mouth and held by the trainer off camera would cause Ed to move his lips as if he were speaking. In the real world, so far birds (like with most animal training, teaching a parrot or cockatoo to talk involves a combination of repetition and reward), dogs, cats, and even an elephant have been reported to mimic human speech. Not to be outdone, the current trend is, of course, Internet driven: people showing YouTube videos of animals "talking" to their owners—but in reality, it just *sounds* as if they are saying human words … if you stretch your imagination.

lier than cats. In fact, dogs came to work closely with humans, relying on them for food and shelter, while cats maintained a more distant relationship, ridding ships and farms of mice (in other words, relying on their own instincts to hunt and catch food). A dog plays fetch because it is an extension of the cooperative working-hunting relationship she or he has with humans, while the cat is more independent and has no such relationship. Then again, this could be debated by many cat owners though—many have cats that actually do like to fetch like a dog.

Can horses really do math?

At the end of the nineteenth century, a performing horse in Germany known as Clever (or Kluge) Hans was able to tap out the answers to mathematical problems written on a chalkboard. Hans would use his right forefoot to indicate the single digits (0–9) and his left forefoot for the tens place (10, 20, 30, etc.). His amazing performances continued for a number of years until the German comparative biologist and psychologist Oskar Pfungst (1874–1932) was able to show that Hans was simply counting until his questioner indicated (subconsciously) that Hans had reached the correct sum. Even though the horse was not actually performing calculations, his ability to observe and respond to subtle changes in human behavior is still quite noteworthy.

Glossary

Abiotic—Pertaining to the non-living, including temperature, water, wind, rocks, and soil.

Acid—A chemical that releases a hydrogen ion in solution with water, as opposed to a base.

Acid Rain—When the rain has a lower pH (less than 5.6), often the result of air pollutants from combustion of fossil fuels.

Adaptation—Any structural, biochemical, or behavioral characteristic of an organism that helps it survive potentially harsh conditions in its environment.

Adenine—One of the nitrogenous bases found in DNA and RNA molecules.

Adenosine Triphosphate (ATP)—The organic compound that stores respiratory energy for transport from one part of a cell to another; it is in the form of a chemical bond.

Allele—A pair of genes that exist at the same location on a pair of homologous chromosomes; they exert parallel control over the same genetic trait.

Amino acid—An organic compound that is the major component unit of proteins.

Angiosperms—Another name for flowering plants.

Animal—One of the (some say five) kingdoms in the classification of organisms; they are usually defined as multicellular organisms whose cells are not bounded by cell walls—and they do not exhibit photosynthesis.

Antibody—A chemical substance that an organism's body produces in the presence of a specific antigen; they are produced by B lymphocytes.

Antigen—Any chemical substance (mostly proteins) that is recognized by the immune system as an "invader" in the body of an organism; in the majority of cases, it is neutralized by a specific antibody.

Apoptosis—Also known as programmed cell death, it is a naturally occurring process in most multicellular organisms.

Asexual reproduction—When new organisms are produced from a single parent organism.

Bacteriocytes—In the relationship between some insects and their endosymbiont, the insects house the resident bacteria in special cells called bacteriocytes.

Bacteriomes—The special structures in the bacteriocytes that house the endosymbiont bacteria.

Base—A chemical that releases a hydroxyl ion in solution with water, as opposed to acid.

Binary fission—When a cell divides by mitosis followed by cytoplasmic division.

Biome—A geographical grouping of similar ecosystems; it is usually named for the major flora (plant-life) in a region.

Budding—A type of asexual reproduction in which the organism produces a "copy" of itself by mitosis—or division—into what resembles a bud.

Capillary—The smallest of the blood vessels that connects an artery to a vein, and through which all absorption into the blood occurs.

Carbohydrate—An organic compound made of carbon, hydrogen, and oxygen in a 1 to 2 to 1 (1:2:1) ratio.

Cartilage—A connective tissue found in many parts of the body; it is flexible and is most common in the embryonic stages of development.

Catalyst—A substance that slows down or speeds up a chemical reaction.

Cell membrane—Cell membranes define and compartmentalize space, regulate the flow of materials, detect external signals, and mediate interactions between cells.

Centrioles—Centrioles are found only in animal cells; they both assemble and organize long, hollow cylinders of protein known as microtubules, and are also known as microtubule-organizing centers (MTOCs).

Chemosynthesis—When certain bacteria use energy from chemical oxidation to convert inorganic raw materials into organic food molecules.

Clot—A structure that essentially plugs a ruptured blood vessel; it is usually the result of enzyme-controlled reactions from a cut or wound.

Coenzyme—Most of the time, a chemical substance that aides the action of a particular enzyme.

Cotyledon—The part of a plant seed—or embryo—that is a source of nutrition for a young, resulting seedling before the plant begins photosynthesis.

Cyclosis—The movement of cell fluid, or cytoplasm, within a cell's interior.

Cyton—The "cell body" of a neuron that generates the nerve impulses in the body.

Cytosine—One of the nitrogenous bases found in both DNA and RNA molecules.

Diffusion—A form of transport by which a soluble solution is passively absorbed or released by a cell or many cells.

Embryo—The earliest stage of an organism's development following fertilization.

Endergonic—A process that absorbs energy.

Endosymbiont—An organism that depends on another organism living in its cells; for example, bacteria living in specialized cells of an insect. If the insect host cannot survive without its bacteria, the bacteria are called endosymbionts.

Enzyme—An organic catalyst that controls the metabolism rate of certain functions; they are usually proteins.

Epidermis—The outermost layer of cells of an organism—usually in reference to plants and animals.

Essential amino acid—An amino acid that cannot be synthesized by the human body but must be ingested.

Fatty acid—The organic molecules that is a component of certain lipids.

Fauna—A "collective" name for animal species.

Flora—A "collective" name for plant species.

Food chain—The relationship between organisms that explains how nutritional relationships and food energy are passed from one to another; the usual sequence is herbivore to carnivore (or omnivore) to decomposer.

Gamete—A special reproductive cell produced by sexually reproducing organisms.

Gametogenesis—The process in which cells divide and form gametes.

Gene—A unit of heredity; it is also a portion of a chromosome that is responsible for the inheritance of a genetic trait.

Gene pool—The total of the inheritable genes for traits in a given reproducing (sexually) population.

Genome—The complete complement of DNA of organisms that provide the "operating instructions" for everything the organisms needs to do to survive and reproduce. A genome is the complete set of genes inherited from one's parents. Genome sizes vary from one species to another.

Glycerol—An organic compound that is a component of certain lipids.

Guanine—One of the nitrogenous bases found in DNA and RNA molecules.

Habitat—The specific environment or area in which an organism (or many organisms) live.

Hermaphrodite—An organism (animal) that has both male and female gametes.

Hormone—Chemicals that are the product of an endocrine gland; they are usually responsible for regulating cell metabolism.

Hypocotyl—The part of a developing plant embryo that becomes the lower part of the stem and root systems.

Interferon—A protein produced by a virus-infected cell that halts the multiplication of the virus; it is said to be important in the fight against human cancer.

Lipid—An organic compound that is made of carbon, hydrogen, and oxygen, but the hydrogen and oxygen are not in a 2 to 1 (2:1) ratio; most of them are a glycerol and three fatty acids.

Lymph—The intercellular fluid that passes through the lymph vessels (the series of branching tubes that collect the intercellular fluid from the tissues and redistributes it as lymph).

Lysosomes—Organelles that are only found in animal cells; they are membrane-bound sacs containing digestive enzymes that break down macromolecules such as proteins, carbohydrates, fats, and nucleic acids.

Monosaccharide—A type of carbohydrate that is also called a "simple sugar"; they all have the molecular formula of $C_6H_{22}O_6$.

Mucus—The somewhat-slimy, protein-rich mixture that is found coating and moistening the respiratory tract surfaces, such as the inside of the human nose.

Mutation—When genetic material is altered—either a chromosome or gene—in any organism.

Neuron—The cells that specifically carry the nerve impulses throughout an organism's body.

Nucleic acid—The organic compound that is made of repeating units of a nucleotide.

Nucleolus—The organelle within a cell's nucleus that is responsible for protein synthesis.

Nucleotide—The repeating unit that makes up the nucleic acid polymer.

Nucleus—The organelle in a cell that most importantly contains the cell's genetic information in the form of chromosomes.

Organelle—The small structures within a cell—each with a special function, such as the mitochondria and nucleus.

Osmosis—The action of water being absorbed or released by a cell.

Oxyhemoglobin—A form of hemoglobin that is loosely bound to hemoglobin in the blood; its primary task is to transport oxygen throughout the body.

Phagocyte—A type of white blood cell in the body that engulfs and destroys invading, harmful bacteria.

Population—The entire membership of a particular species in a specific area over a certain period.

Predation—This refers not only to one animal eating another animal, but also can mean animals eating plants.

Ribose—A five-carbon sugar that is part of the nucleotides of RNA molecules only.

Ribosome—A specific organelle in an organism's cell that is the site of protein synthesis.

Saprophyte—Organisms that get their nutrition from the decomposing of dead plant and animal tissues, such as fungi and bacteria.

Sessile—When an organism does not "move," such as barnacles, they are called sessile.

Spore—A special asexual reproductive cell that is produced by certain plants, such as certain ferns.

Stroma—Small "openings" in a plant leaf, in the chloroplast, in which carbon reactions take place.

Succession—In an established ecological area, it is the situation in which that community is replaced by another—until the last (called climax) community is established.

Tendon—A form of connective tissue that attaches the skeletal muscles to the bone.

Tendonitis—In humans, it is where the junction between the tendon and bone becomes irritated—and even inflamed—usually from repeated actions, such as tendonitis of the elbow from over-practicing a tennis swing.

Thymine—One of the nitrogenous bases found only in DNA.

Tropism—A growing plant's response to an environmental stimulus (such as how a plant grows in response to gravity).

Uracil—The nitrogenous base that is part of the nucleotides of RNA molecules only.

Vaccination—In order to stimulate a person's immune system, an inoculation of a dead or weakened organism associated with the disease is given; this allows the body to build up antibodies to help fight off the disease if a person is exposed again.

Vasodilation—The enlargement of a body's blood vessels that increases the blood supply.

Viroids—The small fragments of nucleic acid (RNA) without a protein coat; they are usually associated with plant diseases and are several thousand times smaller than a virus.

Zygote—When gametes are fused in sexual reproduction, the result is a single diploid cell called a zygote; it is also the term synonymous with a fertilized egg.

Further Reading

(Note: The following is only a minute fraction of the biology book offerings—recent and not-so-recent. The best idea is the pick a topic you're interested in, and use that keyword in your favorite search engine—or even use the keyword in your library or bookstore search engine. Here are just a few we found enjoyable.)

BOOKS

Angrist, Misha. *Here Is a Human Being: At the Dawn of Personal Genomics*, Harper Collins Publishing, 2010.

Ball, Edward. *The Genetic Strand: Exploring a Family History through DNA*, Simon & Schuster, 2007.

Barnes-Svarney, Patricia, and Thomas E., Svarney. *The Oryx Guide to Natural History: The Earth and All Its Inhabitants*, Oryx Press, 1999.

Bryson, Bill. *A Short History of Nearly Everything*, Broadway Books, 2003.

Carroll, Sean B. *Endless Forms Most Beautiful: The New Science of Evo Devo and the Making of the Animal Kingdom*, illustrated by Jamie W. Carroll and Josh P. Klaiss, W.W. Norton and Company, 2006.

Church, George M., and Ed Regis. *Regenesis: How Synthetic Biology Will Reinvent Nature and Ourselves*, Basic Books, 2012.

Conniff, Richard. *The Species Seekers: Heroes, Fools, and the Mad Pursuit of Life on Earth*, W. W. Norton and Company, 2010.

De Waal, Frans, *Our Inner Ape: A Leading Primatologist Explains Why We Are Who We Are*, Riverhead Trade, 2006.

Gould, Stephen Jay. *The Panda's Thumb: More Reflections in Natural History*, W. W. Norton and Company, 1992.

Hazen, Robert M. *Genesis: The Scientific Quest for Life's Origins*, Joseph Henry Press, 2005.

Heinrich, Bernd. *Winter World: The Ingenuity of Animal Survival*, Harper Perennial, 2004.

Hine, Robert, editor. *Oxford Dictionary of Biology,* 6th edition, Oxford University Press, 2008.

Hölldobler, Bert, and Edward O. Wilson. *The Superorganism: The Beauty, Elegance, and Strangeness of Insect Societies*, W. W. Norton and Company, 2008.

Macdonald, David, editor. *The Princeton Encyclopedia of Mammals*, Princeton University Press, 2009.

Morton, Oliver. *Eating the Sun: How Plants Power the Planet*, Harper, 2008.

Nouvian, Claire. *The Deep: The Extraordinary Creatures of the Abyss*, University of Chicago Press, 2007.

Perrin, William, editor. *Encyclopedia of Marine Mammals,* 2nd edition, Elsevier Publishing, 2009.

Pollan, Michael. *The Botany of Desire: A Plant's-Eye View of the World*, Random House, 2002.

Resh, Vincent, and Ring T. Cardé. *Encyclopedia of Insects*, Academic Press, 2003.

Roach, Mary. *Gulp: Adventures on the Alimentary Canal*, W. W. Norton and Company, 2013.

Sachs, Jessica Snyder. *Good Germs, Bad Germs: Health and Survival in a Bacterial World*, Hill and Wang, 2007.

Stewart, Amy. *Wicked Bugs: The Louse That Conquered Napoleon's Army & Other Diabolical Insects*, Algonquin Books, 2011.

Tudge, Colin. *The Tree: A Natural History of What Trees Are, How They Live & Why They Matter*, Crown, 2006.

Wells, Spencer. *The Journey of Man: A Genetic Odyssey*, Random House, 2002.

Wilson, Edward O. *The Future of Life*, Abacus, 2002.

Wynnem, Clive. *Do Animals Think?*, Princeton University Press, 2006.

Zimmer, Carl. *A Planet of Viruses*, University of Chicago Press, 2011.

WEBSITES

(Note: Similar to books—there are thousands, if not hundreds of thousands, of websites that feature biology—from those offered at the high school level and college biology departments to government agencies and biological organizations. Here are only a few of our favorites; and by the time you read this, there will no doubt be more! Also note: the addresses for websites change frequently; we regret any inconvenience an incorrect address may cause.)

American Botanical Council (http://abc.herbalgram.org/): For information about herbs— medicinal, culinary, and otherwise.

American Chemical Society (http://portal.acs.org/portal/acs/corg/content): Filled with information not only on chemistry but on the bio-chemical connection.

American Fern Society (http://amerfernsoc.org/): In existence for over a hundred years, the society is concerned with the study, preservation, and education about ferns around the world.

BioInteractive (http://www.hhmi.org/biointeractive/): This interactive biology site is sponsored by the Howard Hughes Medical Institute.

Biology Online (http://www.biology-online.org/): "Answers all your biology questions" and is an easy site to navigate.

Biology Project (http://www.biology.arizona.edu/): The University of Arizona's page that includes updated resources about biology.

Botanical Society of America (http://www.botany.org/): Follow the latest about plants.

Cells Alive! (http://www.cellsalive.com/): Is a great site to learn more about the intricacies of the cell—plant, animal, and in between.

The Centers for Disease Control and Prevention (http://www.cdc.gov/): This is a government organization site that has a good deal of information about human biology.

Climate Central (http://www.climatecentral.org/): On this site, an independent group of scientists and reporters covers the latest about climate—especially climate change.

EnviroLink Network (http://www.envirolink.org/): Has information about the environment and online resources to check out.

The Environmental Protection Agency (http://www.epa.gov): This site has links to home and the environment, including the latest rules and regulations concerning the environment.

Everyday Mysteries (http://www.loc.gov/rr/scitech/mysteries/): Library of Congress page that has links to biology facts.

The Franklin Institute (http://www.fi.edu/learn/heart/index.html): All about the human heart.

The Genetics Information Center (http://www.kumc.edu/gec/): is at the University of Kansas Medical Center site that is a gateway to other genetic sites.

The Human Connectome Project (http://www.humanconnectomeproject.org/): Funded by the National Institutes of Health, this site features amazing details about the human brain, including the first 3-D digital images of the brain.

Human Genome Project (http://www.ornl.gov/sci/techresources/Human_Genome/home .shtml): Information put out by the U.S. Department of Energy.

Institute for Genome Studies (http://www.igs.umaryland.edu/): The latest research on genome studies from the University of Maryland's institute, this is part of the university's School of Medicine.

The Internet Pathology Laboratory of Medical Education (http://library.med.utah.edu/Web Path/webpath.html): This Mercer University School of Medicine site has some great illustrations of biological items, such as cells and actual photos of livers.

Koko (http://www.koko.org): This is the site that features one of the oldest "educated gorillas."

Microbe World (http://www.microbeworld.org): Just like it reads: all about the microbes in the world around us.

Mycological Society of America (http://www.msafungi.org/): All about fungi.

NASA (http://ipv6.nasa.gov/): Often has information about life—especially the speculation of life on other planets and in the universe.

National Center for Biotechnology Information (http://www.ncbi.nlm.nih.gov/): The National Institutes of Health's National Center for Biotechnology Information advances science and health by providing access to biomedical and genomic information—but it's also interesting to research biological information.

The National Center for Science Education (http://ncse.com/): A great site for teachers and students interested in evolution and climate science.

National Human Genome Research Institute (http://www.genome.gov/): The site of the government's extensive research on genes.

National Institutes of Health (http://www.nih.gov): Has many connections to biological topics.

The New York Botanical Gardens (http://www.nybg.org/): A great site for plant information and even some virtual walks through their gardens.

North American Mycological Association (http://namyco.org/): A useful site for information about fungi.

Plant Facts (http://plantfacts.osu.edu/): A great plant facts site from the University of Ohio—complete with information, images, and videos.

The Royal Horticultural Society (http://www.rhs.org.uk/): This United Kingdom organization has some of the best links and offerings about plants.

The Smithsonian Institute (http://www.si.edu): Well-known for its science expertise, including biology.

Society for the Study of Evolution (http://cms.gogrid.evolutionsociety.org/): Started in 1946, this organization offers information and publications about evolution.

University of California's Museum of Paleontology (http://evolution.berkeley.edu/): A site from the offers the science and history of evolution.

USDA Agricultural Service (http://planthardiness.ars.usda.gov/): Check out the latest plant hardiness map from the U.S. Department of Agriculture—for all gardeners to see which plants grow best in your area. They also have plant fact sheets at http://plants.usda.gov/java/factSheet.

U.S. Fish and Wildlife Service (http://www.fws.gov/): The is a government agency that protects wildlife and offers information about conservation efforts in the United States.

U.S. Food and Drug Administration (http://www.fda.gov): The government's site has information about what we eat and human health.

Virology Journal (http://www.virology.net/): All the Virology on the Web, this site "seeks to be the best single site for Virology information on the Internet."

The World Environmental Organization (http://www.world.org/): Offers links to sites across the globe, including information about conservation and animals, trees, and other concerns.

NEWS SITES

Genetic Engineering and Biotech News (http://www.genengnews.com/): This group provides some of the latest information in genetics.

International Science Grid (http://www.isgtw.org/): Offers coverage of the sciences around the world, including biology.

PhysOrg (http://phys.org/): This web-based science, research and technology news service has some up-to-date information about biology.

Science Daily (http://www.sciencedaily.com/): Find out about the newest studies in biology through this site.

Science Recorder News (http://www.sciencerecorder.com/): Includes up-to-date coverage of biology.

Scientific American (http:/www.scientificamerican.com): The magazine often has topics about biology—human and otherwise.

EXTRA

For those of you interested in the writings of Charles Darwin, the following lists some of his best published works. You can often find these archived in certain libraries or on the Internet:

- *Journal of Researches into the Geology and Natural History of the Various Countries Visited by HMS "Beagle" under the Command of Capt. FitzRoy, R.N., from 1832 to 1836* (1839)

- *Geological Observations on Coral Reefs, Volcanic Islands, and on South America: Being the Geology of the Voyage of the "Beagle," under the Command of Capt. FitzRoy, during the Years 1832–36* (1846)

- *A Monograph on the Sub-Class Cirripedia* (1851–1854)

- *A Monograph on the Fossil Lepadidae, or, Pedunculated Cirripedes of Great Britain* (1851)

- *A Monograph on the Fossil Balanidae and Verrucidae of Great Britain* (1854)

- *On the Origin of Species by Means of Natural Selection; or, the Preservation of Favoured Races in the Struggle for Life* (1859)

- *On the Various Contrivances by Which British and Foreign Orchids Are Fertilised by Insects, and on the Good Effects of Intercrossing* (1861)

- *The Movements and Habits of Climbing Plants* (1865)

- *The Variation of Animals and Plants under Domestication* (1868)

- *The Descent of Man, and Selection in Relation to Sex* (1871)

- *The Expression of the Emotions in Man and Animals* (1872)

- *Insectivorous Plants* (1876)

- *The Effects of Cross and Self Fertilisation in the Vegetable Kingdom* (1876)

- *The Different Forms of Flowers on Plants of the Same Species* (1877)

- *The Power of Movement in Plants* (1880)

- *The Formation of Vegetable Mould, through the Action of Worms, with Observations on Their Habits* (1881)

- *The Movements and Habits of Climbing Plants* (1882)

Index

Note: (ill.) indicates photos and illustrations.

440

441

H

457

468

471

Also from Visible Ink Press

The Handy African American History Answer Book
by Jessie Carnie Smith
ISBN: 978-1-57859-452-8

The Handy American History Answer Book
by David Hudson
ISBN: 978-1-57859-471-9

The Handy Anatomy Answer Book
by James Bobick and Naomi Balaban
ISBN: 978-1-57859-190-9

The Handy Answer Book for Kids (and Parents), 2nd edition
by Gina Misiroglu
ISBN: 978-1-57859-219-7

The Handy Art History Answer Book
by Madelynn Dickerson
ISBN: 978-1-57859-417-7

The Handy Astronomy Answer Book, 3rd edition
by Charles Liu
ISBN: 978-1-57859-190-9

The Handy Bible Answer Book
by Jennifer R. Prince
ISBN: 978-1-57859-478-8

The Handy Chemistry Answer Book
by Ian C. Stewart and Justin P. Lamont
ISBN: 978-1-57859-374-3

The Handy Civil War Answer Book
by Samuel Willard Crompton
ISBN: 978-1-57859-476-4

The Handy Dinosaur Answer Book, 2nd edition
by Patricia Barnes-Svarney and Thomas E. Svarney
ISBN: 978-1-57859-218-0

The Handy Geography Answer Book, 2nd edition
by Paul A. Tucci
ISBN: 978-1-57859-215-9

The Handy Geology Answer Book
by Patricia Barnes-Svarney and Thomas E. Svarney
ISBN: 978-1-57859-156-5

The Handy History Answer Book, 3rd edition
by David L. Hudson, Jr.
ISBN: 978-1-57859-372-9

The Handy Investing Answer Book
by Paul A. Tucci
ISBN: 978-1-57859-486-3

The Handy Law Answer Book
by David L. Hudson Jr.
ISBN: 978-1-57859-217-3

The Handy Math Answer Book, 2nd edition
by Patricia Barnes-Svarney and Thomas E. Svarney
ISBN: 978-1-57859-373-6

The Handy Mythology Answer Book
by David A. Leeming, Ph.D.
ISBN: 978-1-57859-475-7

The Handy Ocean Answer Book
by Patricia Barnes-Svarney and Thomas E. Svarney
ISBN: 978-1-57859-063-6

The Handy Personal Finance Answer Book
by Paul A. Tucci
ISBN: 978-1-57859-322-4

The Handy Philosophy Answer Book
by Naomi Zack
ISBN: 978-1-57859-226-5

The Handy Physics Answer Book, 2nd edition
By Paul W. Zitzewitz, Ph.D.
ISBN: 978-1-57859-305-7

The Handy Politics Answer Book
by Gina Misiroglu
ISBN: 978-1-57859-139-8

The Handy Presidents Answer Book, 2nd edition
by David L. Hudson
ISB N: 978-1-57859-317-0

The Handy Psychology Answer Book
by Lisa J. Cohen
ISBN: 978-1-57859-223-4

The Handy Religion Answer Book, 2nd edition
by John Renard
ISBN: 978-1-57859-379-8

The Handy Science Answer Book®, 4th edition
by The Science and Technology Department Carnegie Library of Pittsburgh, James E. Bobick, and Naomi E. Balaban
ISBN: 978-1-57859-140-4

The Handy Sports Answer Book
by Kevin Hillstrom, Laurie Hillstrom, and Roger Matuz
ISBN: 978-1-57859-075-9

The Handy Supreme Court Answer Book
by David L Hudson, Jr.
ISBN: 978-1-57859-196-1

The Handy Weather Answer Book, 2nd edition
by Kevin S. Hile
ISBN: 978-1-57859-221-0

Please visit the "Handy" series website at www.handyanswers.com.